INTERNATIONAL CENTRE FOR MECHANICAL SCIENCES

COURSES AND LECTURES - No. 277

# THEORETICAL ACOUSTICS AND NUMERICAL TECHNIQUES

EDITED BY

P. FILIPPI

LABORATOIRE DE MECANIQUE ET D'ACOUSTIQUE

MARSEILLE

SPRINGER - VERLAG WIEN - NEW YORK

Printed in Italy



ISBN 3-211-81786-7 Springer Verlag Wien-New York
ISBN 0-387-81786-7 Springer Verlag New York -Wien

## CONTENTS

# PREFACE

Though Acoustics is a part of continuous media mechanics, the mathematical methods recently developed in solid mechanics are almost never used by acousticians. This is, of course, due to a lack of scientific effort on noise and sound mathematical problems. But the main reason is that the difficulties encountered with the wave equation are strongly different from those which appear in solid mechanics. And the very convenient mathematical tools have been developed during the past fifteen years only.

## 1/ ACOUSTICS AND CLASSICAL MATHEMATICS

Let, first, have a brief survey of the mathematical problems appearing in Acoustics. The time-dependant governing equation is of hyerbolice type (a much less simple case than the parabolic type ), and unbounded domains must be considered as soon as environmental acoustics or under-water propagation are concerned.

Because of the difficulty to solve the wave equation, and because lots of noise and sound sources are periodic (or can be considered as periodic ), the Helmholtz equation is more frequently used. If the propagation domain is bounded, resonnances appear. If the propagation domain is unbounded, the total energy involved is unbounded, too. For these reasons, the use of the classical variational techniques is much less easy than for the heat equation, static solid mechanics, or incompressible fluid dynamics. Another difficulty is that as soon as energy is lost within the boundaries - and this generally the case - the operators involved are not self-adjoint : consequently, the powerful spectral theory does not apply in its classical form.

Another type of difficulty will appear if acoustic energy can propagate in the boundaries, and this is often the case. Such boundaries are known by acousticians as "non locally reacting surfaces " : their effect is described by an integral relationship between the pressure and the normal velocity. Such non-local boundary conditions cannot be accounted for by the classical theories.

Nevertheless, during the first half of this century, lots of progress have been made in Acoustical Engineering. And when the large computers appear, numerical techniques have been developed.

First of all, analytical solutions of particular diffraction problems have been established. Room acoustics has been studied for simple geometrical configurations. The basic tools are the spatial Fourier transform and the separation of variables. Then, for less simple geometrical data, pertubation techniques have been used.

Another classical method is to use asymptotic expansions with respect to the distance or the frequency, or other characteristic parameters. The Geometrical Theory of Diffraction belongs to this category ; though they are based on considerations which seem satisfactory to the physicist, the results are not always proved.

In the last fifteen years, boundary integral equations have been used in acoustic diffraction (and, simultaneously, in electromagnetism ). More or less simple numerical procedures have been adopted, but their convergence has been proved recently, only. At the same period, finite elements methods have been succesfully used for solving problems in bounded domains ; for unbounded domains, these methods appear to be less efficient than the boundary integral equations method.

## 2/ THE MODERN MATHEMATICAL ANALYSIS

As far as the data (boundary surfaces, source distribution, space characteristic parameters of the physical medium, ... )are described by sufficiently regular functions, and if local boundary conditions are considered, the existence and uniqueness theorems of the solution can be proved, using very classical mathematics. But, in many practical cases, the necessary regularity hypothesis are not fulfilled : the boundaries can have corners ; the actual sources are efficiently described by distributions (think of multipole sources encountered in jet noise description ). Non local boundary conditions are of practical interest : this is the case of domains bounded by vibrating structures ; another example is provided by under water sediments in which a sound wave can propagate. For harmonic time dependant problems, the best adapted theory is certainly that of pseudo-differential operators and Poisson pseudo-kernels (existence and uniqueness of the solution, eigenmodes, regularity theorems, wedge conditions, ..., are easily proved ). Furthermore, these recent mathematical tools enables to prove the convergence of the numerical techniques used to solve acoustic boundary integral equations. Moreover, it can be expected that new boundary finite elements , well adapted to acoustic problems, will be defined and proved to provide a faster convergence.

During the last ten years, the Fourier Integral Operators theory has been developed to study hyperbolic partial differential equations. The typical example is the wave equation. It is particularly efficient for wave-fronts propagation problems. Moreover, it provides high frequency asymptotical results : as a consequence, some results of the Geometrical Theory of Diffraction can now be justified, or proved to be somewhat incorrect. It can be expected that, for the wave equation, the Fourier Integral Operators theory will provide to the physicist as useful results as those derived from the pseudo-differential operators theory.

## 3/ CONTENTS OF THE PRESENT COURSE

The linearisation of the fluid dynamics equations leading to classical acoustics equations will be recalled briefly in the next section of this preface. A short bibliography of the basic papers and books is given at the end.

The first chapter is devoted to acoustic boundary integral equations : it is first shown how to establish them. Then, a simple numerical technique is described and illustrated by various examples. An "experimental" convergence is shown by comparing the numerical solution either to analytical results, or to model experiments.

In the second chapter, the finite elements method is presented. Here again, the numerical technique is described from an engineering point of view.

The third chapter deals with propagation problems above layered media. It is an example of the use of the spatial Fourier transform. The results presented are quite new. It is shown that various representations of the solution can be obtained : the first two ones recieve a simple physical interpretation, while the third one is well adapted to numerical computation. Moreover, the method here developed is very general and can be applied to a wide class of problems of wave propagation above layerd media, or within a layered medium as well (think of waves in shallow water bounded by stratified sediments ).

The convergence of the numerical solutions of boundary integral equations is studied in the fourth chapter. Two main techniques are presented and compared : it is shown that the collocation method is much more efficient than the techniqhe based on the variational formulation of the problem.This is satisfactory for the physicists who have mainly used this simple procedure.

Chapter five is devoted to an introduction to Fourier Integral Operators, Pseudo-Differential Operators, and Poisson Pseudo-Kernels. The basis of these theories is presented in the scope of acoustic problems. As an example, the boundary integral equation encountered in the diffraction by a hard infinitely thin screen is established, and the so-called wedge conditions are proved.

The sixth chapter illustrates the interest of the Fourier Integral Operators. A problem of singularities propagation and high frequency asymptotics is solved.

The seventh and last chapters devoted to interface problems in acoustics, and in electromagnetism as well. Though some of the results are quite reasonable from a physical point of view, it is shown that their mathematical proof cannot be established without the theories presented in chapter five.

## 4/ THE EQUATIONS OF LINEAR ACOUSTICS

An acoustical motion is a pertubation of a fluid motion. So, the different acoustic wave equations derive from the Navier-Stokes equations. The conservation equations of the mechanics of continuous media can be established in two ways. One way is to describe the modifications of an infinitely small part of the medium by exterior forces. The other way is to express that various quantities (mass, momentum, energy ) are conserved within any arbitrary domain. The advantage of this last method is that the necessary regularity assumptions on the unknown functions are weaker.

4-1 The conservation equations :

Let S be an isolated physical medium : here, "isolated" means that no mass sources exist. Let $\rho(t,X)$ be the volumic mass of the medium at a point X, and time t. Consider, at time t, a volume $\Omega$ of S. If a motion exists, the total mass M(t) of the particles contained in $\Omega$ does not change (although the domain $\Omega$ changes, of course ). Let U(t,X) stand for the particle velocity at the point X, and time t. The mass conservation relationship :

$$\frac{dM(t)}{dt} = \frac{d}{dt}\int_{\Omega} \rho(t,X)\, dv = 0$$

leads to :

$$(1) \qquad \int_{\Omega}\left\{\frac{\partial\rho(t,X)}{\partial t} + \operatorname{div}\left[\rho(t,X)\, U(t,X)\right]\right\}dv = 0$$

From a mathematical point of view, this expression only requires that the quantities involved are locally integrable. If they are defined everywhere, since (1) must be true for any $\Omega$, the mass conservation equation becomes :

$$(2) \qquad \frac{\partial\rho(t,X)}{\partial t} + \operatorname{div}\left[\rho(t,X)\, U(t,X)\right] = 0 .$$

The second equation is the generalisation of the classical law of dynamics , $f=m\gamma$, which relates the accelaration $\gamma$ of a point mass m to the force f acting on it. In fact, if U is the velocity, an equivalent form is $f=d(mU)/dt$, which shows that the time variation of the momentum mU is equal to the exterior force acting on the point mass. For a domain $\Omega$, the total momentum is expressed by :

$$Q = \int \rho(t,X)\ U(t,X)\ dv\ .$$

Two kinds of forces can act on $\Omega$ : volume forces, with density $F_i$; and surface forces, with density $\Sigma$ applied on $\partial\Omega$, the boundary of $\Omega$. These surface forces are exerted by the particles of the domain $(S-\Omega)$ on $\Omega$. Physical and mathematical considerations lead to express the vector $\Sigma$ as the product of a symetrical third-order tensor $\sigma$ and the outward unit vector n normal to $\partial\Omega$, i.e. :

$$\Sigma_i = \sigma_{ij}\ n_j \qquad i=1,2,3$$

in cartesian co-ordinates. The tensor $\sigma$ is called the stress tensor. The integral form of the momentum conservation equation is then :

$$(3) \qquad \frac{d}{dt}\int_{\Omega} \rho\ U_i\ dv - \int_{\partial\Omega} \sigma_{ij}\ n_j\ ds = \int_{\Omega} F_i\ dv \qquad i=1,2,3.$$

Using Ostrogradskhi theorem, and since equation (3) is valid for any domain $\Omega$, one gets :

$$(4) \qquad \frac{\partial(\rho U_i)}{\partial t} + (\rho U_i U_j)_{,j} - \sigma_{ij,j} = F_i \qquad i=1,2,3\ ;$$

it is the differential form of the momentum conservation equation.

At the present step, it is useful to specify that the physical system under consideration is a newtonian fluid. Experience has shown that, for this kind of fluid, the stress tensor is given by :

$$(5) \qquad \begin{cases} \sigma_{ij} = -\ p\ \delta_{ij} + \lambda\ D_{kk}\ \delta_{ij} + 2\ \mu\ D_{ij} \\ D_{ij} = \frac{1}{2}\ (\ U_{i,j} + U_{j,i}\ ) \end{cases}$$

Then a new variable appears : the pressure p. The parameters $\lambda$ and $\mu$ are the viscosity coefficients. We are now left with four scalar equations (eq. 2 and 4 ), and five unknown scalar functions $\rho$, $U_i$ (i=1,2,3), and p.

The third relationship to be introduced, without specifying more the physical medium, is the energy conservation equation. It involves heat exchanges. Let r be the volumic density of heat rate provided to the system by external sources ; and q the heat flow vector which describes the heat exchanges between $\Omega$ and S-$\Omega$. Finally, let e denote the internal energy density of the medium. The integral form of the energy conservation equation is :

$$(6) \quad \frac{d}{dt}\int_{\Omega} \rho \left( e + \frac{1}{2} U.U \right) = \int_{\Omega} ( F.U + r ) \, dv + \int_{\partial\Omega} ( \Sigma.U + q.n ) \, ds$$

Using Ostrogradskhi theorem, the surface integral is transformed into a volume integral, which leads to the partial differential equation :

$$(7) \quad \frac{\partial}{\partial t}\left[ \rho \left( e + \frac{1}{2} U_i U_i \right) \right] + \left[ \rho U_i \left( e + \frac{1}{2} U_j U_j \right) - U_j \sigma_{ji} - q_i \right]_{,i} =$$

$$= F_i U_i - r$$

This equation introduces four additional unknowns : e, and $q_j$ (j=1,2,3 ).

## 4-2 The equations of thermodynamics :

In fact the vector function q is related to a scalar function, the temperature T, by the heat conduction equation :

$$(8) \quad q = -k(T) \nabla T$$

in which k(T) is the conduction factor.

To complete the set of equations, the thermo-mechanical behaviour of the fluid must be introduced. To this end, the pressure p, the temperature T, and the internal energy density e are related together by introducing a last thermodynamical variable, the specific entropy s. Experimental results have shown that the thermodynamical state of a fluid is completely determined by the knowledge of two thermodynamical variables, only. If the density $\rho$ and the specific entropy s are chosen, then the specific internal energy e is a function of s and $1/\rho$, and can be determined experimentally :

(9) $e = f(s, 1/\rho)$

The remaining variables p and T are related to s and $\rho$ by :

(10) $p = \left.\frac{\partial e}{\partial(1/\rho)}\right|_s \quad , \quad T = \left.\frac{\partial e}{\partial s}\right|_{1/\rho} \quad ;$

that is : T is the partial derivative of e with respect to s, for constant $1/\rho$ ; and p is the partial derivative of e with respect to $1/\rho$, for constant s.

4-3 The perfect gas equation :

A perfect gas is a Newtonian fluid which satisfies :

a/ Mariotte's law : the ratio $p/\rho$ is a function of the temperature T only;

b/ Joule's law : the specific internal energy e is a function of the temperature T only.

As a consequence, it can be shown that :

(11) $\frac{p}{\rho} = R\,T$

where R is the perfect gas constant. Then, the constant volume specific heat $C_v$, and the constant pressure specific heat $C_p$ are given by :

(12) $C_v = \frac{de}{dt} \quad , \quad C_p = R + C_v \quad .$

In air, the parameters $C_V$ and $C_P$, and their ratio $\gamma = C_p/C_v$ are constant. Thus, the pressure p can be related to the density $\rho$ and the specific entropy s by :

(13) $p = \rho^\gamma \exp(s/C_v)$

Finally, the last important thermodynamical variable to be defined is the sound speed ; it is given by :

$$(14) \quad c^2 = \frac{\partial p}{\partial \rho} = \gamma \rho^{\gamma-1} \exp(s/C_v)$$

4-4 The equations of linear acoustics :

Let us assume that a known fluid motion, described by the quantities $\rho^0$, $p^0$, $U^0$, $T^0$, etc, is "perturbed" by an additional "small" force f. The state variables will become :

$$\rho^0+\delta \;;\; p^0+\Pi \;;\; U^0+u \;;\; T^0+t \;;\; \text{etc...}$$

Obviously, the perturbations $\delta$, $\Pi$, u, t, ... satisfy non-linear equations. But, if the perturbating force f is "small" enough, it can be expected that the perturbations remain correctly described when all their non-linear combinations are neglected in the equations. Such a perturbation can be called an "acoustic wave", and acoustics appears as an asymptotic theory.

There is a very wide variety of linearised equations, depending on the assumptions made on the perturbed motion. A rather complete analysis can be found in papers by P.E. DOAK [4, 5, 6, 7 ], and B.T. CHU and L.S. KOVASZNAY [8]. A simple example is obtained under the following asumptions

a/ there is no heat source : $r = 0$ ;

b/ the flow is isentropic: $s$ = constant.

Thus, the mass and momentum equations are sufficient to describe the pressure $\Pi$ and the momentum w, perturbations of the pressure $p^0$ and the momentum $W^0$ :

$$(15) \quad \frac{1}{c_0^{\,2}} \frac{\partial \Pi}{\partial t} + \operatorname{div} w = 0$$

$$(16) \quad \frac{\partial w_i}{\partial t} + \left\{ W_i^0\left(\frac{w_j}{\rho_0} - \frac{W_j^0}{\rho_0^2 c_0^2}\Pi\right) + w_i \frac{W_j^0}{\rho_0} \right\}_{,j} + \left\{\Pi\delta_{ij} - \lambda d_{kk}\delta_{ij} - 2\mu d_{ij}\right\}_{,j} = f_i \quad , \; i=1,2,3 \;;$$

with :

$$d_{ij} = \frac{1}{2}\left\{\left(\frac{w_i}{\rho_0} - \frac{W_i^0}{\rho_0^2 c_0^2}\Pi\right)_{,j} + \left(\frac{w_j}{\rho_0} - \frac{W_i^0}{\rho_0^2 c_0^2}\Pi\right)_{,i}\right\}$$

$$w = \delta\, U^0 + \rho^0\, u$$

By differentiating equation (16) with respect to time, and replacing $\partial\Pi/\partial t$ by its expression derived from (15), one gets three partial differential equations for the components of $w$.

4-5 The d'Alembert equation and the Helmholtz equation :

They describe the acoustic motion of a non-viscous fluid initially at rest, and excited by a potential force, that is :

$$W^0 = 0 \quad \text{and} \quad f = -\nabla\psi$$

In that case, an equation governing the acoustic pressure $\Pi$ is easily obtained :

$$(17) \quad \frac{1}{c_0{}^2}\frac{\partial^2\Pi}{\partial t^2} - \Delta\Pi = \Delta\psi$$

If another quantity $\phi$ , the velocity potential, is introduced :

$$u = -\nabla\phi$$

equation (17) becomes :

$$(17') \quad \frac{1}{c_0{}^2}\frac{\partial^2\phi}{\partial t^2} - \Delta\phi = \frac{1}{c_0{}^2}\frac{\partial\psi}{\partial t}$$

Equation (17) is known as the d'Alembert equation. If a time harmonic excitation $-\Psi e^{-i\omega t}$ is considered, the Helmholtz equation

$$(18) \quad \left(\Delta + \frac{\omega^2}{c_0{}^2}\right)\Pi = \Delta\Psi$$

is obtained.

All this course is devoted to this simple forms of acoustic equations. Nevertheless, the main ideas of the mathematical analysis presented here enable to get analytical, and even numerical results for more complex acoustical motions.

I am pleased to thank the International Center for Mechanical Sciences who has offered me the opportunity to organise this summer school on Acoustics.

I am very grateful to my colleagues who have accepted to present lectures, and to prepare very carefully the texts which are published in this book.

Paul Filippi

Marseille, October 1983.

REFERENCES

[1] L. LANDAU and E. LIFCHITZ, 1971, *Mécanique des fluides*. Moscow, Editions Mir.

[2] P. GERMAIN, 1973, *Cours de mécanique des milieux continus*. Masson et Cie., Editeurs.

[3] P.M. MORSE and K.U. INGARD, 1968, *Theoretical Acoustics*. Mc Graw-Hill Book Company, New York.

[4] P.E. DOAK, 1965, *Journal of Sound and Vibration* 2, 53-73. Analysis of internally generated sound in continuous materials : (I) Inhomogeneous acoustic wave equations.

[5] P.E. DOAK, 1971, *Journal of Sound and Vibration* 19, 211-225. On the interpedence between acoustic and turbulent fluctuating motion in moving fluid.

[6] P.E. DOAK, 1972, *Journal of Sound and Vibration* 25, 263-335. Analysis of internally generated sound in continuous materials : (II) A critical review of the conceptual adequacy and physical scopes of existing theories of aerodynamic noise, with special reference to supersonic jet noise.

[7] P.E. DOAK, 1973, *Journal of Sound and Vibration* 26, 91-120. Analysis of internally generated sound in continuous materials : (III) The momentum potential field description of fluctuating fluid motion as a basis for a unified theory of internally generated sound.

[8] B.T. CHU and L.S. KOVASZNAY, 1958, *Journal of Fluid Mechanics* 3, 494-514. Non-linear interactions in a viscous heat-conducting compressible gas.

[9] M.J. LIGHTHILL, 1972, *Journal of Sound and Vibration* 24, 471-492. The fourth annual fairey lecture : the propagation of sound through moving fluids.

# INTEGRAL EQUATIONS IN ACOUSTICS

**Paul J.T. FILIPPI**
**Laboratoire de Mécanique et d'Acoustique**
**BP 71 – 31, chemin Joseph Aiguier**
**13277 Marseille cedex 9**

## I. - INTRODUCTION

The aim of this course is twofold. First, the integral equations of linear acoustics are established for both interior and exterior problems. The integral representation of the diffracted field has several advantages : a/ the regularity theorems of the solution are easily obtained using the theories of "pseudo-differential operators" [1, 2] and "Poisson pseudo-kernels"[3, 4] ; b/ it is probably the most convenient formulation when no-local boundary conditions are involved ; c/ numerical methods provide analytical approximations of the total field which are very useful for exterior problems (far-field diffraction patterns are easily obtained, constant level curves can be drawn,...). Another significant result (which is not established here) concerns the so-called "edge-conditions" which appear when the propagation domain has a non-regular boundary, or more, when the diffracting obstacle is an infinitely thin screen. Such boundaries or obstacles can be considered as the limit of a sequence of regular boundaries or no-zero thickness regular obstacles. It can be shown that the corresponding sequence of solutions has an unique limit which belongs to a functional space, the proporties of which depend on the boundary irregularities. The edge conditions are included in the definition of this functional space. The fundamental ideas of the modern symbolic calculus of the pseudo-differential operators theory were already described in the book "Multidimensional singular integral equations" by S.G. MIKHLIN [5]. But the method used by this author is rather complicated, and the proofs must be established for each particular case. The recent theories are of a great generality and the basic results, useful in acoustics, are very simple.

The second aim of this paper is to gather different examples of numerical experiments which have been published during the past fifteen years. They deal with interior problems and exterior ones, as well. The numerical method used is generally very simple (collocation technique), but, nevertheless appears to be efficient enough for engineering purposes. Recently, different authors have used more sophisticated methods (surface finite elements techniques) : there is, of course, a loss of simplicity, but the convergence rapidity is much increased.

In the remainder of this introductory section, it is shown how a boundary value problem can be replaced by a system of integral equations. Then, some of the inherant difficulties are pointed out.

I.1. - BOUNDARY VALUE PROBLEMS AND LAYER POTENTIALS REPRESENTATION OF THE SOLUTION :

Let $\Omega$ be a bounded or unbounded domain of the $\mathbb{R}^n$ space ($n = 2$ or $3$). Its boundary $\Gamma$ is assumed to be a $C^\infty$ closed surface, with unit normal vector $n$ pointing out of $\Omega$.

Let $\mathcal{L}$ be a partial differential elliptic operator of order $2m$ with indefinitely differentiable coefficients. One seeks the solution $u(M)$ of the following equation

$$\mathcal{L}\, u(M) = f(M)$$

where $f(M)$ is a function (or more generally, a distribution) compactly supported in $\Omega$ (i.e., $f$ is zero outside a bounded domain strictly contained in $\Omega$).

Let $\partial_n^{(j)} u(M)$ be the succesive normal derivatives of $u(M)$ defined for $j \leqslant 2m - 1$. The influence of $\Gamma$ is described by $m$ boundary operators $\ell_i$, leading to $m$ boundary conditions

$$\ell_i(u, \partial_n u, \ldots, \partial_n^{(2m-1)} u)\,(P) = 0, \quad P \in \Gamma, \quad i = 1,2,\ldots,m$$

Classically, the $\ell_i$ are partial differential operators with $C^\infty$ coefficients (locally reacting boundaries). But a more general class of boundary operators is provided by integro-differential operators.

They described extended reaction boundaries, as it is the case when $\Gamma$ is the surface limiting a second medium in which the field $u$ can propagate.

If $\Omega$ is an unbounded domain, a Sommerfeld condition at infinity must be added to ensure the uniqueness of $u$.

From a physical point of view, this condition expresses the conservation of energy principle from which the partial differential equation is derived. Two other equivalent conditions can be used : the limit amplitude principle, and the finite amplitude principle.

Let $G_Q(M)$ be the elementary kernel of ${}^t\mathcal{L}$, the formal adjoint of defined by :

$$< \mathcal{L}\, v, w > \; = \; < v, {}^t\mathcal{L}\, w >$$

where $<\cdot\,,\cdot>$ stands for the duality product between the distributions space $\mathcal{D}'(\mathbb{R}^n)$ and the $C^\infty$- compactly supported functions space $\mathcal{D}(\mathbb{R}^n)$ (for the definitions, see [6]). $G_Q(M)$ satisfies the equation :

$${}^t\mathcal{L}\, G_Q(M) = \delta_Q(M)$$

where M is considered as a parameter, and the derivatives are taken with respect to Q. Furthermore, the function $M \to G_Q(M)$ is assumed to satisfy the Sommerfeld conditions. Obviously, $G_Q(M)$ represents the free-field response to a point isotropic source $\delta_Q(M)$, located at the point Q.

Let now define the distribution $\mu_j \otimes \partial_n^j \delta_\Gamma$ by :

$$< \mu_j \otimes \partial_n^j \delta_\Gamma , \varphi > = (-1)^j \int_\Gamma \mu_j \partial_n^j \varphi \, d\Gamma$$

where $\mu_j$ is a function defined on $\Gamma$, and $\varphi \in \mathscr{D}(R^n)$. It will be called a layer of order j.

An integral representation of u(M) is provided by :

$$u(M) = < f(Q), G_Q(M) > + \sum_{j=0}^{q} < [\mu_j \otimes \partial_n^j \delta_\Gamma^j](Q), G_Q(M) >$$

If a sufficient regularity of f and the $\mu_j$ is assumed, this last expression takes the following form :

$$u(M) = \int_\Omega f(Q)\, G_Q(M)\, dQ + \sum_{j=0}^{q} (-1)^j \int_\Gamma \mu_j(Q)\, \partial_{n(Q)}^j G_Q(M)\, dQ$$

The q+1 functions $\mu_j$ have to be determined by the m boundary conditions. Consequently it is certainly necessary to have $q+1 \geqslant m$.

## I.2. - INTEGRAL EQUATIONS

Let now express that the integral representations of the function u(M) satisfies the boundary conditions ; one gets :

$$\sum_{j=0}^{q} \ell_i \{ \lim_{M\in\Omega \to P\in\Gamma} \partial_{n(P)}^k < [\mu_j \otimes \partial_n^j \delta_\Gamma](Q), G_Q(M) > \} =$$

$$- \ell_i \{ \lim_{M\in\Omega \to P\in\Gamma} \partial_{n(P)}^k < f(Q), G_Q(M) > \}, \quad i = 1,2,\ldots,m .$$

It can be proved that the limits involved exist and are unique. Furthermore, the right-hand members are equal to :

$$- \ell_i \{ \partial_{n(P)}^k < f(Q), G_Q(P) > \}$$

due to the assumption that the support of f is strictly included in $\Omega$. But, a layer potential can have discontinuous derivatives, and its derivatives cannot always be expressed by a Rieman integral (Cauchy's principal values and non integrable kernels can be involved). Consequently, even if expressions of the form

$$\partial_{n(P)}^k \int_\Gamma \mu_j(Q)\, \partial_{n(Q)}^j G_Q(P)\, dQ , \quad P \in \Gamma$$

are meaningful, they are often different from

$$\lim_{M\in\Omega \to P\in\Gamma} \partial_{n(P)}^k \int_\Gamma \mu_j(Q)\, \partial_{n(Q)}^j G_Q(M)\, dQ .$$

So, the expression of the boundary conditions requires some care.

The existence and uniqueness conditions of the solutions $\mu_j$ of the boundary integral equations system are provided by the Fredholm alternative. This implies first that (q+1-m) additional relationships between the $\mu_j$ are arbitrarily choosen to get as many equations as unknown functions. The non-uniqueness of such a representation can be used to get a system of integral boundary equations which can be called "equivalent" to the initial boundary value problem : i.e., if u exists and is unique, the integral equations system has one and only one solution ; if the boundary value problem has n eigenfunctions, there exist n and only n independant sets of solutions $\mu_j$.

The last general remark to be made concerns the regularity of the solution u of the differential system, and that of the $\mu_j$. Using elementary results of the theories of pseudo-differential operators and Poisson pseudo-kernels, the following result is obtained : if $\Gamma$ is a $C^\infty$ surface and if $\omega$ is the sources' support strictly contained in $\Omega$, then u is a $C^\infty$ function in $\Omega - \omega$, and the $\mu_j$ are $C^\infty$ functions on $\Gamma$.

Section II is devoted to the establishment of integral representation of the solutions of the Helmholtz equation for interior and exterior problems. The necessary mathematical justifications are developped in the lectures by M. Durand.

Section III deals with the numerical solutions of the integral equations of scalar diffraction. Here again, the mathematical aspect is not developped ; the readear must refer to the lectures by W. Wendland and E. Stephan. Various typical examples are given, and the numerical techniques used by their authors are detailed.

## II. - THE SCALAR HELMHOLTZ EQUATION AND RELATED INTEGRAL EQUATIONS

The propagation of harmonic waves in a homogeneous isotropic fluid is governed by the Helmholtz equation. This is the simplest form of the linearized Navier-Stokes equations, as described in the lectures by M. Howe (for a simplest but less general, derivation of the equations, the readear can refer to [7] ). Various integral representations of the scattered or diffracted field are introduced, and the corresponding integral equations recalled. Emphasis is given to some delicate cases, such as exterior problems (the integral equations having eigenfrequencies, their solutions are not always unique) or diffraction by a perfectly hard thin screen (integral equation with non integrable kernel).

### II.1. - HELMHOLTZ EQUATION AND LOCAL BOUNDARY CONDITIONS :

#### *II.1.1. - BOUNDED DOMAINS :*

Let consider, in space $\mathbb{R}^n$ , a bounded domain $\Omega$ . Its boundary $\Gamma$ is assumed to be regular ( $C^\infty$ for simplicity), and has an unit normal vector n pointing out to $\complement\bar{\Omega}$ , the space complementary of $\bar{\Omega}$ . Let f be a regular function, defined for all M in $\Omega$ ; Trf and $\mathrm{Tr}\,\partial_n f$ are the functions defined on $\Gamma$ by the following limits :

(1) $$\mathrm{Tr}\, f(P) = \lim_{M\in\Omega \to P\in\Gamma} f(M)$$

(2) $$\mathrm{Tr}\,\partial_n f(P) = \lim_{M\in\Omega \to P\in\Gamma} n(P)\,.\,\mathrm{grad}\, f(M)$$

We consider the boundary value problem : find $\varphi$ satisfying

$$\begin{cases} (\Delta + k^2)\varphi = S & \forall M \in \Omega \quad (3) \\ \alpha\,\mathrm{Tr}\,\varphi + \beta\,\mathrm{Tr}\,\partial_n\varphi = 0 & \forall M \in \Gamma \quad (4) \end{cases}$$

Where S is any function (or, more generally, distribution) compactly supported in $\Omega$ ; $\alpha$ and $\beta$ are $C^\infty$ functions defined on $\Gamma$ .

*Theorem : Let $k_q^2$ be the eigenvalues of the Laplace operator with respect to the domain $\Omega$ and the boundary condition (4).*

*a/ There exist a denumbrable infinite sequence of such $k_q^2$ ; for each $k_q^2$ , the system (3-4) has no solution if $S \not\equiv 0$, and a finite number of linearly* *independant solutions,* $\varphi_j^q$ *uniquely determined up to a multiplicative constant. The* $\varphi_j^q$ *are called the eigenfunctions of the system (3-4).*

*b/ If* $k^2 \neq k_q^2$*, the system (3-4) has one and only one solution, for any source distribution S.*

To establish this theorem, the Green formula representation of the solution can be used : it is shown that the corresponding boundary integral equation is of Fredholm type.

*II.1.2. - UNBOUNDED DOMAINS AND UNIQUENESS CONDITIONS :*

If $\Omega$ is an unbounded domain, it is intuitive that the system (3-4) is not sufficient to determine a unique solution.

Indeed, physical experiments show that no energy is reflected from points at infinity. This must be expressed mathematically. There are three equivalent "radiation" conditions.

*a/ - The Sommerfeld condition*. It describes the asymptotic behaviour of the solution at infinity [8, 9, 10]. Let r be the distance from M to the coordinates origin. If the harmonic time dependance is $e^{-i\omega t}$, one seeks $\varphi_-$ satisfying the relationships :

$$(5) \quad \begin{cases} \lim_{r \to \infty} \varphi_- = \mathcal{O}(r^{-\frac{n-1}{2}}) \\ \lim_{r \to \infty} [\partial_r \varphi_- - ik\varphi_-] = o(r^{-\frac{n-1}{2}}) \end{cases}$$

For a time dependance $e^{+i\omega t}$, conditions (5) are replaced by :

$$(6) \quad \lim_{r \to \infty} \varphi_+ = \mathcal{O}(r^{-\frac{n-1}{2}}) \qquad \lim_{r \to \infty} [\partial_r \varphi_+ + ik\varphi_+] = o(r^{-\frac{n-1}{2}})$$

Conditions (5) and (6) determined uniquely $\varphi_-$ and $\varphi_+$.

*b/ - The limit amplitude principle*. Let $\psi_-$ (t, M) be the unique solution of the initial boundary value problem :

$$(7)\quad \begin{cases} (\Delta - \frac{1}{c^2}\partial^2_{tt})\ \psi_-(t,M) = Y(t)\ S(M)\, e^{-i\omega t}\ , \quad \omega^2 = c^2k^2 \\ \psi_- = \partial_t \psi_- = 0 \quad \text{for} \quad t < 0 \\ Y(t) = 0 \ \text{for} \ t < 0\ , \ = 1 \ \text{for} \ t > 0 \end{cases}$$

which satisfies the boundary condition (4). There exists a unique function $\varphi_-$ defined by :

$$(8)\quad \varphi_-(M) = \lim_{t \to \infty} \psi_-(t,M)\, e^{i\omega t}$$

Replacing equations (7) by

$$(9)\quad \begin{cases} (\Delta - \frac{1}{c^2}\partial^2_{tt})\ \psi_+(t,M) = Y(t)\ S(M)\, e^{i\omega t} \\ \psi_+ = \partial_t \psi_+ = 0 \quad \text{for} \quad t < 0 \end{cases}$$

A unique limit function $\varphi_+$ can be defined by :

$$(10)\quad \varphi_+(M) = \lim_{t \to \infty} \psi_+(t,M)\, e^{-i\omega t}$$

*c/ - <u>The limit absorption principle</u>.* Define, for $\varepsilon < 0$ , the functions $\varphi_{+\varepsilon}$ and $\varphi_{-\varepsilon}$ which are the unique bounded solutions of the system (3-4) in which $k^2$ is replaced by $(k + i\varepsilon)^2$ and $(k - i\varepsilon)^2$, respectively. It is shown that :

$$(11)\quad \varphi_- = \lim_{\varepsilon \to 0} \varphi_{-\varepsilon}$$

$$(12)\quad \varphi_+ = \lim_{\varepsilon \to 0} \varphi_{+\varepsilon}$$

exist and are unique.

It is now possible to state the following existence and uniqueness theorem.

*<u>Theorem</u> : The boundary value problem (3-4) has one and only one solution satisfying either of the three equivalent radiation conditions (5), (8) or (10). Another unique solution is defined by either of the three equivalent radiation conditions (6), (10) or (12).*

If the ratio $\alpha/\beta$ is real (or if $\alpha$ and $\beta$ are real), the two different solutions are complex conjugate functions.

### *II.1.3. - PHYSICAL INTERPRETATION OF THE RADIATION CONDITIONS :*

The meaning of the limit amplitude principle is obvious. A physical perturbation is always created a given finite time.

When the excitation is a periodic function of time for $t > 0$, it is intuitive that the response will tend to a time peirodic limit. A solution of the Helmholtz equation is such a limit (if this limit exists). But, the mathematical expression of this solutions depends of the mathematical form which has been adopted for describing the excitation. Nevertheless, it must be noticed that the physical quantity described by the two different mathematical expressions is unique. Indeed, if $\varphi_{\pm}$ is a complex pressure amplitude, the physical pressure is given by :

$$p(M,t) = \mathrm{Re}(\varphi_{-}\, e^{-i\omega t}) = \mathrm{Re}(\varphi_{+}\, e^{+i\omega t}) .$$

The limit amplitude principle corresponds to an idealization of the physical propagation medium. Experiments have shown that waves propagating in a fluid are progressively damped with the distance. The damping factor being generally small, it can be ignored as far as distances involved are not too large. The "damped" Helmholtz equation has two mathematical solutions : one is increasing exponentially at infinity ; the other one, which describes the physical phenomenon, decreases exponentially. Depending of the time dependance description ($e^{-i\omega t}$ or $e^{+i\omega t}$), the damping factor $\varepsilon$ is positive or negative. This leads to the two limits defined by equations (11) or (12).

The interpretation of the Sommerfeld conditions is less obvious. In fact, it must be recalled that the wave equation is derived from the principle of energy conservation. As a consequence, the mean energy flux, over one period, through a spherical surface of radius $r \to \infty$, must be positive and finite. This is expressed by equations (5) or (6), depending on the assumed time dependance. A simple proof is given in reference [8], for example.

## II. 2. - NON LOCAL BOUNDARY CONDITIONS, AN EXAMPLE.

Let $\Omega$ be a bounded domain of space $\mathbb{R}^3$, and $\Gamma$ its boundary (assumed to be $C^\infty$) with exterior unit normal $n$. $\Omega$ is occupied by a perfect homogeneous fluid, characterized by a density $\rho$ and a sound velocity $c$ ($\rho$ and $c$ are real). Assume that $\complement\bar{\Omega}$ is occupied by an isotropic and homogeneous porous medium characterized by a complex density $\rho'$ and a complex sound speed $c'$.

The harmonic ($e^{-i\omega t}$) sound pressure $\varphi$ in $\Omega$, and the sound pressure $\varphi'$ in $\complement\bar{\Omega}$ satisfy the system of equations :

$$(13)\quad \begin{cases} (\Delta + k^2)\varphi = & \text{for } M \in \Omega, \quad k^2 = \omega^2/c^2 \\ (\Delta + k'^2)\varphi' = 0 & \text{for } M \in \Omega, \quad k'^2 = \omega^2/c'^2 \\ \left.\begin{array}{l} Tr\varphi = Tr\varphi' \\ Tr\partial_n \varphi/\rho = Tr\partial_n \varphi'/\rho' \end{array}\right\} & \text{for } M \in \Gamma \\ \lim_{r \to \infty} |\varphi'| < \infty & \end{cases}$$

Using the results established in the following subsections, it is seen that $\varphi'$ can be expressed by :

$$(14)\qquad \varphi'(M) = -\int_\Gamma \{ Tr\varphi'(P)\, \partial_{n(P)} \frac{e^{ik'r(M,P)}}{4\Pi r(M,P)} - Tr\partial_n \varphi'(P) \frac{e^{ik'r(M,P)}}{4\Pi r(M,P)} \}\, dP$$

Because of the continuity relationships, this expression becomes :

$$(15)\qquad \varphi'(M) = -\int_\Gamma \{Tr\varphi(P)\, \partial_{n(P)} \frac{e^{ik'r(M,P)}}{4\Pi r(M,P)} - \frac{\rho'}{\rho} Tr\partial_n \varphi(P) \frac{e^{ik'r(P,M)}}{4\Pi r(P,M)} \}\, dP$$

It is obvious now that the function $\varphi$ satisfies the following pseudo-differential boundary value problem :

$$(16)\quad \begin{array}{c} (\Delta + k^2)\,\varphi = S \quad \text{for} \quad M \in \Omega \\ Tr\varphi(M) + Tr \int_\Gamma \{Tr\varphi(P)\, \partial_{n(P)} \frac{e^{ik'r(M,P)}}{4\Pi r(M,P)} - \frac{\rho'}{\rho} Tr\partial_n \varphi(P) \frac{e^{ik'r(M,P)}}{4\Pi r(M,P)} \}\, dP = 0 \quad \text{for } M \in \Gamma \end{array}$$

It is shown [11] , that system (16) has one and only one solution for any real k (the eigenfrequencies are complex). System (13) has, equally, a unique solution $\{ \varphi , \varphi' \}$ , for real k. This type of problems is not studied here, but the general methods developped in the following sections can be applied.

## II.3. - ELEMENTARY SOLUTION ; LAYER POTENTIALS ; INTEGRAL EQUATIONS.

In what follows, the time dependance $e^{-i\omega t}$ is always assumed. Using Fourier transformation, and the limit amplitude principle, the free-field Green function of the Helmholtz equation is determined. Then, two classes of functions satisfying the homogeneous Helmholtz equation in an open domain, are defined : the simple layer potentials, and the double layer potentials. The last part of this section is devoted to the integral equations derived from the representation of the scattered or diffracted field with layer potentials.

### *II.3.1. - ELEMENTARY SOLUTIONS AND ELEMENTARY KERNELS OF THE HELMHOLTZ EQUATION :*

A function (or more precisely a distribution) G is an elementary solution of the Helmholtz equation if it satisfies [6] :

$$(\Delta + k^2)\, G = \delta \tag{17}$$

where $\delta$ is the Dirac measure. Physically, G represents the sound field due to a point isotropic source at the coordinates origin. But G is not uniquely determined: in fact, it depends on the conditions which are imposed on the boundary of the domain in which equation (17) is valid. If the propagation domain is the whole space ( $\mathbb{R}^2$ or $\mathbb{R}^3$ ), and if a radiation condition is imposed, then G is unique, and represents the free-field radiation of a point isotropic source located at origin.

Similarly, an elementary kernel of the Helmholtz operator is a distribution G(M,S) of two variables, solution of [6] :

$$(\Delta + k^2)\, G(M,S) = \delta_S(M) \tag{18}$$

If equation (18) holds in the whole space, and a radiation condition is imposed , then G(M,S) is unique. It represents the sound field at M due to point isotropic source located at S. It is easily seen that the reciprocity principle applies, and that G(M, S) can be interpreted as the radiation at S of a point source located at M (symetry of the $\delta$ measure).

To solve (17), the Fourier transform can be used. If f is an integrable function, its transform $\hat{f}$ is defined by :

$$\hat{f}(\xi_1,\xi_2,\xi_3) = \int_{-\infty}^{+\infty}\int_{-\infty}^{+\infty}\int_{-\infty}^{+\infty} f(X_1,X_2,X_3)\, e^{2i\pi(X_1\xi_1 + X_2\xi_2 + X_3\xi_3)}\, dX_1\, dX_2\, dX_3 \,.$$

It is easily seen that $\hat{G}$ is given by :

(19) $$\hat{G} = \frac{1}{-4\Pi^2 \xi^2 + k^2} \quad , \quad \xi^2 = \xi_1^2 + \xi_2^2 + \xi_3^2$$

This equation being defined for any real $\xi$ if $k^2$ is complex. To use the limit amplitude principle, (19) is considered as the limit, for $\varepsilon \longrightarrow 0$, of :

(19') $$\hat{G}_\varepsilon = \frac{1}{-4\Pi^2 \xi^2 + (k + i\varepsilon)^2} \quad , \quad \varepsilon > 0$$

The function $\hat{G}_\varepsilon$ depending of $\xi^2$ only, it is shown [6], that its inverse Fourier transform is given by the integral :

(20) $$G = \frac{2\Pi}{r^{(n-2)/2}} \int_0^\infty \frac{1}{-4\Pi^2 \xi^2 + (k + i\varepsilon)^2} J_{\frac{n-2}{2}} (2\Pi\xi r) \, \xi^{\frac{2}{n}} \, d\xi$$

where $J_{\frac{n-2}{2}} (2\Pi\xi r)$ is the Bessel function of order $(n-2)/2$.

A simple transformation shows that (20) is equivalent to :

(21) $$G = \frac{\Pi}{r^{(n-2)/2}} \int_{+\infty e^{i\Pi}}^{+\infty} \frac{1}{-4\Pi^2 \xi^2 + (k + i\varepsilon)^2} \xi^{n/2} H^{(1)}_{(n-2)/2} (2\Pi\xi r) \, d\xi$$

The integration can now be performed by the residue method, using the contour

$$\{ -R \leqslant \xi \leqslant +R \; ; \; \xi = Re^{i\theta} \, , \; 0 \leqslant \theta \leqslant \Pi \; ; \; R \rightarrow \infty \}$$

and we are left with :

(22) $$\begin{cases} G = -\dfrac{e^{ikr}}{4\Pi r} & \text{in } \mathbb{R}^3 \\[2ex] G = -\dfrac{i}{4} H_0^{(1)} (kr) & \text{in } \mathbb{R}^2 \end{cases}$$

where $r$ is the radial coordinate of M. The free-field elementary kernel G(S,M) is given by (22) with $r$ standing for the distance between S and M.

*II.3.2. - SIMPLE AND DOUBLE LAYER POTENTIAL :*

First of all, the expression of the Laplacian of a discontinuous function must be defined in the distribution sense. Let $\Gamma$ be a closed $C^{\infty}$- surface (or curve) deviding the space $\mathbb{R}^3$ (or $\mathbb{R}^2$) into two regions $\Omega$, and $\complement\bar{\Omega}$. Its unit normal vector is choosen to point out to $\complement\bar{\Omega}$. Let $\varphi$ be a function defined in the whole space but on $\Gamma$, twotimes differentiable in $\Omega$ and $\complement\bar{\Omega}$; furthermore, it is assumed that the following limits exist :

(23)
$$\begin{cases} Tr^{+}\varphi = \lim\limits_{M\in \complement\Omega \to P\in\Gamma} \varphi(M) \\ Tr^{-}\varphi = \lim\limits_{M\in\Omega \to P\in\Gamma} \varphi(M) \end{cases}$$

(24)
$$\begin{cases} Tr^{+}\partial_n\varphi = \lim\limits_{M\in \complement\Omega \to P\in\Gamma} \partial_{n(P)}\varphi(M) \\ Tr^{-}\partial_n\varphi = \lim\limits_{M\in\Omega \to P\in\Gamma} \partial_{n(P)}\varphi(M) \\ \partial_{n(P)}\varphi(M) = n(P)\cdot \text{grad}\,\varphi(M) \end{cases}$$

That is, $\varphi$ and $\partial_n\varphi$ are functions which have a discontinuity on $\Gamma$. It is shown that, when the Laplace operator is applied to such a function, the result is not a function but a distribution. Denoting by $\{\Delta\varphi\}$ the discontinuous function, equal to the Laplacian of $\varphi$ in $\Omega$ and $\complement\bar{\Omega}$, but undefined on $\Gamma$, the distribution $\Delta\varphi$ is defined by :

(25)
$$\Delta\varphi = \{\Delta\varphi\} + (Tr^{+}\varphi - Tr^{-}\varphi)\otimes\delta'_{\Gamma} + (Tr^{+}\partial_n\varphi - Tr^{-}\partial_n\varphi)\otimes\delta_{\Gamma}.$$

The distributions supported by $\Gamma$ appearing in equation (25) are defined by :

(26)
$$<\mu\otimes\delta_{\Gamma}, v> = \int_{\Gamma}\mu\, Tr v\, d\Gamma$$

(27)
$$<\nu\otimes\delta'_{\Gamma}, v> = -\int_{\Gamma}\nu\, Tr\,\partial_n v\, d\Gamma$$

where $v$ is a function in $\mathscr{D}$ ($C^{\infty}$- compactly supported function on $\mathbb{R}^3$).

Assume now that $\varphi$ satisfies the homogeneous Helmholtz equation in $\Omega$ as well as in $\complement\bar{\Omega}$ . Equation (25) yields to :

(28) $$(\Delta + k^2)\varphi = (Tr^+\varphi - Tr^-\varphi) \otimes \delta'_\Gamma + (Tr^+\partial_n\varphi - Tr^-\partial_n\varphi) \otimes \delta_\Gamma$$

As a distribution defined in the whole space, $\varphi$ satisfies a non-homogeneous Helmholtz equation. The second term of the right-hand side

$$\mu \otimes \delta_\Gamma = (Tr^+\partial_n\varphi - Tr^-\partial_n\varphi)\otimes\delta_\Gamma$$

represents a layer of simple sources, the density of which is $\mu$ . The first term

$$\nu \otimes \delta'_\Gamma = (Tr^+\varphi - Tr^-\varphi) \otimes \delta'_\Gamma$$

is a layer of dipole sources with density $\nu$ . The free- field radiation of these kinds of sources can be expressed by surface integrals involving the free-field elementary kernel of the Helmholtz operator (if $\mu$ and $\nu$ are sufficiently regular)

*a/ The simple layer potential* $\varphi_1$ is the solution of the non-homogeneous equation :

(29) $$(\Delta + k^2)\varphi_1 = \mu \otimes \delta_\Gamma$$

Using the fundamental property of elementary kernels [6] , one gets :

(30) $$\varphi_1(M) = < G(M,P)\,,\,[\mu\otimes\delta_\Gamma](P) > = \int_\Gamma \mu(P)\,G(M,P)\,dP\,.$$

The last part of this equality implies that $\mu(P)\,G(M,P)$ is integrable.

From equations (25) and (29), it is obvious that $\varphi_1$ is a continuous function, but that its normal derivative has a discontinuity given by :

(31) $$Tr^+\partial_n\varphi_1 - Tr^-\partial_n\varphi_1 = \mu\,.$$

Furthermore, the following equalities are easily proved :

(32) $$\left\{\begin{aligned} Tr^+\partial_n\varphi_1(P_0) &= \frac{\mu(P_0)}{2} + \int_\Gamma \mu(P)\,\partial_{n(P_0)}G(P_0,P)\,dP \\ Tr^-\partial_n\varphi_1(P_0) &= -\frac{\mu(P_0)}{2} + \int_\Gamma \mu(P)\,\partial_{n(P_0)}G(P_0,P)\,dP \end{aligned}\right.$$

*b/ The double layer potential* $\varphi_2$ is the solution of the non-homogeneous equation :

(33) $$(\Delta + k^2)\varphi_2 = \nu \otimes \delta'_\Gamma$$

The solution of which is :

(34) $$\varphi_2(M) = < G(M,P)\,,\,[\nu \otimes \delta'_\Gamma](P) > = -\int_\Gamma \nu(P)\,\partial_{n(P)}\,G(M,P)\,dP\,.$$

Here again, the integrability of the function $\nu(P)\,\partial_{n(P)}\,G(M,P)$ ensures the validity of the last part of this equality.

This function has a discontinuity given by :

(35) $$Tr^+\varphi_2 - Tr^-\varphi_2 = \nu$$

and the following equalities hold :

(36) $$\begin{cases} Tr^+\varphi(P_0) = \dfrac{\nu(P_0)}{2} - \displaystyle\int_\Gamma \nu(P)\,\partial_{n(P)}\,G(P_0,P)\,dP\,. \\[2ex] Tr^-\varphi(P_0) = -\dfrac{\nu(P_0)}{2} - \displaystyle\int_\Gamma \nu(P)\,\partial_n(P)\,G(P_0,P)\,dP\,. \end{cases}$$

The normal derivative of $\varphi_2$ is continuous, and is defined by :

(37) $$Tr\varphi_2(P_0) = \lim_{M\in\Omega \to P_0\in\Gamma} \int_\Gamma \nu(P)\,\partial_{n(P_0)}\,\partial_{n(P)}\,G(M,P)\,dP\,.$$

The limit is proved to exist though the kernel $\partial_{n(P_0)}\,\partial_{n(P)}\,G(P_0,P)$ has a non-integrable singularity. Consequently, the function defined by (37) cannot be expressed by a convergent Rieman integral. It is shown in [12] and [13] that, if $\Gamma$ is a closed surface of $\mathbb{R}^3$, and if $\nu$ is a $C^1$ – function, the function $Tr\,\partial_n\varphi_2$ can be expresses by the following cauchy principal value integral :

(38) $$Tr\,\partial_n\varphi_2(P_0) = -\,V.p.\int_\Gamma [\nu(P) - \nu(P_0)]\,\partial_{n(P_0)}\,\partial_{n(P)}\,\frac{1}{4\Pi r(P_0,P)}\,dP$$
$$-\int_\Gamma \nu(P)\,\partial_{n(P_0)}\,\partial_{n(P)}\,[G(P_0,P) - \frac{1}{4\Pi r(P_0,P)}]\,dP$$

Otherwise, the first integral in (38) is not convergent. A logarithmic singularity appears, which can be extracted analytically, leaving a convergent integral. Nevertheless, despite the non-integrability of the kernel in (37), the numerical calculation of such a limit can be performed, using rather elementary approximation procedures as shown in section III.

*II.3.3. - INTEGRAL EQUATIONS :*

For the sake of simplicity, we will restrain our study to the Dirichlet and Neumann problems.

Let $\Gamma$ be a closed surface in the $\mathbb{R}^3$ space (or a closed curve in the $\mathbb{R}^2$ space), the inside of which is $\Omega$, and the outside $\complement\bar{\Omega}$. One looks for a solution $\varphi$ of the non-homogeneous Helmholtz equation :

(39) $$(\Delta + k^2)\varphi = S$$

in which S is compactly supported distribution of sources and k is a real wave number. The boundary conditions under consideration here are:

(40) $$\begin{cases} Tr\varphi = 0 & \text{Dirichlet problem} \\ Tr\partial_n\varphi = 0 & \text{Neumann problem} \end{cases}$$

In these expressions, the symbol $Tr$ stands for $Tr^+$ or $Tr^-$, depending on wether an exterior or an interior problem is looked at. The exterior and interior Dirichlet problems are denoted $D_e$ and $D_i$ respectively, while the exterior and interior Neumann ones are refered to as $N_e$ and $N_i$ respectively. For exterior problems, a radiation condition must be added (equation (5), (8), or (11)). Let G(M,M') be the Helmholtz elementary kernel, uniquelly determined by the radiation conditions. One seeks a solution $\varphi$ of the form :

(41) $$\begin{cases} \varphi(M) = \varphi_0(M) + \langle G(M,P), [\alpha\mu\otimes\delta_\Gamma + \beta\nu\otimes\delta'_\Gamma]\, P\rangle \\ \varphi_0(M) = \langle G(M,P), S(P)\rangle \end{cases}$$

where $\alpha$ and $\beta$ are constants. More explicitely, the solution $\varphi$ is written as :

(42) $$\varphi(M) = \varphi_0(M) + \int_\Gamma \{\alpha\mu(P)\, G(M,P) - \beta\nu(P)\,\partial_{n(P)} G(M,P)\}\, dP$$

The functions $\mu$ and $\nu$ must be determined by the boundary condition.

It is first of all obvious that the only boundary condition is not sufficient to ensure a possible uniqueness of the functions couple $\{\mu,\nu\}$ .

An arbitrary relationship between $\mu$ and $\nu$ must be added ; the simplest one $\mu = \nu$ , will be adopted, leading finally to the field representation

(43) $$\varphi(M) = \varphi_0(M) + \int_\Gamma \mu(P)\,\{\alpha G(M,P) - \beta\,\partial_{n(P)} G(M,P)\}\,dP$$

To the four boundary value problems, correspond four integral equations, namely :

For the $D_i$ problem :

(44) $$\alpha \int_\Gamma \mu(P)\,G(P_0,P)\,dP - \beta\,\frac{\mu(P_0)}{2} - \beta \int_\Gamma \mu(P)\,\partial_{n(P)} G(P_0,P)\,dP = -\varphi_0(P_0) \qquad \forall P_0 \in \Gamma$$

For the $D_e$ problem :

(45) $$\alpha \int_\Gamma \mu(P)\,G(P_0,P)\,dP + \beta\,\frac{\mu(P_0)}{2} - \beta \int_\Gamma \mu(P)\,\partial_{n(P)} G(P_0,P)\,dP = -\varphi_0(P_0) \qquad \forall P_0 \in \Gamma$$

For the $N_i$ problem :

(46) $$-\alpha\,\frac{\mu(P_0)}{2} + \alpha \int \mu(P)\,\partial_{n(P_0)} G(P_0,P)\,dP - \beta\,\mathrm{Tr} \int \mu(P)\,\partial_{n(P_0)}\partial_{n(P)} G(P_0,P)\,dP = -\partial_{n(P_0)}\varphi_0(P_0) \qquad \forall P_0 \in \Gamma$$

For the $N_e$ problem :

(47) $$+\alpha\,\frac{\mu(P_0)}{2} + \int_\Gamma \mu(P)\,\partial_{n(P_0)} G(P_0,P)\,dP - \beta\,\mathrm{Tr} \int_\Gamma \mu(P)\,\partial_{n(P_0)}\,\partial_{n(P)} G(P_0,P)\,dP = -\partial_{n(P_0)}\varphi_0(P_0) \qquad \forall P_0 \in \Gamma$$

These equations are of Fredholm type, and one has the following

<u>*Theorem*</u> : *a/ For the interior problems* $D_i$ *and* $N_i$ *equations (44) and (46) have one and only one solution, whatever* $\alpha$ *and* $\beta$ *are unless k belongs to the denumbrable set of eigenvalues of the corresponding boundary pseudo-differential operator. If k is an eigenvalue, then the homogeneous integral equation has a finite number of linearly independant eigenfunctions uniquely determined up to an arbitrary multiplicative constant. To each eigensolution of the integral equation, corresponds an eigenfunction for the initial boundary value problem.*

*b/ For the exterior problems* $D_e$ *and* $N_e$ *, equations (45) and (47) have one and only one solution for any real k, if the ratio* $\alpha/\beta$ *has a non-zero imaginary part.*

The proof of this theorem is not given here. Nevertheless, let us give a rapid explanation of the condition imposed to the ratio $\alpha/\beta$ in part b of the theorem. Consider, for example, equation (45). In fact, it is the transposed of the equation obtained for an interior problem satisfying the mixed boundary condition

$$\alpha \mathrm{Tr}\varphi - \beta \mathrm{Tr}\partial_n \varphi = 0$$

the solution of which being represented with a simple layer potential. Consequently, equation (45) has the eigenvalues of this interior problem, which are real if $\alpha/\beta$ is real, and complex if $\mathrm{Im}(\alpha/\beta) \neq 0$.

### *II.3.4. - THE GREEN REPRESENTATION :*

Let go back to the representation (42) of the solution. Another way to get non-independant layer densities $\alpha\mu$ and $\beta\nu$ is to choose them so that expression (42) is identically zero outside the propagation domain.

For the interior problems, this is obtained with :

$$\alpha\mu = -\mathrm{Tr}^- \partial_n \varphi \,, \quad \beta\nu = -\mathrm{Tr}^- \varphi \,,$$

leading to :

$$\varphi(M) = \varphi_0(M) - \int_\Gamma \{\mathrm{Tr}^- \partial_n \varphi(P)\, G(M,P) - \mathrm{Tr}^- \varphi(P)\, \partial_{n(P)} G(M,P)\}\, dP \,, \; M \in \Omega \,. \tag{48}$$

Indeed, the function (4 8) has the discontinuity

$$\mathrm{Tr}^+\varphi - \mathrm{Tr}^-\varphi = -\mathrm{Tr}^-\varphi \,,$$

and its normal derivative has the discontinuity

$$\mathrm{Tr}^+\partial_n\varphi - \mathrm{Tr}^-\partial_n\varphi = -\mathrm{Tr}^-\partial_n\varphi .$$

This shows that $\mathrm{Tr}^+\varphi$ and $\mathrm{Tr}^+\partial_n\varphi$ are zero, and, as a consequence, $\varphi$ is identically zero in $\Omega$

Similarly, the Green representation of the solution of an exterior problem is given by :

$$\varphi(M) = \varphi_0(M) + \int_\Gamma \{\mathrm{Tr}^+ \partial_n \varphi(P)\, G(M,P) - \mathrm{Tr}^+ \varphi(P)\, \partial_{n(P)} G(M,P)\}\, dP \,. \tag{49}$$

But this expression is not convenient when numerical computations have to be done. Indeed, for both the $D_e$ and $N_e$ problems, the integral equation derived from (49) has real eigenvalues, though it has always at least one solution leading to a unique field function.

ORDER NO: N7276311
ORDER DATE: 08/02/84

THEORETICAL ACOUSTICS AND NUMERICAL TECHNIQUES, ED. BY P.FILIPPI. N T A S

INTL. CENTRE FOR MEC

SPRINGER VLG 1983 1 VOLS
23.20 PAPER

0-387-81786-7
N7-276311

QTY ORDERED: 001
QTY SHIPPED: 001
C482139

OF AT SAN ANI 12997000XI XTSA

## II.4. - INTEGRAL REPRESENTATION OF THE FIELD DIFFRACTED BY AN INFINITELY THIN SCREEN.

Let us consider an obstacle having one of its dimensions very small compared to the other ones, and assume that this thickness is small compared to the wavelength.

It is reasonable to guess that, from a mathematical (and even numerical) point of view, this small thickness can be neglected in some way. It can be expected that the corresponding diffracted field is equal, up to a negligible error, to the field diffracted by an idealized screen of zero thickness.

Let $\Gamma_\varepsilon$ be a sequence of non zero thickness regular obstacles such that the limit, for $\varepsilon \to 0$, , of the $\Gamma_\varepsilon$ is a surface $\Gamma$ (or a curve segment in $\mathbb{R}^2$) with a unit normal vector n (see fig.II 1 ). A sequence $\varphi_\varepsilon$ of uniquely determined sound field functions is defined by :

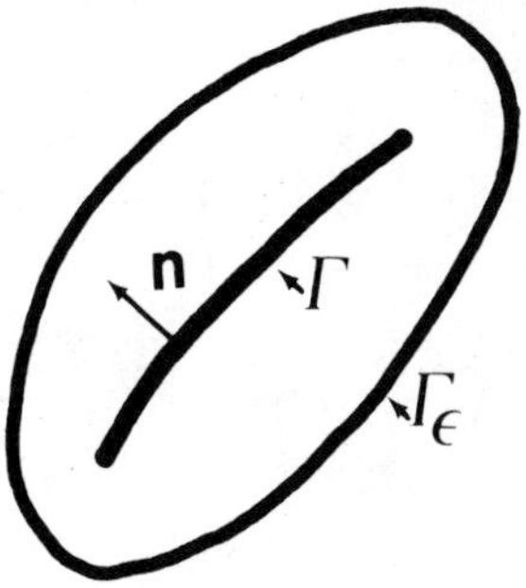


Figure II. 1.

(50) $\begin{cases} (\Delta + k^2)\varphi_\varepsilon = S & \text{in the exterior of } \Gamma_\varepsilon, \\ \text{boundary condition on } \Gamma_\varepsilon, \\ \text{radiation condition.} \end{cases}$

For the seek of simplicity, Dirichlet or Neumann conditions will be considered only. The limit diffraction problem is: find $\varphi$ such that :

(51) $\begin{cases} (\Delta + k^2)\varphi = S & \text{in } \Gamma, \\ \text{Dirichlet or Neumann condition on } \Gamma, \\ \text{radiation condition.} \end{cases}$

The only possible integral representation of $\varphi$ is obviously :

(52) $$\varphi(M) = \langle G(M,P), S(P) \rangle + \langle G(M,P), [\mu \otimes \delta_\Gamma](P) \rangle$$

for the Dirichlet problem

(53) $$\varphi(M) = \langle G(M,P), S(P) \rangle + \langle G(M,P), [\mu \otimes \delta'_\Gamma](P) \rangle$$

for the Neumann problem.

Indeed, the Dirichlet condition implies that $\varphi$ is a continuous function. Consequently, the integral representation of the diffracted field can involve a simple layer potential only. The same kind of remark shows that, for the Neumann problem, a double layer potential is needed. Using equations (52) and (53), the following integral equations are obtained :

$$\int_{\Gamma} \mu(P)\, G(P_0, P)\, dP = - \operatorname{Tr} \varphi_0(P_0) \qquad \forall P_0 \in \Gamma \tag{54}$$

(Dirichlet problem)

$$\operatorname{Tr} \int_{\Gamma} \mu(P)\, \partial_{n(P_0)} \partial_{n(P)} G(P_0, P)\, dP = - \operatorname{Tr} \partial_{n(P_0)} \varphi_0(P_0) \quad \forall P_0 \in \Gamma \tag{55}$$

(Neumann problem).

The uniqueness of the solution of equations (54) or (55) is ensured by using edge conditions, as is done in ref [14]. Another way is to prove that the sequence $\varphi_\varepsilon$ converges to a function $\varphi$ satisfying the system (51). This method has several advantages. First, the infinitely thin screen is considered as the saymptotic limit of a physical screen, the thickness of which is small compared to the wavelength. Second, the functional space, which $\mu$ belongs to, is determined when the convergence of the $\varphi_\varepsilon$ is obtained. Third, the functional spaces in which equations (54) and (55) have one and only one solution are derived from the regularity properties of $\varphi$. These delicate mathematical developments are given in the lectures by M. Durand, and will not be reproduced here. The results are as follows :

*a/- For the Dirichlet problem, equation (54) has a unique solution* $\mu$, *which is a* $C^\infty$*- function everywhere on* $\Gamma$ *but at the edge, and has an integrable singularity at the edge of* $\Gamma$.

*b/- For the Neumann problem, equation (55) has a unique solution* $\mu$, *which is a* $C^\infty$*-function everywhere on* $\Gamma$ *but at the edge, and is zero at the edge of* $\Gamma$.

These properties are quite sufficient to justify the numerical method used in section III.4.

## III. - NUMERICAL SOLUTION OF THE INTEGRAL EQUATIONS ASSOCIATED WITH THE SCALAR HELMHOLTZ EQUATION

The present section is devoted to various examples which have been solved by the so-called collocation method. This is the simplest approximation. The layer density is approximated by a N-steps function . The same kind of approximations made on the second member of the integral equation leads to solve a NxN linear system of algebraîc equations. The accuracy of the results so obtained is quite sufficient as far as the distance between the observation point and the boundary is not too small.

In the preceeding section, three classes of problems have been pointed out : a/ the interior problems for which forced oscillations as well as eigenmodes occur ; b/ the exterior problems for which the associated integral equations can have real eigenwavenumbers ; c/ the diffraction by an infinitely thin screen which can lead to boundary integral equations with a non integrable kernel.

For these three classes of problems, numerical solutions are presented. The results are compared either to analycal solutions, or to experiments.

### III.1. - THE COLLOCATION METHOD SCHEME :

In the preceeding sections, it has been shown that an integral equation of the form

$$K\,\mu = f \qquad (56)$$

can be associated to any boundary value problem of steady waves acoustics. The operator K can be either an integral operator

$$\mu \rightarrow K\,\mu = \int_\Gamma K(P_0, P)\,\mu(P)\,dP \qquad (57)$$

or the sum of such an integral operator and the identity

$$\mu \rightarrow K\,\mu = \mu(P_0) + \int_\Gamma K(P_0, P)\,\mu(P)\,dP\,. \qquad (58)$$

The integrals involved are either Rieman or Cauchy principal value integrals, or must be defined as a limit when non integrable kernels are involved.

Equations of the form (56) are shown to have a unique solution which belongs to a Sobolev space $H^s(\Gamma)$ for any second member f in a Sobolev space $H^{s'}(\Gamma)$ , the difference (s - s') being determined by the operator K . This implies that solving an equation of the form (56) is implicitely a variational problem, and that the variational theories apply.

Whatever the operator K is, the same approximation scheme is adopted. Let devide the boundary $\Gamma$ into N elements $\Gamma_j$ (j = 1, 2, ...N). The center of each element $\Gamma_j$ is $P_j$. The areas (or lengths in $\mathbb{R}^2$) of the $\Gamma_j$ have magnitudes of the same order.

Numerical experiments have shown that the linear dimensions of the $\Gamma_j$ must be less than $\lambda/6$, one sixth of the wavelength.

In the collocation method, the function $\mu$ is approximated by a N-steps function equal to a constant $\mu_j$ on each $\Gamma_j$. Denoting again by $\mu_j$ the function which is zero everywhere on $\Gamma$ but $\Gamma_j$, the first member of equation (56) is approximated by :

$$K\mu(P_0) \cong \sum_{j=1}^{N} K\mu_j(P_0) \tag{59}$$

The second step consists in approximating the second member $f(P_0)$ and the functions $K\mu_j(P_0)$ in the same way :

$$\begin{cases} f(P_0) \cong f(P_i) \quad \text{for} \quad P_0 \in \Gamma_i, \quad i = 1,2,\ldots,N \\ K\mu_j(P_0) \cong K\mu_j(P_i) \quad \text{for} \quad P_0 \in \Gamma_i, \quad i = 1,2,\ldots,N. \end{cases} \tag{60}$$

The equation (56) is so approximated by a linear system of algebraïc equations :

$$\sum_{j=1}^{N} K\mu_j(P_i) = f(P_i), \quad i = 1,2,\ldots,N. \tag{61}$$

But the functions $K\mu_j(P_i)$ involve an integral over $\Gamma_j$ of the kernel function $\tilde{K}(P_i, P)$. In general this integral cannot be calculated analyticaly and must be estimated numericaly. Three techniques are used :

a/ If $i \neq j$, the kernel $\tilde{K}(P_i, P)$ is regular within $\Gamma_j$, and the simplest approximation

$$\mu_j \int_{\Gamma_j} \tilde{K}(P_i, P)\, dP \cong \mu_j K(P_i, P_j) \times \text{area of } \Gamma_j \tag{62}$$

can be adopted.

b/ If $i = j$, because of the singularity of the kernel, $\Gamma_j$ is replaced by a simpler surface, and an analytical integration is performed. For example, in $\mathbb{R}^2$, $\Gamma_j$ is replaced by a disk, tangent to $\Gamma_j$ in $P_j$, and having the same area.

c/ If such approximations are not accurate enough, the element $\Gamma_j$ is first devided into Q subelements $\Gamma_j^q$. Then, on each $\Gamma_j^q$, the former approximations are used. It seems that 4 to 9 subelements are in general sufficient.

## III.2. - EIGENFREQUENCIES AND FORCED OSCILLATIONS OF BOUNDED DOMAINS :

In the two examples here proposed, both eigenfrequencies and forced oscillations are computed. The first domain considered is the interior of a disk ; this allows an analytical solution which the numerical results are compared to. The second domain is the space region between the exterior of a spherical obstacle and the interior of ellipsoïdal surface ; the numerical results are compared to experimental ones.

### *III.2.1. - TWO-DIMENSIONAL INTERIOR PROBLEM : COMPARISON OF THE NUMERICAL SOLUTION TO THE ANALYTICAL ONE [15] :*

Let $\Omega$ be the interior of a disk of radius a and centered in 0 (see fig. III-1).

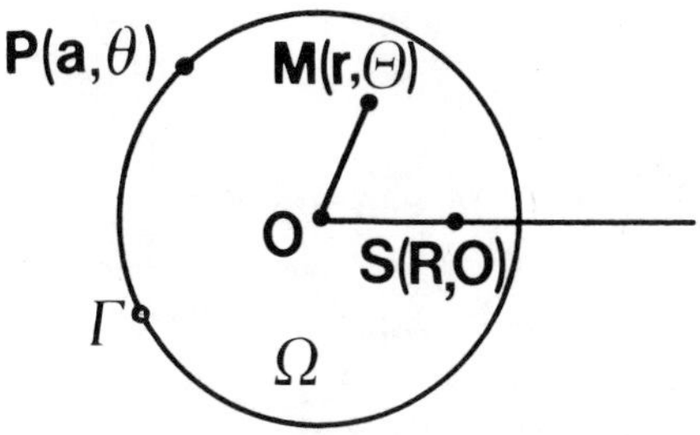


Figure III.1. - Geometry of the plane circular domain.

A point isotropic source is located at S, with cylindrical co-ordinates (R, 0). The co-ordinates of a point $M \in \Omega$ are r and $\Theta$ ; those of a point $P \in \Gamma$ are a and $\theta$. The acoustical field $\varphi$ satisfies the Helmholtz equation :

$$(\Delta + k^2)\,\varphi = \delta_S \qquad \forall M \in \Omega . \tag{63}$$

And a Neumann boundary condition

$$\mathrm{Tr}\,\partial_n \varphi = 0 \qquad \forall P \in \Gamma \tag{64}$$

is assumed, in which **n** is the outer unit normal vector. If the diffracted field is represented by a simple layer potential, one has :

$$\varphi(M) = -\frac{i}{4} H_0\,[k\,d(M,S)] - \frac{i}{4}\int_\Gamma \mu(P)\, H_0[k\,d(M,P)]\,dP \tag{65}$$

where $H_0 = J_0 + i Y_0$ is the Hankel function of the first kind. The corresponding boundary integral equation is :

(66) $$-\frac{\mu(P_0)}{2} - \frac{i}{4}\int_0^{2\Pi} \partial_{r_0} H_0[k\, d(P_0, P)]\, \mu(P)\, a\, d\theta = -\frac{i}{4}\, \partial_{r_0} H_0[k\, d(P_0, S)]$$

$$\forall P_0 \in \Gamma$$

with :

$$P = (a, \theta)\,, \quad P_0 = (r_0 \to a, \theta_0)\,.$$

*a/ Analytical solution :*

A classical calculation shows that $\mu(P)$ can be represented by a Fourier series :

(67) $$\mu(P) = -\sum_{m=-\infty}^{+\infty} \frac{J_m(kR)\, H'_m(ka)}{2\Pi a\, J'_m(ka)\, H_m(ka)}\, e^{im\theta}$$

This yields to a series representation of the diffracted field :

(68) $$\varphi(M) = -\frac{i}{4} H_0[k\, d(M, S)] + \frac{i}{4}\sum_{m=-\infty}^{+\infty} \frac{H'_m(ka)}{J'_m(ka)}\, J_m(kR)\, J_m(kr)\, e^{im\Theta}$$

These series are defined for all values of $k$ but those for which $J'_m(ka)$ is zero. For fixed values of $a$, $R$, and $r$, the series (68), considered as a function of $\Theta$, in convergent for the $L^2(0, 2\Pi)$ norm.

*b/ Numerical solution :*

The circle $\Gamma$ is divided into N equal arcs $\Gamma_j$, with central points $P_j$. The following linear system is solved :

(69-a) $$\sum_{j=1}^{N} A_{ij}\, \mu_j = f_i\,, \quad i = 1,2,\ldots, N$$

(69-b) $$A_{ij\neq i} = -ikH_1(kd_{ij}) \cos(\vec{d}_{ij}, \vec{n}_i)\, \Gamma_j\,, \quad \vec{d}_{ij} = \overrightarrow{P_i P_j}\,, \quad d_{ij} = \| \vec{d}_{ij} \|\,.$$

(69-c) $$A_{ii} = 2 - \frac{i\Pi\Gamma_i}{4a}\{S_0(Z)\, H_1(Z) - H_0(Z)\, S_1(Z)\}_{z=k\Gamma_i/2}$$

(69-d) $$f_i = ikH_1(k\rho_i)\cos(\vec{\rho}_i, \vec{n}_i)\,, \quad \vec{\rho}_i = \overrightarrow{SP_i}\,, \quad \rho_i = \| \vec{\rho}_i \|\,.$$

Equation (69-c) is obtained by analytical integration of the integral equation kernel over the arc $\Gamma_i$ ; and the Struve functions $S_0(Z)$ and $S_1(Z)$ appear [16]. The corresponding approximation of the total field is :

$$\varphi(M) \cong -\frac{i}{4} H_0[k\,d(M,S)] - \frac{i}{4}\sum_{j=1}^{N} \mu_j H_0[k\,d(M,P_j)]\,\Gamma_j . \tag{70}$$

Though the problem has a symetry axis, it has been ignored by the author of ref. [15] to get a more general check of the efficiency of his technique.

*c/ Numerical and analytical results* :

First, the approximate eigenwavenumbers are determined : for that purpose the modulus of the determinant of the matrix $\{A_{ij}\}$ is computed for $0 \leqslant ka \leqslant 10$. The values of k such that a minimum is reached are taken as approximations of the successive first aigenwavenumbers. The number of elements $\Gamma_j$ is N = 20 and N = 40.

The second calculation concerns forced oscillations due to a point source located at the point (R = a/2;0).

Three approximation orders have been used : N = 20, 40 and 80. Results are presented for ka = 4,5 and ka = 18. In each case, the sound field is calculated along the diameters defined by $\theta = 0$, and $\theta = \Pi/6$.

*d/ Discussion of the results* :

Table III.1 shows that the eigenwavenumbers obtained with N = 40 are accurate enough for engineering purposes : indeed, the relative error is less than $4 \cdot 10^{-5}$ for all the first sixteen eigenwavenumbers computed here.

Figures III-2(a) and III-2(b) show that the N = 40 approximation is sufficient for ka = 4.5 . But, for ka = 18, the N = 80 approximation is required to get an equivalent accuracy (figures III-2(c) and III-2(d)). It must be remarked that the accuracy here obtained is higher than that required by physicists : indeed, logarithmic scales used in acoustics allow much stronger relative errors.

### *III.2.2. - THREE DIMENSIONAL INTERIOR PROBLEM : COMPARISON OF THE NUMERICAL RESULTS TO EXPERIMENTAL ONES [17]* :

The domain $\Omega$ is represented in fig. III-3: $\Sigma$ is an oblate spheroidal surface with diameters a and a/2 ; $\sigma$ is a spherical surface with radius 0.4a, and centered at the ellipsoïd center. The Neumann boundary condition is assumed on $\Sigma$ , and on $\sigma$ .

Table III-1.
Exact ($v_+$) and approximate ($v_a$) eigenwavenumbers

| | N = 20 | | N = 40 | |
|---|---|---|---|---|
| $v_+$ | $v_a$ | $\frac{\Delta v}{v}$ | $v_a$ | $\frac{\Delta v}{v}$ |
| 1.84118 | 1.84105 | 7.06 (-5) | 1.841165 | 0.815(-5) |
| 3.05424 | 3.05393 | 1.015(-4) | 3.05419 | 1.64 (-5) |
| 3.83171 | 3.83138 | 8.61 (-5) | 3.83175 | 1.04 (-5) |
| 4.20119 | 4.20052 | 1.57 (-4) | 4.20110 | 2.14 (-5) |
| 5.31755 | 5.31650 | 1.97 (-4) | 5.31738 | 3.13 (-5) |
| 5.33144 | 5.33083 | 1.14 (-4) | 5.33139 | 0.938(-5) |
| 6.41562 | 6.41404 | 2.46 (-4) | 6.41579 | 2.74 (-5) |
| 6.70613 | 6.70516 | 1.44 (-4) | 6.70600 | 1.94 (-5) |
| 7.01559 | 7.01462 | 1.38 (-4) | 7.01542 | 2.42 (-5) |
| 7.50127 | 7.49900 | 3.03 (-4) | 7.50097 | 4.00 (-5) |
| 8.01524 | 8.01379 | 1.81 (-4) | 8.01503 | 2.62 (-5) |
| 8.53632 | 8.53486 | 1.71 (-4) | 8.53609 | 2.69 (-5) |
| 8.57784 | 8.57501 | 3.30 (-4) | 8.57748 | 4.20 (-5) |
| 9.28240 | 9.28053 | 2.01 (-4) | 9.28209 | 3.35 (-5) |
| 9.64742 | 9.64389 | 3.66 (-4) | 9.64695 | 4.87 (-5) |
| 9.96947 | 9.96748 | 2.00 (-4) | 9.96917 | 3.01 (-5) |

Figure III-2. Exact and approximate results for the field in the circular plane domain.

——— Exact solution

Numerical approximations : ● N=20; ✱ N=40; ★ N=80.

$\theta$ is the angle between the diameter passing through the source and the diameter along which the acoustic field is computed.

(a) - ka = 45 ; $\theta = 0; \pi$.

(b) - ka = 45 ; $\theta = \pm\pi/6; \pi \pm \pi/6$.

(c) - ka = 18 ; $\theta = 0; \pi$.

(d) - ka = 18 ; $\theta = \pm\pi/6; \pi \pm \pi/6$.

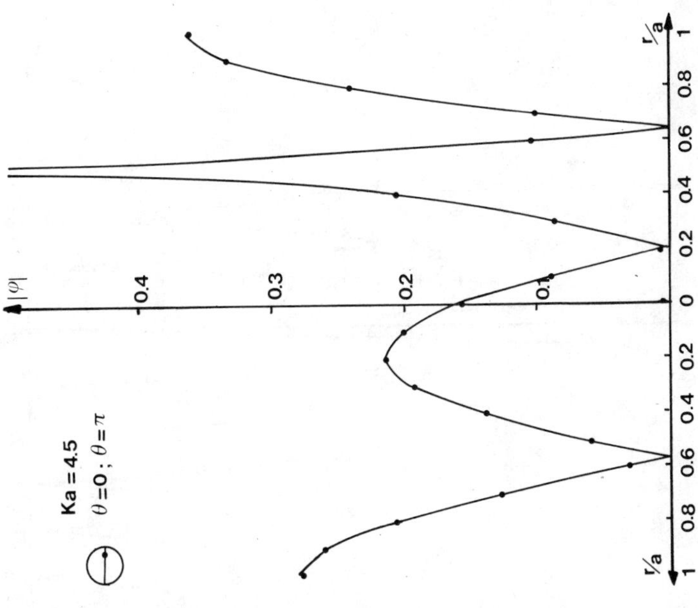


Fig. III - 2 (a)

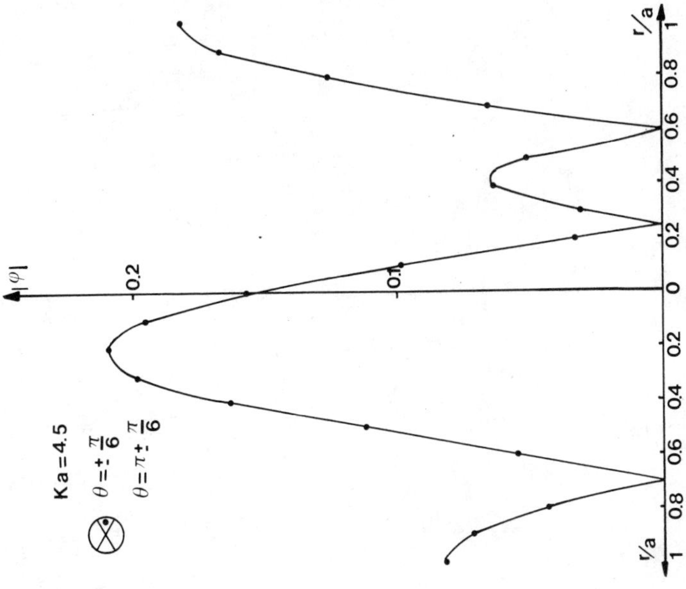


Fig. III - 2 (b)

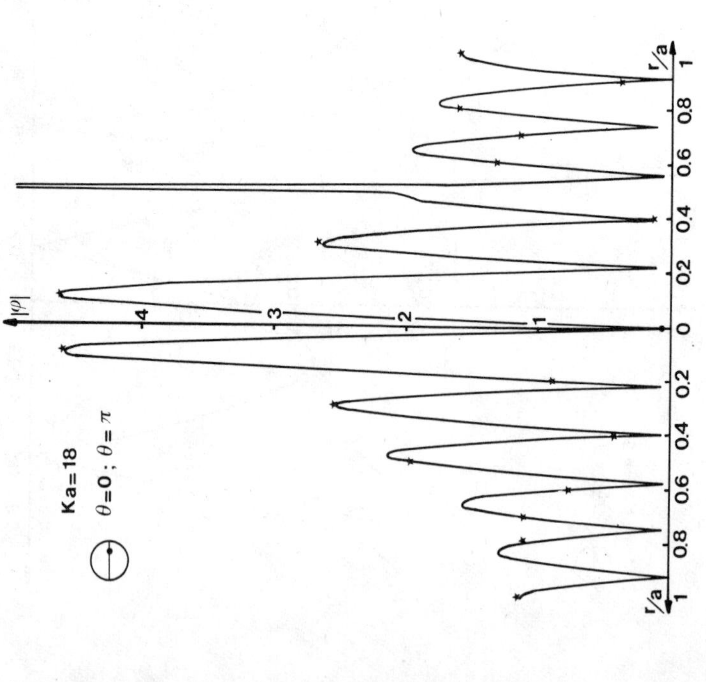


Fig. III - 2 (c)

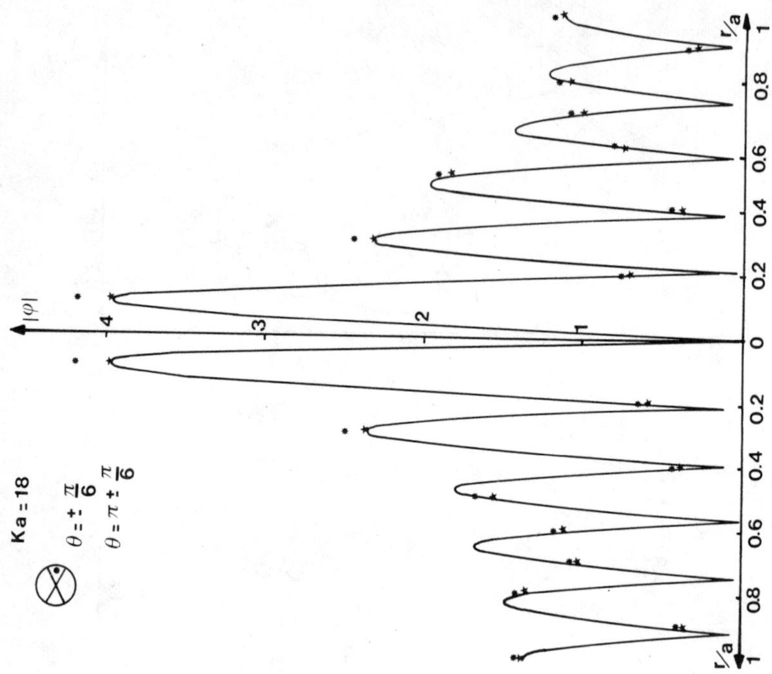


Fig. III - 2 (d)

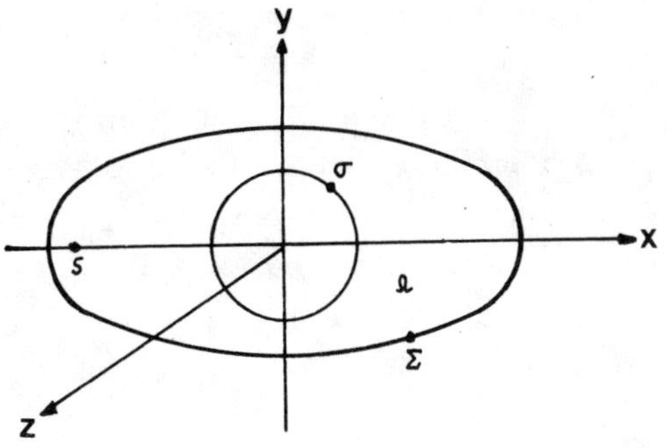


- Figure III-3 - Geometry of the multiply connected domain.

A point isotropic source is located at one focus S of the ellipsoïd. For such a problem, no analytical solution is known.

The diffracted field is described by a simple layer potential, with density $\mu$ on $\Sigma$, and $\nu$ on $\sigma$. These densities are solution of a system of two integral equations, namely :

(71-a)
$$-\frac{\mu(P)}{2} + \int_\Sigma \mu(P')\,\partial_{n(P)} G(P,P')\,dP' + \int_\sigma \nu(Q')\,\partial_{n(P)} G(P,Q')\,dQ' = = -\,\partial_{n(P)}\,\varphi_0(P)\,, \qquad \forall P \in \Sigma$$

(71-b)
$$\int_\Sigma \mu(P')\,\partial_{n(Q)} G(Q,P'/dP' - \frac{\nu(Q)}{2} + \int_\sigma \nu(Q')\,\partial_{n(Q)} G(Q,Q')\,dQ' = = -\,\partial_{n(Q)}\,\varphi_0(Q)\,, \qquad \forall Q \in \sigma$$

where :
$$G(M,P) = -\frac{e^{ikr(M,P)}}{4\Pi r(M,P)}\,, \quad \varphi_0(M) = -\frac{e^{ikr(S,M)}}{4\Pi r(S,M)}\,;$$

n(P) and n(Q) are unit outgoing normal vectors.

The geometrical data have an axis of symmetry. Thus, the functions $\mu$ and $\nu$ depend on the x co-ordinate only ; this simplification is taken into account. The surface $\Sigma$ is divided into 80 annular elements $\Sigma_i$ on each of which $\mu$ is approximated by a constant $\mu_i$ (i = 1,2, ...80).

The surface $\sigma$ is divided into 32 annular elements $\sigma_i$ and $\nu$ is approximated by a set of constants $\nu_j$ (j = 1, 2, ...32). Each annular surface element $\Sigma_i$ and $\sigma_i$ is divided into 80 subelements $\Sigma_{ik}$ and $\sigma_{ik}$, and the following approximation is adopted in equation (71-a) :

$$\int_\Sigma \mu(P') \partial_{n(P_i)} G(P_i, P') dP' \cong \sum_{\substack{j \neq i \\ j=1}}^{80} \mu_j \sum_{k=1}^{80} \partial_{n(P_i)} G(P_i, P'_{jk}) \Sigma_{jk} + A_{ii} \mu_i$$

$$A_{ii} = \sum_{k=2}^{80} \partial_{n(P_i)} G(P_i, P'_{jk}) \Sigma_{jk} + \partial_{n(P_i)} G(P_i, P_{i1}) \Sigma_{i1}$$

$\Sigma_{jk}$(j,k = 1,2,...80) = areas of the 80 sub-elements of the 80 annular elements of $\Sigma_j$

$P_i$ (i = 1, 2, ...80) = center of the sub-element (i, 1)

$P_{i1}$ (i = 1, 2, ...80) = center of the commun limit arc of $\Sigma_{i1}$ and $\Sigma_{i2}$

$P'_{jk}$(j, k = 1, 2...80) = center of the subelement (j,k)

$$\int_\sigma \nu(Q') \partial_{n(P_i)} G(P_i, Q') dQ' \cong \sum_{\ell=1}^{80} \nu_j \sum_{j=1}^{32} \partial_{n(P_i)} G(P_i), Q'_{\ell j}) \sigma_{\ell j}$$

$\sigma_{j\ell}$ ($\ell$ = 1, 2...,80 ; j = 1, 2,...32) = areas of the 80 subelements of the 32 annular elements of $\sigma$ ;

$Q'_{\ell j}$ ($\ell$ = 1, 2, ..., 80 ; j = 1, 2,..., 32) = center of the subelement $\sigma_{j\ell}$ .

A similar approximation is used for equation (71-b).

Here again, the eigenfrequencies of the domain and forced oscillations are computed. The results are compared to a modelexperiment in which a = 1 m.

Table III-2 shows that, for the first twelve eigenfrequencies, the relative error does not exceed 2%. In figures III-4 (a), (b), and(c), the computed values (circles) are plotted on continuous recording of the sound level observed along the axis of symmetry. These results show that, despite the complexity of the geometrical data, the method provides a correct prediction of the sound level distribution.

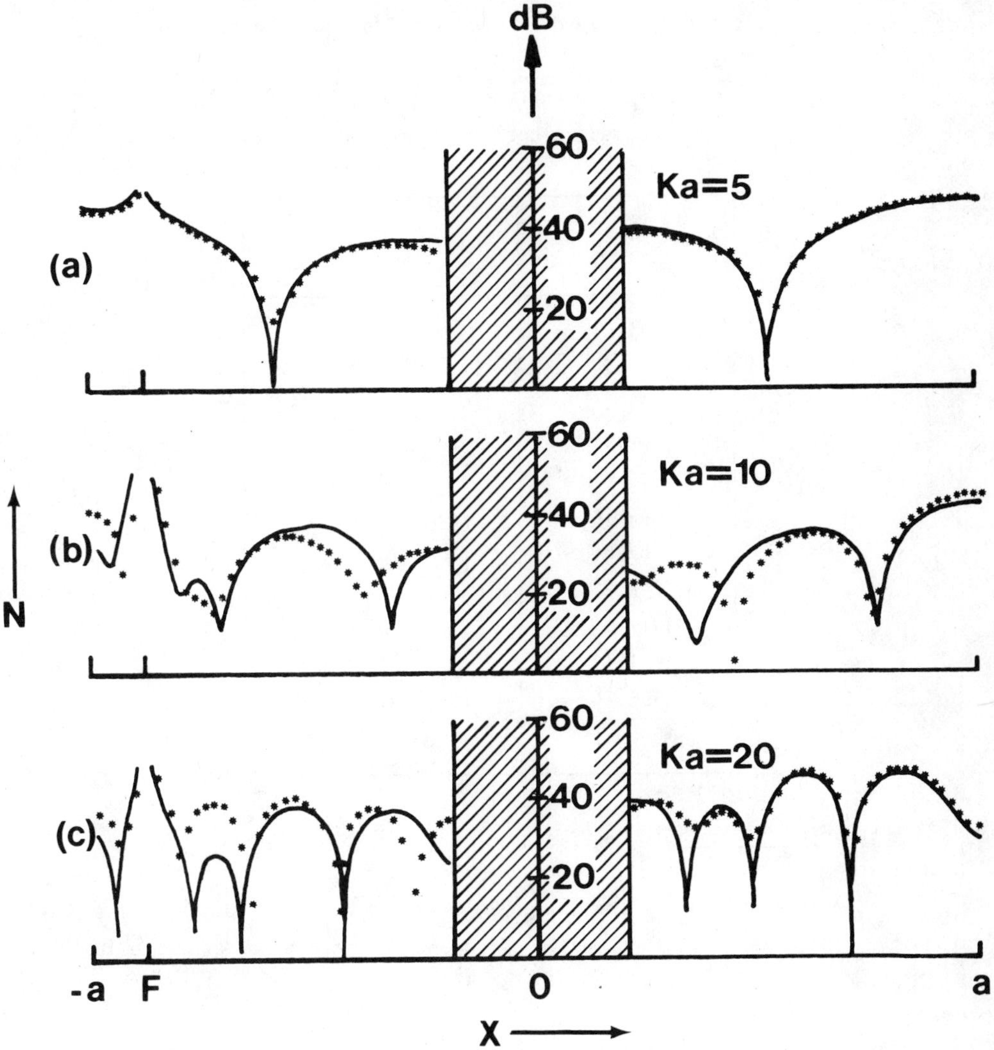


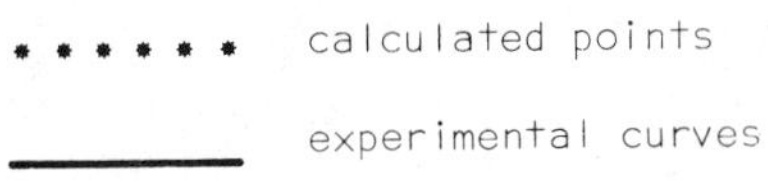


Figure III-4. Sound field within the ellipsoidal room containing a sphere.

Table III-2
Eigenfrequencies of the ellipsoïdal room containing a sphere.

$v_c$:computed values $v_m$:measured values

| $v_c$ (Hz) | $v_m$ (Hz) |
|---|---|
| 111.0 | 113 |
| 201.2 | 204 |
| 277.1 | 279 |
| 364.9 | 364 |
| 438.5 | 441 |
| 642.1 | 640 |
| 782.6 | 784 |
| 849.2 | 854 |
| 912.6 | 907 |
| 931.3 | 933 |
| 996.1 | 990 |
| 1040.6 | 1033 |

III.3. - EXTERIOR PROBLEMS :

The present results are due to Bolomey and Tabbara [18, 19] . The main interest of their paper is to show the efficiency of the use os a simple layer - double layer combination with a complex coefficient to describe the field diffracted by an obstacle.

Let consider, in the space $\mathbb{R}^2$ a plane incident wave $\varphi_0(M) = e^{ikx}$ which is diffracted by a cylindrical obstacle $\Gamma$ of radius a . On $\Gamma$ , the total field $\varphi(M)$ satisfies a Dirichlet condition. The Green representation of $\varphi(M)$

(72) $$\varphi(M) = \varphi_0(M) - \int_\Gamma \mathrm{Tr}\,\partial_n(P')\, G(M,P')\, dP'$$

leads to the integral equation :

(73) $$\frac{\mathrm{Tr}\,\partial_n\,\varphi(P)}{2} + \int_\Gamma \mathrm{Tt}\,\partial_n(P')\,\partial_{n(P)}\, G(P,P')\, dP' = \mathrm{Tr}\,\partial_n\,\varphi_0(P)\ ,\forall P \in \Gamma .$$

If the wave number k is choosen equal to an eigenwavenumber of the interior Dirichlet problem, equation (73) has a non-unique solution. The uniqueness of the solution can be theoretically obtained by adding the condition that expression (72) is identically zero in in the interior of $\Gamma$ (property of the Green representation) This method has been proposed by Schenck [20], and is used by Bolomey and Tabbara to get a numerical approximation of $\varphi(M)$ based on the collocation method.

Then these authors seek the total field in the following form :

$$\varphi(M) = \varphi_0(M) + \int_\Gamma \mu(P') \left[\partial_{n(P')} G(M,P') - iG(M,P')\right] dP' . \tag{74}$$

The corresponding integral equation is :

$$-\frac{\mu(P)}{2} + \int_\Gamma \mu(P') \left[\partial_{n(P')} G(P,P') - iG(P,P')\right] dP' = -\varphi_0(P), \ \forall P \in \Gamma, \tag{75}$$

and it is solved by the collocation method.

In figure III-5, the modulus of the determinant of the algebraïc linear system approximating the integral equation is shown for both methods : for the linear system derived from equation (73), the derterminant presents a sequence of minima, while that derived from equation (74) is a monotonic function ot the wave-number.

The total field is then computed for the eigenwavenumber k = 3.8317/a, and the result obtained is compared to the exact Bessel series solution.

It appears, in figure III-6 that the mixed layer potential representation coïncides with the exact solution within an error less than the drawing error.

Conversely, the Green representation gives a very poor approximation of the solution : the error in the vicinity of the obstacle can be larger than 10 dB.

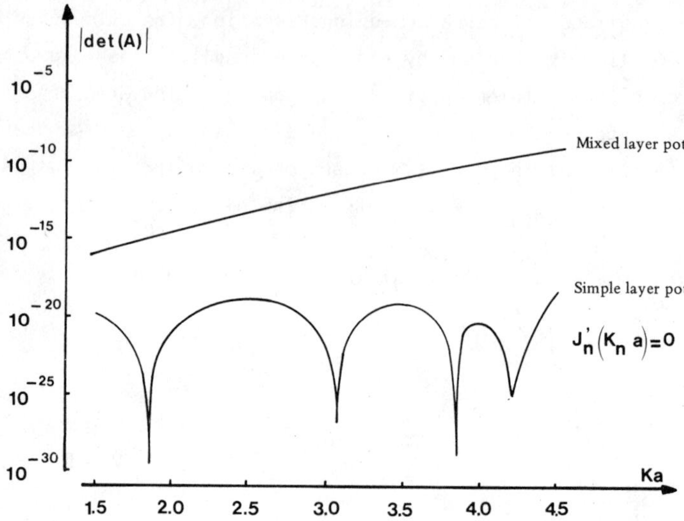


Fig. III - 5. Comparison of the moduli of the determinants of the algebraic systems approximating equations (73) and (74) for, respectively, a simple layer potential and a mixed layer potential.

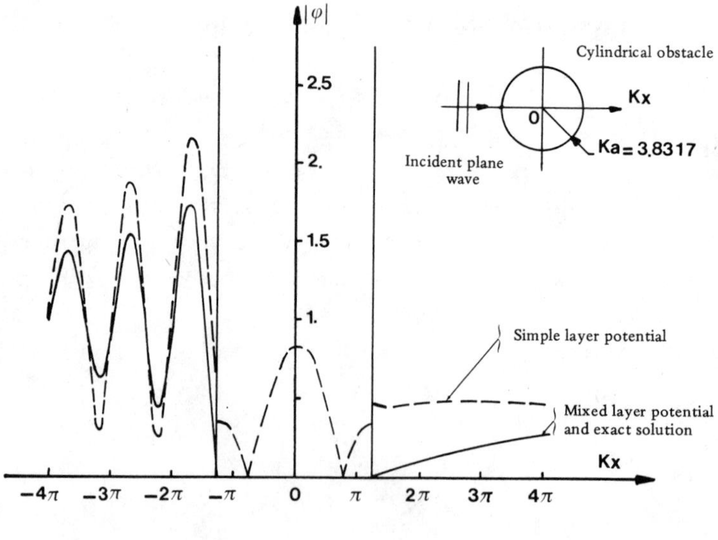


Fig. III - 6. Comparison of the simple layer (— — —) and the mixed layer (——) approximate field of the soft cylindrical obstacle.

## III.4. - DIFFRACTION BY AN INFINITELY THIN AND PERFECTLY HARD SCREEN :

In section II- 4 , it has been pointed out that the only possible integral representation of the sound field by an infinitely thin and perfectly hard screen is provided by a double layer potential. The corresponding integral equation involves a non-integrable kernel. The principal aim if this section is to show that an elementary technique, as the collocation method, enables the computation of numerical approximations.

Three examples are detailed. The first one deals with the diffraction of a plane wave by an infinite strip (two dimensional problem) : the layer density is computed and compared to its exact value obtained by sepration of the variables. In the second example, the incident field is that due to a point isotropic source and the screen is rectangular : the numerical results are compared to field measurements. The aim of the third example is twofold : the effect of an infinite hard strip normal to an infinite hard plane, in presence of line source is examined ; in the first part, the integral equation results are compared to the classical approximations (roughly, the first term of the geometrical theory of diffraction) ; in the second part, it is shown that this simple model can predict correctly the excess attenuation due to a long screen in presence of a line of point sources.

### *III.4.1. - DIFFRACTION OF A PLANE WAVE BY AN INFINITE STRIP [21] :*

Let consider a plane wave $\varphi_0 = e^{iky}$ impinging on an infinite plane strip located in the plane Y = 0, and extending between X = -a and X =+a. The diffracted field being described by a double layer potential, one has to solve the following integral equation :

$$(76)\qquad \mathrm{Tr} \int_{-a}^{+a} \mu(X') \frac{\partial^2 G}{\partial Y \partial Y'} (X, X')\, dX' = -ik, \quad -a \leqslant X \leqslant +a$$

with :

$$\frac{\partial^2 G}{\partial Y \partial Y'} (X, X') = \frac{\partial^2}{\partial Y \partial Y'} \left\{ -\frac{i}{4} H_0 \left[ k \sqrt{(X - X')^2 + (Y - Y')^2} \right] \right\}_{Y'=0}$$

Because of the symmetry of the data with respect to the X = 0 axis, the layer density $\mu(X)$ is an even function.

The interval $[0,a]$ is divided into N segments, the extremity of which are $\{(j-1)a/N, ja/N ; j = 1,2,...N.\}$.

Along each segment, the density $\mu$ is approximated by a constant $\mu_j$ and the integral equation is replaced by a linear system :

(77) $$\sum_{j=1}^{N} \mu_j(A_{jn} + B_{jn}) = - ik , \; n = 1,2, ..., N .$$

with :

$$A_{jn} = \int_{-(j-1)a/N}^{-ja/N} \frac{\partial^2 G}{\partial Y \partial Y'} [(n-\frac{1}{2}) a/N , X'] dX'$$

$$B_{j\neq n n} = \int_{(j-1)a/N}^{ja/N} \frac{\partial^2 G}{\partial Y \partial Y'} [(n-\frac{1}{2}) a/N , X'] dX'$$

$$B_{nn} = \lim_{Y \to 0} \int_{(N-1)a/N}^{na/N} \frac{\partial^2 G}{\partial Y \partial Y'} [M_n , X'] dX'$$

$$M_n = \{(n - 1/2) a/N , Y\}$$

The terms $A_{jn}$ and $B_{j\neq nn}$ are convergent intergals. They are evaluated as follows : the function $\partial_{y'} H_0[k\sqrt{(X_n - X')^2 + (Y - Y')^2}]$ is expanded in terms of cylindrical harmonics centered at $P_j(X_j = (j-1/2)a/N)$, differentiated term by term with respect to $Y$ ; and then, letting $Y = 0$, the first three terms of the resulting series are intergated analitically. To evaluate the coefficient $B_{nn}$, the ascending series of the Hankel function is used : an analytical integration of the first terms is made, and then the limit for $Y \to 0$ is taken.

The comparison between the numerical values $\mu_j$ and the exact function $\mu(X)$ is presented on figures III-7 and III-8, for the two values N = 10 and N = 100. The agreement is quite satisfactory.

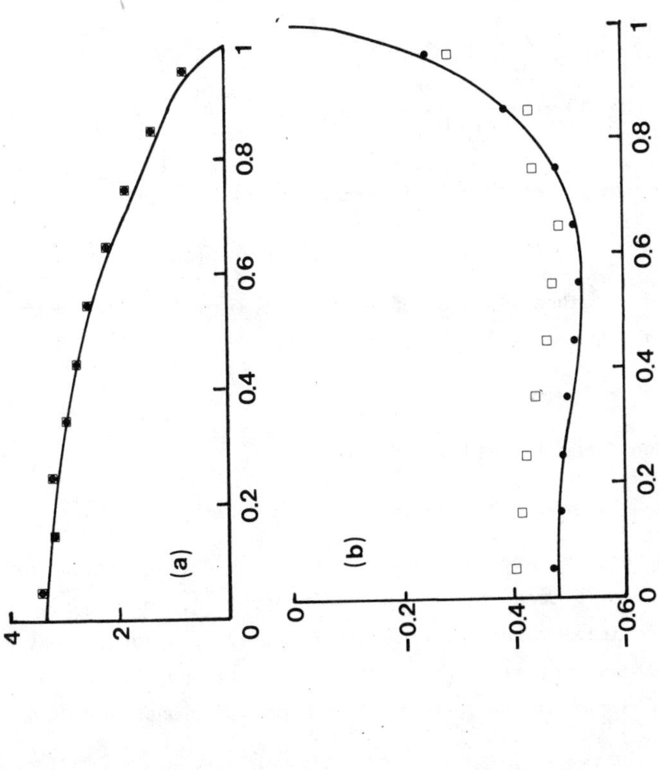


Fig. III - 7. Diffraction by a strip, ka = 2.
(a) Real part of $\mu$
(b) Imaginary part of $\mu$
□ 10 equations
● 100 equations

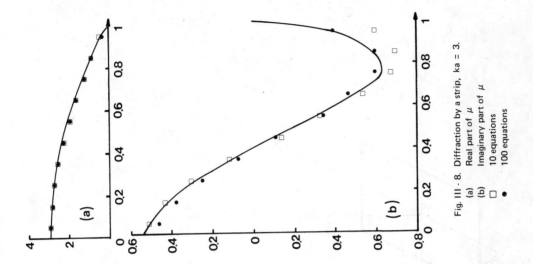

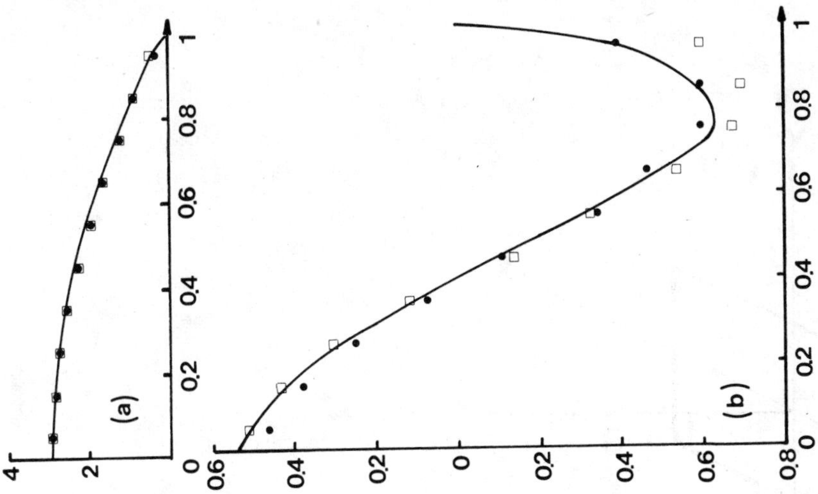


Fig. III - 8. Diffraction by a strip, ka = 3.
(a) Real part of $\mu$
(b) Imaginary part of $\mu$
□ 10 equations
● 100 equations

### III.4.2. - *DIFFRACTION OF A SPERICAL WAVE BY A RECTANGULAR HARD SCREEN [22, 23] :*

Let $\Sigma$ be a plane rectangular screen of dimensions 2a and 2b, lying in the $Z = 0$ plane, as shown on figure III-9 . A point isotropic source S is located on the Z-axis. Assuming a Neumann condition, the double layer potential representation of the diffracted field leads to the integral equation:

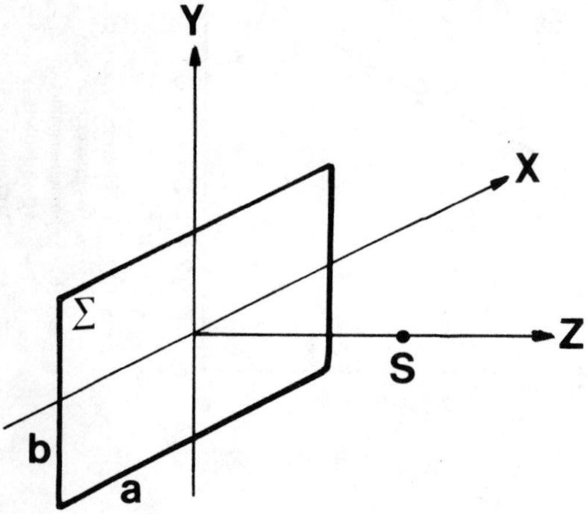


- Figure III-9. Geometry of the rectangular screen.

(78)
$$\mathrm{Tr} \int_{\Sigma} \mu(P') \frac{\partial^2 G(P,P')}{\partial n(P)\,\partial n(P')} dP' = -\mathrm{Tr}\, \partial_{n(P)} \varphi_0(P) \,, \quad \forall P \in \Sigma$$

with :

$$G(M,M') = -\frac{e^{ikr(M,M')}}{4\Pi r(M,M')} \,, \quad \varphi_0(M) = -\frac{e^{ikr(M,S)}}{4\Pi r(M,S)} \,.$$

The symmetry of the geometrical data is introduced and one has :

$$\mu(P) = \mu(X,Y) = \mu(\pm X, \pm Y)$$

To use the collocation method, the screen is divided into identical rectangular elements of area $\sigma$ , and centered at the points :

$$X_r = (r - m - 0.5)\, a/m \,, \quad Y_S = (s - n - 0.5)\, b/n \,.$$

These elements are limited by the lines :

$$X = X_r \pm 0.5\, a/m \,, \quad Y = Y_s \pm 0.5\, b/n \,.$$

The parameters m and n are integers which define the approximation order, and the integers s and r vary in the intervals $[1, 2m]$ and $[1, 2n]$ respectively. So, the total number of elements is 4 mn. By an elementary transformation, the double index (r, s) is replaced by a single one $1 \leqslant i \leqslant 4$ mn. To construct the approximation matrix, two kind of approximations are used :

Figures III-10. Diffraction by a rectangular screen. These figures show the sound level as measured (————) and calculated (••••••••) along a circle, of radius 0.35 m, centered at the point source, and lying in the (x,z) plane.

The screen has the following dimensions :

a = 0.15 m ; b = 0.10 m ; the source screen distance is 0.30 m.

The wavenumbers k correspond to the following wavelenghts :

(a) 0.314 m ; (b) 0.157 m ; (c) 0.105 m.

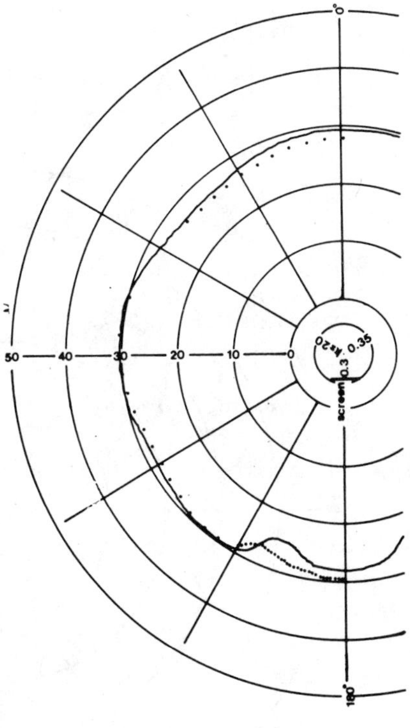


Fig. III - 10 (a)

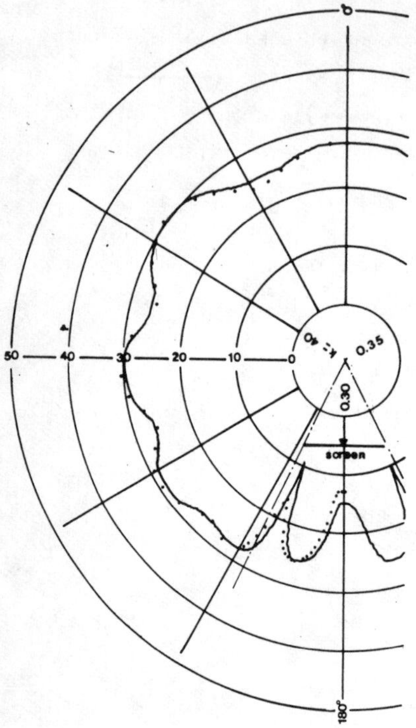


Fig. III - 10 (b)

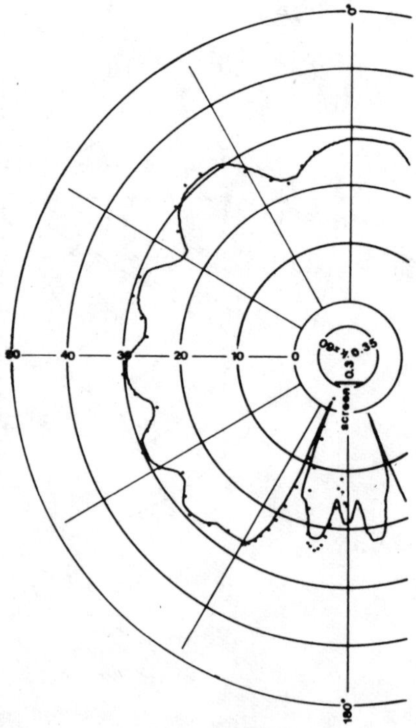


Fig. III - 10 (c)

a/ non diagonal terms $A_{ij}$ : the surface element is divided into 9 equal subelements, and one adopts the approximation

$$\hat{A}_{ij=i} = \frac{1}{9} \sum_{\ell=1}^{9} \sigma \left. \frac{\partial^2 G(M_i, M_{j\ell})}{\partial z\, \partial z'} \right|_{z=z'=0} , \quad \sigma = \frac{1}{9}\frac{ab}{mn} ,$$

where $M_j$ is the center of the $\ell$th subelements of $\sigma_j$ .

b/ diagonal terms $A_{ii}$ : the surface element $\sigma_j$ is replaced by a disk having the same area, and the integral can thus be calculated analytically.

From the $\mu_i$ so obtained, the diffracted field is approximated by :

$$\varphi_d(M) \cong \sum_{i=1}^{4mn} \mu_i \frac{ab}{mn} \left. \frac{\partial G(M, M_i)}{\partial Z'} \right|_{z'=0}$$

where $M_i$ is the point with coordinates $(X_i, Y_i, Z')$.

In figures III-11 (a), (b), and (c), the numerical results are plotted against the recorded experimental curves. It is obvious that the accuracy of the numerical method is good enough for engineering purposes.

### *III.4.3. - EXCESS ATTENUATION OF A NOISE BARRIER :*

Since several years, the noise pollution due to trafic ways has become an important problem.

One of the most commonly used protection is to build up a wall along the way. The most classical method to predict the effect of such a barrier is due to Maekawa [24] .

It is interesting to compare the results obtained by this simple method to those given by the use of an integral representation of the diffracted field ; this is done by Daumas in ref [25] .

Let us consider the two-dimensional problems shown on fig. III-11.

- Figure III-11 -

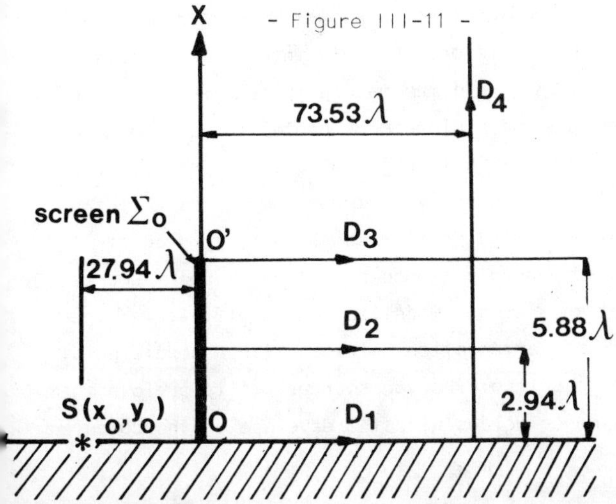


A perfectly hard screen $\Sigma$ is located along the segments $[X = Y = 0 ; X = 0 , Y = h]$ and the plane $Y = 0$ is assumed perfectly hard, too. A point isotropic source is located at $S(X_0 < 0, Y_0 \geqslant 0)$. The resulting sound field $\varphi(M)$ is the restriction to the half-plane $Y > 0$ of the field due to two sources $S(X_0, Y_0)$ and $S'(X_0, -Y_0)$ in presence of a linear screen extending from the point $(X = 0, Y = -h)$ to the point $(X = 0, Y = +h)$.

Let define $\Phi_1$ (M) as the sound field due to the sources S and S' in presence of the perfectly hard half plane ( X = 0, Y $\leqslant$ h) ; and $\Phi_2$ (M), the sound field due to the same sources, but in presence of the perfectly hard half plane ( X = 0, Y $\geqslant$ -h). Let now $\varphi_1$ (M) and $\varphi_2$ (M) be the far high frequency behaviour of $\Phi_1$ and $\Phi_2$ . A classical approximation of $\varphi$ (M) is given by :

$$\varphi(M) \cong \varphi_1(M) + \varphi_2(M) .$$

The quantity of interest for engineering is the excess attenuation defined by :

$$\Delta L = 20 \, Lg \, | \varphi(M) / 2 \varphi_0(M) |$$

where $\varphi_0$ (M) is the incident field.

a/ *Comparison between the classical approximation and the layer potential solution :*

The author of ref [25] solves the integral equation by the collocation method.

The number of points is increased untill the difference between two calculated curves becomes non significant (less than 0.1 dB). On figures III-12 (a), (b) (c), (d), the solid lines represent the layer potential solution which can be considered as "exact", and the dashed lines correspond to the Maekawa's approximation.

The examination of these curves show that the classical approximation provides an optimistic prediction of the screen efficiency.

b/ *The two-dimensional problem as a simple modelization of a long screen :*

From the point of view of environmental noise reduction, it is meaningless to describe the screen effect with precision. The only interesting result is to know wether the envelope of the sound field maxima are lowered when a screen is built up between the noise sources and the region to be protected. Furthermore, in many practical cases , there is a possibility to raise a long screen which can be considered as almost straight. It is consequently, interesting to check if the results obtained on a two-dimensional idealized problem can be used to predict the excess attenuation due to a long screen in presence of a line of point sources. Figures III-12 (a), (b) (c) show that this simple modelization is efficient : the circles represent the experimental excess attenuation obtained with a single point source, while the crosses correspond to a line of seven sources. It is obvious that the envelope of the experimental maxima coïncides with the envelope of the computed curves within 0.2 dB.

Figures III-12 - Screen efficiency

—— Layer potential method

- - - - Classical approximations

●●●● Experimental results for a single point source

★★★ Experimental results for seven point sources

***** Overall efficiency.

a - Screen efficiency along $D_1$

b - Screen efficiency along $D_2$

c - Screen efficiency along $D_3$

d - Screen efficiency along $D_4$.

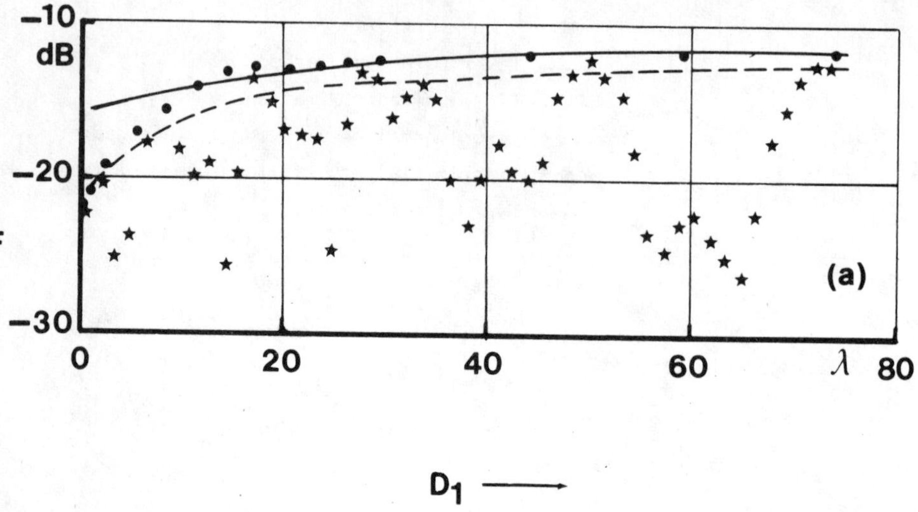


Fig. III - 12 (a)

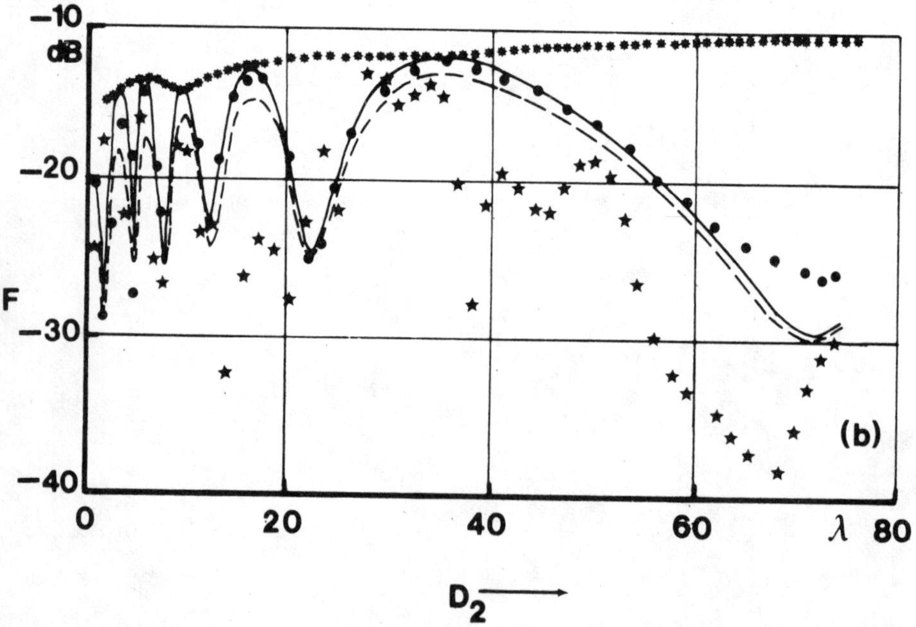


Fig. III - 12 (b)

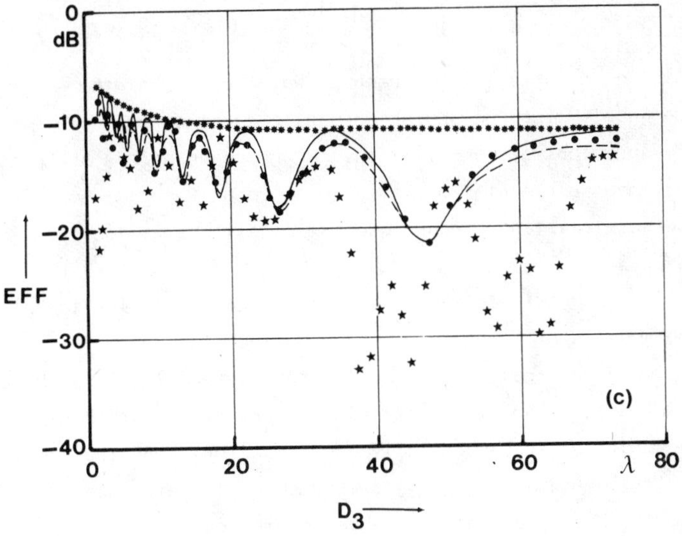


Fig. III - 12 (c)

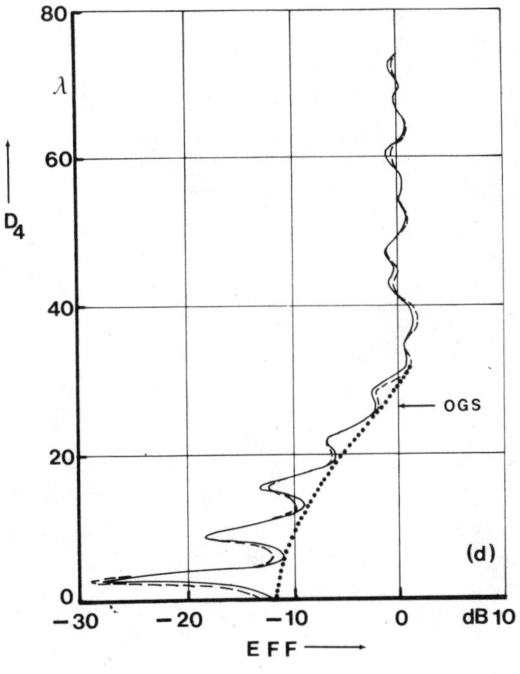


Fig. III - 12 (d)

## IV. - CONCLUSION

In these lectures, it has been shown, first, that the solution of a boundary value problem for the scalar Helmholtz equation can always be represented in terms of layer potentials and that several representations are possible. There exists at least one representation which leads to an integral equation which is subject to the same existence and uniqueness conditions as those of the differential system.

The second result is that the simplest approximation technique, namely the collocation method, is efficient enough for engineering needs. Nevertheless, more sophisticated techniques can provide better results and less expensive computation times. Finite elemnts methods, are certainly more powerfull (in particular, the regularity of the solution can be partially accounted for), and the study of well-adapted approximating functions, as those proposed is reference [26] , will certainly improve the efficiency of integral equation methods.

The final example (given in section III.4.3) shows clearly that for many problems in environmental acoustics, a very simplified mathematical modelization can provide, for a very low cost, the informations needed by engineers.

REFERENCES


1. L. HORMANDER 1968 *American Mathematical Society Proceedings of Symposia in Pure and Applied Mathematics* 10, 138-183. Pseudo-differential and hypoelliptic equations.
2. R. SEELEY 1969 in *C.I.M.E., Pseudo-differential operators, Stresa, 26 Agosto-3 Settembre*. Rome : Edizzioni Cremonese, Topics in pseudo-differential operators.
3. L. BOUTET DE MONVEL 1966 *Journal d'Analyse Mathématique, Jérusalem* 17, 241-304. Comportement d'un opérateur pseudodifférentiel sur une variété à bord.
4. L. BOUTET DE MONVEL 1971 *Acta Mathematica* 126, 11-51. Boundary problems for pseudo-differential operators.
5. S.G. MIKHLIN 1965 *Multidimensional Singular Integral Equations*. Oxford : Pergamon Press.
6. L. SCHWARTZ 1966 *Théorie des distributions*. Paris : Hermann.
7. L. LANDAU and E. LIFCHITZ 1971 *Mécanique des fluides*. Moscow : Editions Mir.
8. I. VEKUA 1968 *New Method for Solving Elliptic Equations*. Amsterdam : North Holland Publishing Company, New York : John Wiley and Sons Inc.
9. B.R. VAINBERG 1966 *Russian Mathematical Surveys* 21, 115-193. Principles of radiation, limit absorption and limit amplitude in the general theory of partial differential equations.
10. C. H. WILCOX 1975 *Lecture Notes in Mathematics : Scattering Theory for the d'Alembert Equation in Exterior Domains*. Berlin-Heidelberg-New York : Springer Verlag.
11. P. FILIPPI 1979 *Journal de Mécanique* 18 (3) 565-591. Problème de transmission pour l'équation de Helmholtz scalaire et problèmes aux limites équivalents : application à la transmission gaz parfait - milieux poreux.
12. M.N. SAYHI, Y. OUSSET and G. VERCHERY 1981 *Journal of Sound and Vibration* 74 (2), 187-204. Solutions of radiation problems by collocation of integral formulations in terms of single and double layer potentials.
13. J.C. NEDELEC 1975 *Ecole Polytechnique, Centre de Mathématiques Appliqués*. Curve finite element methods for the solution of singular integral equation of surfaces in $R^3$.
14. Y. HAYASHI 1973 *Journal of Mathematical Analysis and Applications* 44, 489-530. The Dirichlet problem for the two-dimensional Helmholtz equation for an open boundary.
15. F. CASSOT and G. EXTREMET 1972 *Acustica* 27, 238-245. Détermination numérique du champ sonore et des fréquences propres dans une enceinte circulaire par la méthode de discrétisation.
16. M. ABRAMOVITCH and L.A. STEGUN 1970 *Handbook of Mathematical Tables*. Washington, D.C. : National Bureau of Standards.
17. G. EXTREMET 1970 *Acustica* 23, 307-314. Propagation du son dans une enceinte fermée.
18. CH. BOLOMEY and W. TABBARA 1971 *Journées sur l'Application des potentiels de couches à la mécanique et à la diffraction, Centre de Recherches Physiques de Marseille* (now L.M.A.) , *2 november 1971, note n° 1218*. Sur le couplage entre problèmes complémentaires pour l'équation des ondes.
19. CH. BOLOMEY and W. TABBARA 1973 *Institution of Electrical Engineers Transactions* AP21, 356-363. Numerical aspects of coupling between complementary boundary value problems.
20. H.A. SCHENCK 1968 *Journal of the Acoustical Society of America*, 44, 41-58. Improved integral formulation for acoustic radiation problems.
21. P. FILIPPI and G. DUMERY 1969 *Acustica* 21, 343-350. Etude théorique et numérique de la diffraction par un écran mince.

22. F. CASSOT 1971 *Thèse de spécialité en Acoustique, Marseille 29 octobre 1971.* Contribution à l'étude de la diffraction par un écran mince. (See also Proceedings of the 7th International Congress on Acoustics, Budapest, 1971).
23. F. CASSOT 1975 *Acustica* 34, 64-71. Contribution à l'étude de la diffraction par un écran mince.
24. Z. MAEKAWA 1965 *Mémoires of the Faculty of Engineering, Kobe University* 11 (29). Noise reduction by screens.
25. A. DAUMAS 1978, *Acustica* 40 (4) 213-222. Etude de la diffraction par un écran mince disposé sur le sol.
26. M. SELVA 1977, *Thèse de spécialité en Analyse Numérique, Université de Provence, Marseille.* Sur une équation à noyau singulier issue de la théorie de la diffraction.

COMPLEMENTARY BIBLIOGRAPHY

1. H. LEVINE and J. SCHWINGER 1948, *Physical Review* 74, 958-974. On the theory of diffraction by an aperture in an infinite plane screen, I.
2. H. LEVINE and J. SCHWINGER 1949, *Physical Review* 75, 1423-1432. On the theory of diffraction by an aperture in an infinite plane screen, II.
3. C. J. BOUWKAMP 1954 *Reports on Progress in Physics* 17, 35-100. Diffraction theory.
4. C. MIRANDA 1955 *Equazioni alle derivate parziali di tipo ellitico.* Berlin : Springer.
5. L.I. MUSHKHELISHVILI 1958 *Singular Integral Equations.* Groningen : P. Noordhoff N.V.
6. P. WERNER 1962 *Archives of Rational Mechanics and Analysis* 10, 29-66. Randwertprobleme der mathematischen Akustik.
7. V.D. KUPRADZE 1965 *Potential Methods in Theory of Elasticity.* Jerusalem : Israël Program for Scientific Translations.
8. D. GREENSPAN 1966 *Archives of Rational Mechanics and Analysis,* 23, 288-316. A numerical methof for the exterior Dirichlet problem for the reduced wave equation.
9. T.S. LUU, G. COULMY and J. CORNIGLION 1969 *Association Technique Maritime et Aéronautique (Paris).* Technique des effets élémentaires dans la résolution des problèmes d'hydro-et d'aérodynamique.
10. G.F. ROACH 1970 *Archives of Rational Mechanics and Analysis* 36, 79-88. Approximate Green's function and the solution of related integral equations.
11. T.S. LUU, 1970 *Association Technique Maritime et Aéronautique (Paris).* Calcul de l'hélice marine subcavitante par la méthode de singularité.
12. T.S. LUU, G. COULMY and J. CORNIGLION 1971 *Association Technique, Maritime et Aéronautique.* Etude des écoulements instationnaires des aubes passantes par une théorie non linéaire.
13. T.S. LUU, G. COULMY and J. CORNIGLION 1971 *Association Technique, Maritime et Aéronautique.* Calcul non linéaire de l'écoulement à potentiel autour d'une aile d'envergure finie de forme arbitraire.
14. A.J. BURTON and G.F. MILLER 1971 *Proceedings of the Royal Society, London,* A323, 201-210. The application of integral equation methods to the numerical solution of some exterior boundary problems.
15. J. VIVOLI 1972 *Thèse, Marseille No.A.O. 7868.* Vibrations des plaques et potentiels de couches.
16. D.S. JONES 1972 *Journal of Sound and Vibration* 20, 71-78. Diffraction theory : a brief introductory review.
17. J. VIVOLI and P. FILIPPI 1974 *Journal of the Acoustical Society of America* 53, 562-567. Eigenfrequencies of thin plates and layer potentials.
18. D.S. JONES 1974 *Quaterly Journal of Mechanics and Applied Mechanics* XXVII, 129-142. Integral equations for the exterior acoustic problem.
19. R.E. KLEINMAN and G.F. ROACH 1974 *SIAM Review* 16, 214-236. Boundary integral equations for the three-dimensional Helmholtz equation.

20. J.C. NEDELEC 1976 *Computational Methods in Applied Mechanics and Engineering* 8, 61-80. Curved finite element methods for the solution of singular integral equations in $R^3$.
21. J. GIROIRE 1976 *Ecole Polytechnique, Centre de Mathématiques Appliquées.* Formulation variationnelle par équations intégrales d problèmes aux limites extérieurs.
22. L.M. DELVES and J. WALSH 1974 *Numerical Solution of Integral Equations.* Oxford : University Press.
23. R.E. KLEINMAN and W.L. WENDLAND 1977 *Journal of Mathematical Analysis and Applications* 57, (1), 170-202. On Neumann's method for the exterior Neumann problem for the Helmholtz equation.
24. G.C. HSIAO and W.L. WENDLAND 1977, *Journal of Mathematical Analysis and Applications* 58 (3), 449-481. A finite element method for some integral equations of the first kind.
25. G.C. HSIAO, P. KOPP and W.L. WENDLAND 1980, *Computing* 25, 89-130. A Galerkin collocation method for some integral equations of the first kind.
26. G.C. HSIAO and W.L. WENDALAND 1981, *JOurnal of Integral Equations* 3, 135-299. The Aubin-Nitsche lemma for integral equations.

# FINITE ELEMENT TECHNIQUES FOR ACOUSTICS

M. Petyt
Institute of Sound and Vibration Research
University of Southampton, England

Finite element techniques were first developed for analysing complex, engineering structures. Once the method had been given a firm mathematical foundation, it was only natural that it should be used for analysing other physical problems which could be represented by partial differential equations. The field of acoustics has been no exception.

There are three main types of acoustic problems to which the method has been applied. The first is the determination of the acoustic pressures in a cavity which is enclosed by either rigid or flexible walls. Also the noise radiated by vibrating structures, which are immersed in an infinite acoustic medium, can also be predicted. However, this type of problem can be analysed more efficiently by boundary element methods. Finally, finite element methods have been used to predict the propagation of acoustic waves down variable ducts without and with mean flow.

The following chapters describe the application of finite element techniques to irregular shaped cavities with both rigid and flexible walls and also duct acoustics.

# 1. FINITE ELEMENT ANALYSIS OF IRREGULAR SHAPED CAVITIES WITH ACOUSTICALLY HARD WALLS

## 1.1 Introduction

The method used to determine the acoustic characteristics of a cavity depends upon the size of the cavity in relation to the wavelength of sound. If the wavelength is considerably less than the cavity dimensions, then a statistical approach is used since there are many acoustic resonances. When the wavelength of sound is greater than about one third of the shortest dimension of the cavity, a modal analysis is preferred because the basic assumptions of the statistical approach are no longer valid.

The natural frequencies and modes of cavities of simple shape, such as rectangular hexahedra and cylinders, can be obtained analytically. However, if the cavity is of complex shape, such as the compartments of passenger vehicles, then the modal characteristics can only be found using numerical methods.

## 1.2 Equations of motion and boundary conditions

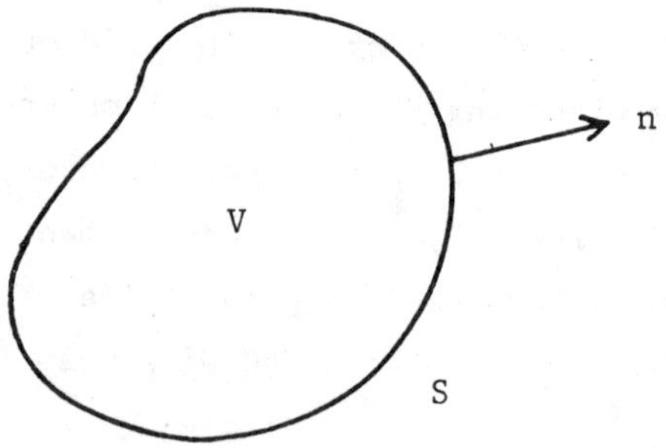


Fig. 1.1 Irregular shaped cavity

Consider a cavity of volume V enclosed by a surface S, which contains an inviscid, compressible fluid. Within V the acoustic pressure, p, must satisfy the Helmholtz equation

$$\nabla^2 p + (\omega/c_o)^2 p = 0 \tag{1.1}$$

where $\omega$ is the frequency of vibration and $c_o$ the speed of sound. If the surface S is acoustically hard, then the fluid particle velocity normal to the surface is zero, giving

$$\frac{\partial p}{\partial n} = 0 \tag{1.2}$$

where n is the outward normal to the surface.

The natural frequencies and modes of the cavity are obtained by solving equation (1.1) subject to the boundary conditions (1.2).

1.3 Approximate methods of solving Helmholtz's Equation

There are several methods of obtaining approximate solutions to differential equations. A number of these are referred to, collectively, as weighted residual techniques. These methods seek a solution in the form

$$p = \sum_{j=1}^{n} \phi_j (x,y,z) \alpha_j \tag{1.3}$$

where the $\phi_j(x,y,z)$ are prescribed functions of position (x.y.z), which are linearly independent, and the $\alpha_j$ are unknown parameters.

Introducing (1.3) into the left hand sides of (1.1) and (1.2) will produce errors, that is

$$\nabla^2 p + (\omega/c_o)^2 p = \varepsilon_1 \neq 0 \quad \text{in } V \tag{1.4}$$

and

$$\frac{\partial p}{\partial n} = \varepsilon_2 \neq 0 \quad \text{on } S \tag{1.5}$$

The parameters $\alpha_j$ are determined by requiring that the weighted average of the errors is zero, that is

$$\int_V (\nabla^2 p + (\omega/c_o)^2 p)\ \delta p\ dV - \int_S \frac{\partial p}{\partial n}\ \delta p\ dS = 0 \qquad (1.6)$$

In this equation a particular type of weight function, $\delta p$, has been used which is given by

$$\delta p = \sum_{j=1}^{n} \phi_j\ \delta\alpha_j \qquad (1.7)$$

This is the same type of weight function that is used in the Galerkin method. However, unlike the Galerkin method, equation (1.6) does not require the prescribed functions, $\phi_j$, in (1.3) to satisfy the boundary conditions (1.2).

Applying Green's Theorem to the first term in equation (1.6) gives

$$-\int_V \nabla p\ .\ \nabla(\delta p)\ dV + \int_V (\omega/c_o)^2\ p\ \delta p\ dV = 0 \qquad (1.8)$$

This equation can be written in the alternative form

$$\delta \int_V \frac{1}{2}\ [(\nabla p)^2 - (\omega/c_o)^2\ p^2]\ dV = 0 \qquad (1.9)$$

Equations (1.1) and (1.2) have now been replaced by an equivalent variational principle (1.9). Substituting (1.3) into (1.9) gives a set of linear equations for the parameters $\alpha_j$. This procedure is known as the Rayleigh-Ritz method. Convergence to the true solution is obtained as the number of terms in (1.3) is increased, provided the prescribed functions, $\phi_j$, satisfy the following conditions:

(1) are linearly independent

(2) are continuous

and (3) form a complete series

Note that the first term in (1.9) is proportional to the kinetic energy of the fluid whilst the second one is proportional to the strain energy of the fluid.

## 1.4 Finite element method

One of the disadvantages of the Rayleigh-Ritz method for solving the Helmholtz equation is that it is difficult to construct suitable approximating functions for irregular shaped cavities. One way of overcoming this is to use the finite element method. This method provides an automatic procedure for constructing the approximating functions. This technique will initially be demonstrated by considering a one-dimensional cavity. It will then be extended to two and three dimensional cavities.

## 1.5 One dimensional column

Consider a one-dimensional column, of constant cross section A and length L, bounded by rigid ends. The cross sectional dimensions are assumed to be small compared with an acoustic wavelength.

The prescribed functions, to be used in the Rayleigh-Ritz method, are constructed in the following manner:

a) Select a set of reference or 'node' points.

b) Associate with each node point a given number of degrees of freedom.

c) Construct a set of functions such that each one corresponds to a unit value of one degree of freedom, the others being taken as zero.

This procedure is illustrated for a one dimensional column in Figure 1.2. In this Figure five node points have been selected at equal intervals. The region between each node is referred to as an 'element'. Since only the functions are required to be continuous, then only the pressure need be taken to be a degree of freedom at each node. In this case the assumed functions are as illustrated in Figure 1.2.

It can be seen that the pressure variation over each element is zero except for two cases, the number being equal to the number of nodes

(2) multiplied by the number of degrees of freedom at each node (1) for a single element. These two pressure variations are identical for each element. Therefore, it will be simpler to evaluate the integrals in the variational expression (1.9) for each element and then add the contribution from the elements together.

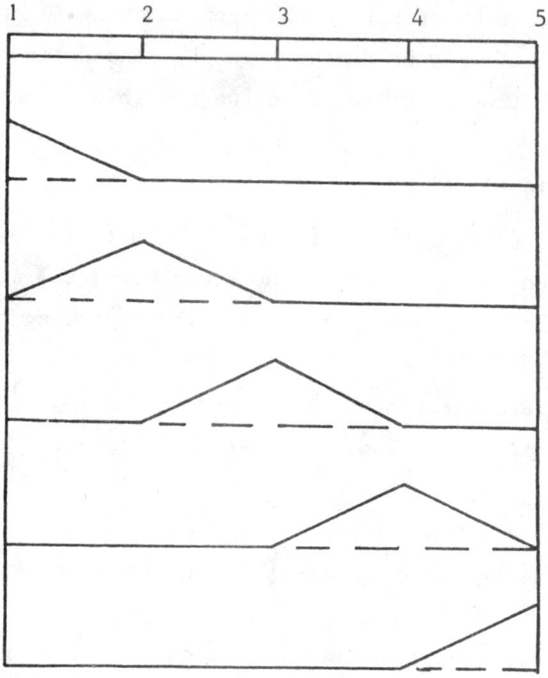


Fig. 1.2 Prescribed functions for a one-dimensional column

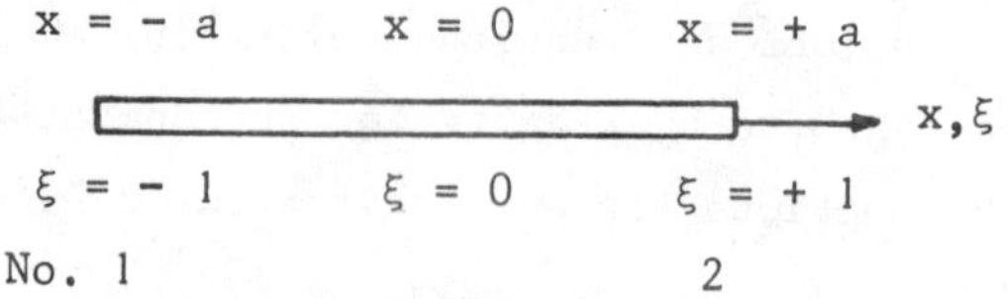


Fig. 1.3 Geometry of a single element

The geometry of a single element of length 2a is shown in Figure 1.3. Both a dimensional coordinate x and a non-dimensional one $\xi = x/a$ are shown. The pressure variation for this element is given by

$$p = \frac{1}{2}(1-\xi)\, p_1 + \frac{1}{2}(1+\xi)\, p_2 \tag{1.10}$$

where $p_1$, $p_2$ are the pressures at nodes 1 and 2.

This expression can be written in matrix form as follows:

$$p = \lfloor \tfrac{1}{2}(1-\xi) \;\; \tfrac{1}{2}(1+\xi) \rfloor \begin{bmatrix} p_1 \\ p_2 \end{bmatrix} = \lfloor N(\xi) \rfloor_e \{p\}_e \tag{1.11}$$

Substituting (1.11) into the integrals in (1.9) gives, for a single element,

$$\int_{V_e} (\nabla p)^2 \, dV = \int_{-a}^{+a} \left(\frac{\partial p}{\partial x}\right)^2 A \, dx = \{p\}_e^T [k]_e \{p\}_e \tag{1.12}$$

and

$$\int_{V_e} \frac{1}{c_o^2} p^2 dV = \int_{-a}^{+a} p^2 \frac{A}{c_o^2} \, dx = \{p\}_e^T [m]_e \{p\}_e \tag{1.13}$$

where

$$[k]_e = \int_{-1}^{+1} \frac{A}{a} \lfloor N'(\xi) \rfloor_e^T \lfloor N'(\xi) \rfloor_e \, d\xi = \frac{A}{a} \begin{bmatrix} 1/2 & -1/2 \\ -1/2 & 1/2 \end{bmatrix} \tag{1.14}$$

and

$$[m]_e = \int_{-1}^{+1} \frac{Aa}{c_o^2} \lfloor N(\xi) \rfloor_e^T \lfloor N(\xi) \rfloor_e \, d\xi = \frac{Aa}{c_o^2} \begin{bmatrix} 2/3 & 1/3 \\ 1/3 & 2/3 \end{bmatrix} \tag{1.15}$$

The next step is to relate the degrees of freedom of a single element to the set of degrees of freedom for the complete system, $\{p\}$. For element e

$$\{p\}_e = [a]_e \{p\} \tag{1.16}$$

where

$$\{p\}^T = \lfloor p_1 \; p_2 \; p_3 \; p_4 \; p_5 \rfloor \tag{1.17}$$

It can easily be verified that each element of $[a]_e$ is either unity or zero.

Substituting the transformation (1.16) into (1.12) and (1.13) and summing over all elements gives

$$\int_V (\nabla p)^2 \, dV = \{p\}^T \sum_{e=1}^{4} [a]_e^T [k]_e [a]_e \cdot \{p\} = \{p\}^T [K_a]\{p\} \tag{1.18}$$

and

$$\int \frac{1}{c_o^2} p^2 \, dV = \{p\}^T \sum_{e=1}^{4} [a]_e^T [m]_e [a]_e \{p\} = \{p\}^T [M_a]\{p\} \tag{1.19}$$

It will be left to the reader to verify the following results:

$$[K_a] = \frac{A}{2a} \begin{bmatrix} 1 & -1 & & & \\ -1 & 2 & -1 & & \\ & -1 & 2 & -1 & \\ & & -1 & 2 & -1 \\ & & & -1 & 1 \end{bmatrix} \tag{1.20}$$

$$[M_a] = \frac{Aa}{3c_o^2} \begin{bmatrix} 2 & 1 & & & \\ 1 & 4 & 1 & & \\ & 1 & 4 & 1 & \\ & & 1 & 4 & 1 \\ & & & 1 & 2 \end{bmatrix} \tag{1.21}$$

where $a = L/8$.

Equation (1.9) can now be written in the form

$$\delta\left[ \frac{1}{2} \{p\}^T [K_a]\{p\} - \omega^2 \frac{1}{2} \{p\}^T [M_a]\{p\}\right] = 0 \tag{1.22}$$

which gives

$$[\underset{\sim}{K}_a - \omega^2 \underset{\sim}{M}_a] \{p\} = 0 \tag{1.23}$$

Equation (1.23) represents a linear, algebraic eigenvalue problem which can be solved by various standard procedures [1.1, 1.2]. The eigenvalues (which are positive) represent the square of the natural frequencies, and the corresponding eigenvectors give the shape of the modes of vibration.

Note that in practice it is unnecessary to form the matrices $[a]_e$, as defined in (1.16), and carry out the matrix multiplication indicated in (1.18) and (1.19). The matrix product $[a]_e^T\,[k]_e\,[a]_e$ effectively locates the positions in $[K_a]$ to which the elements of $[k]_e$ have to be added. In Figure 1.2 element number e has nodes e and (e+1). Therefore, the two rows and columns of the matrix $[k]_e$ are added into the rows and columns e and (e+1) of the matrix $[K]_a$. This procedure is known as the 'assembly process' and applies to the formation of the matrix $[M_a]$ also.

Notice that both $[K_a]$ and $[M_a]$ as defined in equations (1.20) and (1.21) are symmetric and banded. These properties can be exploited when writing computer programs. All the information displayed in either (1.20) or (1.21) can be stored in 9 locations, whereas 25 locations would be required to store the complete matrix.

In the Rayleigh-Ritz method the accuracy of the solution is increased by increasing the number of prescribed functions in the assumed series. To increase the number of prescribed functions in the finite element method, the number of node points, and therefore the number of elements, is increased. The prescribed function for each element should satisfy the Rayleigh-Ritz convergence criteria. Also these functions should be continuous across element boundaries.

## 1.6 Two dimensional cavities

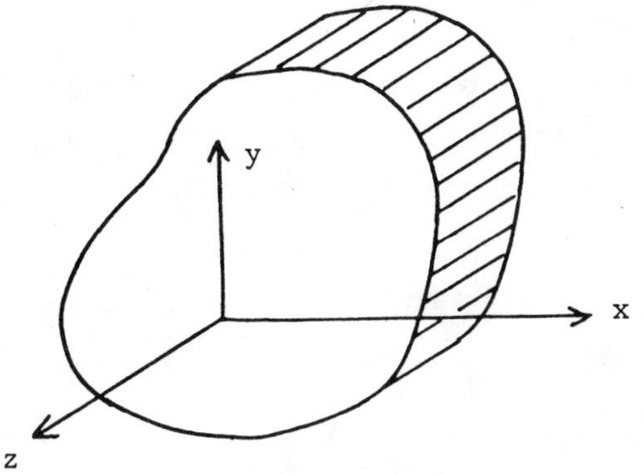


Fig. 1.4 A two dimensional cavity

Some three dimensional cavities have the property that the geometry does not change in one direction. This means that the cavity has a pair of walls which are parallel to one another. Also, all sections parallel to these walls are identical as shown in Figure 1.4. An example of this is the interior of many automobiles. Assuming that $z = 0, L_z$ are the parallel walls, then the solution of equations (1.1) and (1.2) takes the form

$$p(x,y,z) = p(x,y) \cos \frac{n\pi z}{L_z} , \quad n = 0,1,... \tag{1.24}$$

Substituting (1.24) into (1.9) and integrating with respect to z gives

$$\delta \int_A \frac{1}{2} \quad [(\nabla p)^2 - (\omega_1/c_o)^2 \, p^2] dA = 0 \tag{1.25}$$

where A is the cross-sectional area (z = constant) of the cavity, $\nabla_1$ is the two-dimensional form of the operator $\nabla$ and

$$(\omega_1/c_o)^2 = (\omega/c_o)^2 - (n\pi/L_z)^2 \tag{1.26}$$

The variational principle (1.25) can now be solved using two-dimensional finite elements.

The natural frequencies, $\omega$, of the three dimensional cavity are related to the natural frequencies, $\omega_1$, of the two-dimensional cavity as follows

$$\omega^2 = \omega_1{}^2 + (n\pi c_o/L_z)^2 \tag{1.27}$$

Corresponding to each $\omega_1$, there are an infinite number of natural frequencies that have the same pressure variation in the (x,y) plane. However, the pressure varies differently in the z direction, as indicated by equation (1.24). For even values of n, this variation is symmetric about $z = L_z/2$, and for odd values of n, it is antisymmetric.

### 1.6.1 Linear triangular element

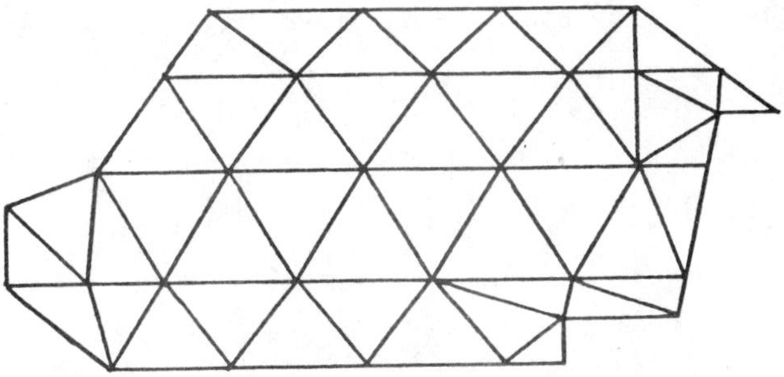

Fig. 1.5 Idealisation of a two dimensional cavity [1.3]

The simplest way of idealising a two-dimensional irregular shaped cavity is to use an assemblage of triangular elements as shown in Fig.1.5.

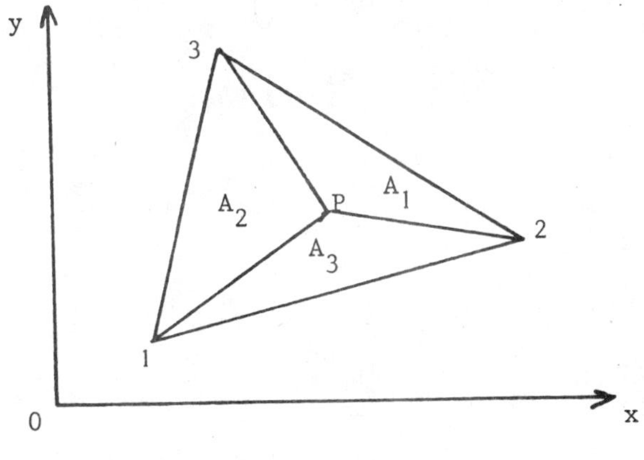


Fig. 1.6 Geometry of a Linear Triangular Element

Figure 1.6 shows a triangular element with three node points, one at each vertex, having a single degree of freedom, the pressure , at each node. The variation in pressure within the element can, therefore, be represented by a polynomial having three terms, that is

$$p = \alpha_1 + \alpha_2 x + \alpha_3 y \qquad (1.28)$$

Evaluating (1.28) at the three nodes and solving the resulting equations for the coefficients $\alpha_1 - \alpha_3$ gives

$$p = \lfloor N(x,y) \rfloor \; \{p\} \qquad (1.29)*$$

with

$$\lfloor N(x,y) \rfloor = \lfloor N_1 \; N_2 \; N_3 \rfloor \; , \; \{p\}^T = \lfloor p_1 \; p_2 \; p_3 \rfloor \qquad (1.30)$$

and

$$N_i = \frac{1}{2A} (A_i^o + a_i x + b_i y) \qquad (1.31)$$

In (1.31)

$$A_i^o = x_j y_\ell - x_\ell y_j , \quad a_i = y_j - y_\ell , \quad b_i = x_\ell - x_j \qquad (1.32)$$

and A is the area of the triangle which is given by

$$A = \frac{1}{2} (a_1 b_2 - a_2 b_1) \qquad (1.33)$$

Substituting (1.29) into the integrals in (1.25) gives the following element matrices

$$[k] = \int_A [B]^T [B] \, dA \qquad (1.34)$$

and

$$[m] = \frac{1}{c_o^2} \int_A \lfloor N \rfloor^T \lfloor N \rfloor \, dA \qquad (1.35)$$

where

$$[B] = \begin{bmatrix} \frac{\partial}{\partial x} \\ \frac{\partial}{\partial y} \end{bmatrix} \lfloor N \rfloor \qquad (1.36)$$

* Subscript e, previously used to denote element matrices, will be omitted for convenience in what follows.

Carrying out the integrations in (1.34) and (1.35) using the functions (1.31) proves to be rather lengthy. A more convenient way of dealing with triangular regions is to use area coordinates. Referring to Figure 1.6, the area coordinates $(L_1, L_2, L_3)$ of the point P are defined as

$$L_1 = A_1/A \ , \quad L_2 = A_2/A \ , \quad L_3 = A_3/A \tag{1.37}$$

where $A_1$, $A_2$, $A_3$ denote the areas of the subtriangles indicated. It can easily be seen that the coordinates of the three vertices are (1,0,0), (0,1,0) and (0,0,1) respectively. The area and Cartesian coordinates are related by

$$L_i = \frac{1}{2A} (A_i^o + a_i x + b_i y) \tag{1.38}$$

Thus, in area coordinates, (1.31) becomes

$$N_i = L_i \tag{1.39}$$

The right hand side of (1.36) is evaluated using the following relationships

$$\frac{\partial N_i}{\partial x} = \sum_{j=1}^{3} \frac{\partial N_i}{\partial L_j} \frac{\partial L_j}{\partial x} \ , \quad \frac{\partial N_i}{\partial y} = \sum_{j=1}^{3} \frac{\partial N_i}{\partial L_j} \frac{\partial L_j}{\partial y} \tag{1.40}$$

Substituting (1.38) and (1.39) into (1.40), it can be seen that $|B|$, as defined by (1.36), is

$$[B] = \frac{1}{2A} \begin{bmatrix} a_1 & a_2 & a_3 \\ b_1 & b_2 & b_3 \end{bmatrix} \tag{1.41}$$

Substituting (1.41) into (1.34) and performing the integration, gives the following expression for the element (i,j) of the matrix $[k]$,

$$k_{ij} = \frac{1}{4A} (a_i a_j + b_i b_j) \tag{1.42}$$

Substituting (1.39) into (1.35), gives the elements of $[m]$ as

$$m_{ij} = \frac{1}{c_o^2} \int_A L_i \, L_j \, dA \tag{1.43}$$

Integrals of this form can be evaluated using the following formula

$$\int_A L_1^m\, L_2^m\, L_3^p \, dA = 2A\ m!\ n!\ p!/(m+n+p+2)! \qquad (1.44)$$

Thus, (1.43) becomes

$$m_{ij} = A/6c_o \ (i=j), \quad A/12c_o \ (i \neq j) \qquad (1.45)$$

The accuracy of the element is not good. Thus, a large number of small elements must be used. This can be overcome by introducing more node points as described in the next section.

1.6.2 Quadratic triangular element

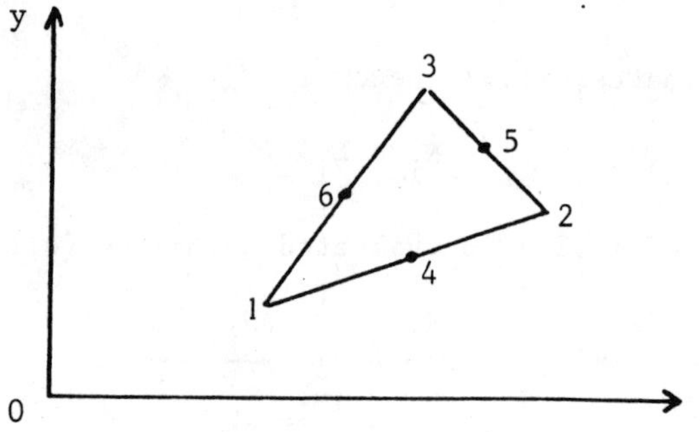


Fig. 1.7 Geometry of a quadratic triangular element

Figure 1.7 shows a triangular element with six node points, one at each vertex and one at the mid point of each side. The variation in pressure within the element can, therefore, be represented by a polynomial having six terms, that is

$$p = \alpha_1 + \alpha_2 x + \alpha_3 y + \alpha_4 x^2 + \alpha_5 xy + \alpha_6 y^2 \qquad (1.46)$$

The coefficients $\alpha_1 - \alpha_6$ can be found by evaluating (1.46) at the six node points. However, this procedure will not be followed as a simpler formulation can be obtained using area coordinates. The variation in pressure, by analogy with (1.29), takes the form

$$p = \sum_{j=1}^{6} N_j \, (L_1, L_2, L_3) \, p_j \qquad (1.47)$$

Remembering that $N_j$ has a unit value at node j and is zero at all other nodes, the functions $N_j$ in (1.47) can be written down by inspection. This gives

$$\begin{aligned} N_1 &= (2L_1 - 1)\,L_1 & N_4 &= 4\,L_1\,L_2 \\ N_2 &= (2L_2 - 1)\,L_2 & N_5 &= 4\,L_2\,L_3 \\ N_3 &= (2L_3 - 1)\,L_3 & N_6 &= 4\,L_3\,L_1 \end{aligned} \qquad (1.48)$$

The element belongs to a family of elements which Zienkiewicz [1.4] has named 'Serendipity' elements.

The matrices [k] and [m], as defined by (1.34) and (1.35) can now be evaluated using (1.48) in the way described in the last section. The result can now be found in reference [1.3].

The cavity shown in Figure 1.5 is analysed in reference [1.3] using this element. The frequencies and modes obtained are compared with experimental results.

### 1.6.3 Rectangular elements

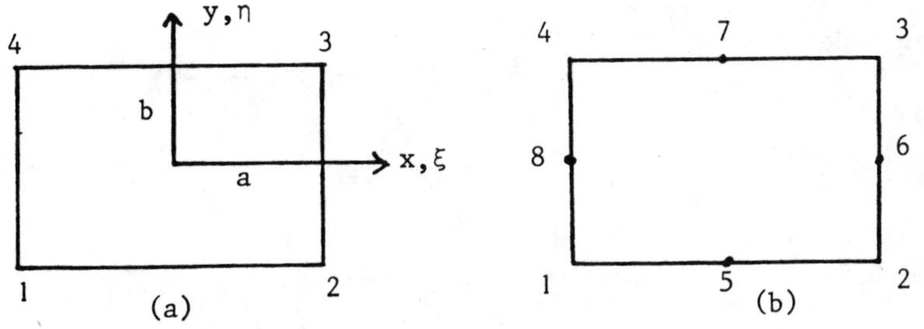


Figure 1.8 Geometry of Rectangular elements (a) Linear (b) Quadratic

The preceding analysis of two-dimensional cavities can be made more efficient by using rectangular elements as much as possible and filling in with triangular elements where the shape of the boundary makes it necessary. Figure 1.8 illustrates two such elements, a linear one with four nodes and a quadratic one with eight nodes. Both these elements are members of the serendipity family, and so the pressure variation can be obtained relatively simply.

Using non-dimensional cartesian coordinates $\xi = x/a$, $\eta = y/b$, as shown in Figure 1.8(a), the pressure variation for the linear element is given by

$$p = \sum_{j=1}^{4} N_j \, p_j \qquad (1.49)$$

where

$$N_j = \frac{1}{4} \, (1+\xi_o)(1+\eta_o) \qquad (1.50)$$

In this expression

$$\xi_o = \xi \, \xi_j \, , \qquad \eta_o = \eta \, \eta_j \qquad (1.51)$$

where $(\xi_j, \eta_j)$ are the non-dimensional coordinates of node j.

The pressure variation for the quadratic element, Figure 1.8(b) is given by

$$p = \sum_{j=1}^{8} N_j \, p_j \qquad (1.52)$$

where for nodes 1-4

$$N_j = \frac{1}{4} \, (1+\xi_o)(1+\eta_o)(\xi_o+\eta_o-1) \qquad (1.53)$$

nodes 5,7

$$N_j = \frac{1}{2} \, (1-\xi^2)(1+\eta_o) \qquad (1.54)$$

and nodes 6,8

$$N_j = \frac{1}{2} \, (1+\xi_o)(1-\eta^2) \qquad (1.55)$$

The matrices [k] and [m], equations (1.34) and (1.35), can be evaluated by the methods given in section 1.6.1.

### 1.6.4 Isoparametric and related elements

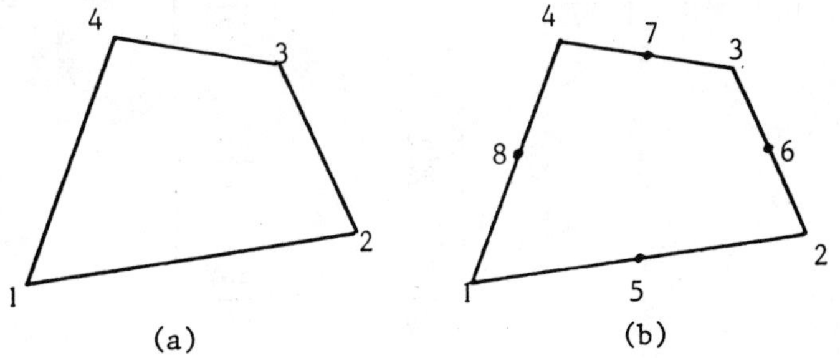


Fig. 1.9 Geometry of Quadrilateral Elements (a) Isoparametric, (b) Sub-parametric.

The elements presented in the last section will be much more versatile if they can be transformed into quadrilateral elements. The elements shown in Figures 1.8(a) and 1.8(b) can be transformed into the quadrilateral elements shown in Figures 1.9(a) and 1.9(b) by means of the relationships:

$$x = \sum_{j=1}^{4} N_j \; (\xi,\eta) \; x_j \; , \qquad y = \sum_{j=1}^{4} N_j \; (\xi,\eta) \; y_j \tag{1.56}$$

where the functions $N_j(\xi,\eta)$ are defined by equations (1.50). The pressure variations are again represented by the functions (1.50) and (1.53)-(1.55).

The element shown in Figure 1.9(a) is an example of an <u>isoparametric</u> element. Such elements use the same functions to define both geometry and the pressure variation. In the case of the element shown in Figure 1.9(b), less nodes are used to define the geometry than in the definition of the pressure variation. Such elements are referred to as <u>sub-parametric</u> elements.

In evaluating the $[B]_e$ matrix, equation (1.36), use should be made of the following relationship

$$\begin{bmatrix} \dfrac{\partial}{\partial x} \\ \\ \dfrac{\partial}{\partial y} \end{bmatrix} = [J]^{-1} \begin{bmatrix} \dfrac{\partial}{\partial \xi} \\ \\ \dfrac{\partial}{\partial \eta} \end{bmatrix} \tag{1.57}$$

where

$$|J| = \begin{bmatrix} \dfrac{\partial x}{\partial \xi} & \dfrac{\partial y}{\partial \xi} \\ \\ \dfrac{\partial x}{\partial \eta} & \dfrac{\partial y}{\partial \eta} \end{bmatrix} \tag{1.58}$$

is the Jacobian matrix of the transformation. This matrix is evaluated using (1.56). The integrals in (1.34) and (1.35) should be transformed to the $(\xi,\eta)$ coordinate using

$$dA = dx\,dy = \det [J]\, d\xi\, d\eta \tag{1.59}$$

resulting in the limits of integration being $\pm 1$.

In most cases the form of the integrand is rather complicated and algebraic integration is either difficult or impossible. This situation is overcome by using numerical integration. The Gauss integration technique is commonly used and a 3 x 3 array of integration points has proved to give accurate results.

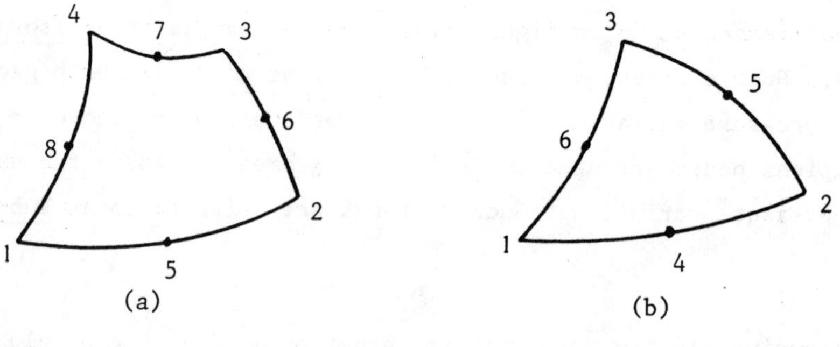


Fig. 1.10 Curvilinear isoparametric elements

The rectangular element in Figure 1.8(b) and the triangular element in Figure 1.7 can be transformed into quadrilateral and triangular elements with curved boundaries as shown in Figure 1.10. These are both isoparametric elements since the same function is used for both geometry and pressure variation. In the case of the quadrilateral element the transformation is given by

$$x = \sum_{j=1}^{8} N_j (\xi,\eta)\, x_j , \qquad y = \sum_{j=1}^{8} N_j (\xi,\eta)\, y_j \tag{1.60}$$

where the functions $N_j(\xi,\eta)$ are defined by equations (1.53) to (1.55). The transformation for the triangular element is

$$x = \sum_{j=1}^{6} N_j (L_1,L_2,L_3)\, x_j , \quad y = \sum_{j=1}^{6} N_j (L_1,L_2,L_3)\, y_j \tag{1.61}$$

where the functions $N_j (L_1,L_2,L_3)$ are defined by equations (1.48).

The integrals in equations (1.34) and (1.35) are evaluated by the procedure outlined above. However, in the case of the triangular element, the $(\xi,\eta)$ coordinates are defined by

$$L_1 = \xi , \quad L_2 = \eta , \quad L_3 = 1 - \xi - \eta \tag{1.62}$$

This gives

$$\frac{\partial}{\partial \xi} = \frac{\partial}{\partial L_1} - \frac{\partial}{\partial L_3} , \quad \frac{\partial}{\partial \eta} = \frac{\partial}{\partial L_2} - \frac{\partial}{\partial L_3} \tag{1.63}$$

Accurate values can be obtained for a triangle by using seven integration points.

## 1.7 Axisymmetric Cavities

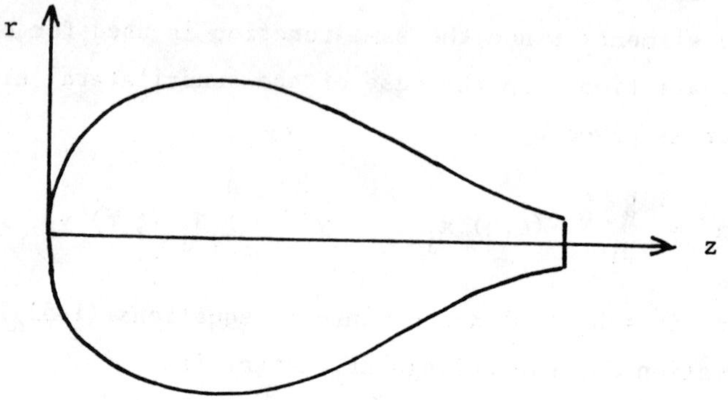


Fig. 1.11 An axisymmetric cavity

Some three dimensional cavities are axisymmetric in shape. The surface of the cavity can, therefore, be generated by rotating a two-dimensional curve about the axis of symmetry. Many fluid containers are of this form. Assuming that the z axis is the axis of symmetry, then the solution of (1.1) in cylindrical coordinates (r, θ, z) takes the form

$$p(r,\ \theta,\ z) = p(z,r)\cos n\,\theta\ , \qquad n = 0,\ 1,\ \ldots \tag{1.64}$$

Substituting (1.64) into (1.9) and integrating with respect to θ gives

$$\delta \int_A \frac{1}{2}\left[(\nabla_1 p)^2 + \left(\frac{n}{r}\right)^2 p^2 - (\omega/c_o)^2\, p^2\right] r\, dA = 0 \tag{1.65}$$

where A is the cross-sectional area (θ = constant) of the cavity and $\nabla_1$ is the two dimensional form of the operator ∇ in the (z,r) plane. Thus, once again, the problem has been reduced to solving a variational principle, namely equation (1.65), by means of two-dimensional finite elements. Notice, however, that in this case it is not convenient to associate the second term in (1.65) with the third term, as was done in (1.25), due to the presence of the factor $1/r^2$. Instead, the second term is associated with the first term.

Any of the element shapes and associated pressure variations described in Section 1.6, can be used in the present case. References [1.5-1.7] use the three-node triangle described in Section 1.6.1, whilst reference [1.8] uses the six-node triangle of Section 1.6.2. Reference [1.5] also uses the four-node quadrilateral of Section 1.6.4, and reference [1.9] uses the eight-node isoparametric quadrilateral, also described in Section 1.6.4.

In all cases the pressure variation in the (z,r) plane can be written in the form

$$p = \lfloor N \rfloor \{p\} \tag{1.66}$$

where $\lfloor N \rfloor$ is a row matrix of functions N(z,r) which describe the variation of pressure over a single element and {p} is a column matrix of nodal pressure values.

Substituting (1.66) into the integrals in (1.65) gives the following element matrices:

$$[k] = \int_A r\ [B]^T [B]\ dA + n^2 \int_A \frac{1}{r} \lfloor N \rfloor^T \lfloor N \rfloor\ dA \tag{1.67}$$

and

$$[m] = \frac{1}{c_o^2} \int_A r\ \lfloor N \rfloor^T \lfloor N \rfloor\ dA \tag{1.68}$$

where

$$[B] = \begin{bmatrix} \dfrac{\partial}{\partial z} \\ \dfrac{\partial}{\partial r} \end{bmatrix} \lfloor N \rfloor \tag{1.69}$$

At first sight, it would appear that the second term of (1.67) needs special attention for elements adjacent to the z-axis, since r=0 there. In fact, reference [1.5] does mention a special trapezoidal element. Some of the references cited only consider the case n=0, and so this term disappears. If numerical integration is used, then the second term of (1.67) will not need special attention provided there are no integration points on r=0.

Reference [1.7] analyses several axisymmetric cavities using the three-node triangle described in Section 1.6.1. The frequencies and modes are compared with experimental measurements in each case.

## 1.8 Three Dimensional Cavities

In order to analyse truly three-dimensional cavities it is necessary to use three-dimensional elements. These can be derived in a similar manner to the two-dimensional elements described in Section 1.6. Reference [1.10] uses both a four-node tetrahedral element and an eight-node rectangular hexahedral element. These are three dimensional versions of the elements described in Sections 1.6.1 and 1.6.3 respectively. Reference [1.11] describes an eight-node, isoparametric hexahedra, whilst reference [1.12] describes a twenty-node, isoparametric hexahedra. These are three-dimensional versions of the two elements shown in Figures 1.9(a) and 1.10 (a) of Section 1.6.4. By way of illustration, the element developed in reference [1.12] will be described.

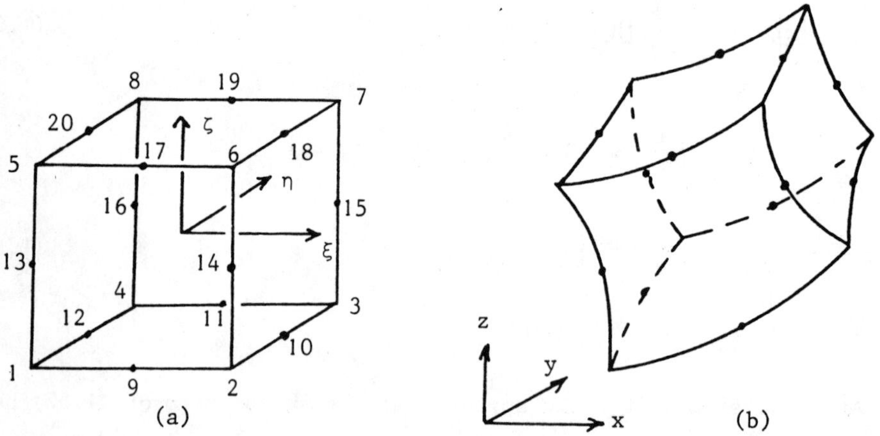


Fig. 1.12 Geometry of a 20-node isoparametric element

The geometry of the element is shown in Figure 1.12. The same functions are used to define both the geometry and pressure variation. Thus,

$$p = \sum_{j=1}^{20} N_j\ (\xi,\eta,\zeta)\ p_j \tag{1.70}$$

$$x = \sum_{j=1}^{20} N_j\ x_j\ , \qquad y = \sum_{j=1}^{20} N_j\ y_j\ , \qquad z = \sum_{j=1}^{20} N_j\ z_j \tag{1.71}$$

The functions $N_j(\xi,\eta,\zeta)$ are defined as follows. For the corner nodes 1 to 8

$$N_j = \frac{1}{8}\ (1+\xi_o)(1+\eta_o)(1+\zeta_o)(\xi_o+\eta_o+\zeta_o-2) \tag{1.72}$$

where $\xi_o$, $\eta_o$ are defined in (1.51) and $\zeta_o = \zeta\,\zeta_j$.
For the mid-edge nodes 9, 11, 17, 19

$$N_j = \frac{1}{4}\ (1-\xi^2)(1+\eta_o)(1+\zeta_o) \tag{1.73}$$

For the mid-edge nodes 10, 12, 18, 20

$$N_j = \frac{1}{4}\ (1+\xi_o)(1-\eta^2)(1+\zeta_o) \tag{1.74}$$

and for the mid-edge nodes 13, 14, 15, 16

$$N_j = \frac{1}{4}\ (1+\xi_o)(1+\eta_o)(1-\zeta^2) \tag{1.75}$$

Substituting (1.70) and (1.71) into the integrals in (1.9) gives the following element matrices

$$[k] = \int_{-1}^{+1}\int_{-1}^{+1}\int_{-1}^{+1} [B]^T\ [B]\ \det\ [J]\ d\xi\ d\eta\ d\zeta \tag{1.76}$$

and

$$[m] = \frac{1}{c_o^2}\int_{-1}^{+1}\int_{-1}^{+1}\int_{-1}^{+1} \lfloor N\rfloor^T\ \lfloor N\rfloor\ \det\ [J]\ d\xi\ d\eta\ d\zeta \tag{1.77}$$

In these expressions we have

$$\lfloor N\rfloor = \lfloor N_1\ N_2\ \ldots\ N_{20}\rfloor \tag{1.78}$$

where the $N_j$ are as defined in (1.72) to (1.75),

$$[J] = \begin{bmatrix} \frac{\partial x}{\partial \xi} & \frac{\partial y}{\partial \xi} & \frac{\partial z}{\partial \xi} \\ \frac{\partial x}{\partial \eta} & \frac{\partial y}{\partial \eta} & \frac{\partial z}{\partial \eta} \\ \frac{\partial x}{\partial \zeta} & \frac{\partial y}{\partial \zeta} & \frac{\partial z}{\partial \zeta} \end{bmatrix} \tag{1.79}$$

and

$$[B] = [J]^{-1} \begin{bmatrix} \frac{\partial}{\partial \xi} \\ \frac{\partial}{\partial \eta} \\ \frac{\partial}{\partial \zeta} \end{bmatrix} \lfloor N \rfloor \tag{1.80}$$

Due to the complexity of the resulting expressions for [k] and [m], they are evaluated using numerical integration. The twenty-seven point integration scheme presented in reference [1.13] was used. This scheme gives the same accuracy as a sixty-four point Gauss integration scheme, but is much more efficient.

Reference [1.12] uses this element to analyse a 1/12th scale model of a light van. The frequencies and modes obtained are compared with experimental measurements.

## 1.9 Other types of elements

All the elements described in the previous sections had only the pressure as nodal degree of freedom. Several elements have more than one degree of freedom at each node.

Reference [1.13] describes a two-dimensional element of triangular shape. It has three nodes with three degrees of freedom at each node, namely, $p$, $\partial p/\partial x$ and $\partial p/\partial y$. Reference [1.14] presents a rectangular element which has the same three degrees of freedom at each of four nodes.

Element matrices are not always formulated in terms of pressure. For example, references [1.6] and [1.15] use a velocity potential, $\phi$. However, the pressure can be obtained using the relationship $p = -\rho_a \partial\phi/\partial t$, where $\rho_a$ is the density of the acoustic medium. The element in reference [1.6] is a linear triangle with $\phi$ as the degree of freedom at each node. Reference [1.15] describes a rectangular hexahedra with $\phi$, $\phi_x$, $\phi_y$, $\phi_z$ as nodal degrees of freedom.

Reference [1.16] uses a displacement potential, also denoted by $\phi$. In this case the pressure is given by the relationship $p = -\rho_a \partial^2\phi/\partial t^2$. The element is an axisymmetric one of rectangular cross-section. It has four nodes with $\phi$, $\partial\phi/\partial z$, $\partial\phi/\partial r$, $\partial^2\phi/\partial z\partial r$, as degrees of freedom.

## 1.10 References


1.1 K.J. BATHE and E.L. WILSON 1976 Numerical Methods in Finite Element Analysis. Englewood Cliffs: Prentice-Hall.

1.2 A. JENNINGS 1977 Matrix Computation for Engineers and Scientists. London: Wiley.

1.3 T.L. RICHARDS and S.K. JHA 1979 Journal of Sound and Vibration 63, 61-72. A simplified finite element for studying acoustic characteristics inside a car cavity.

1.4 O.C. ZIENKIEWICZ 1977 The Finite Element Method in Engineering Science, 3rd Edition. London: McGraw-Hill.

1.5 D.N. HERTING et al 1971 NASA TM X-2378 NASTRAN: Users' Experiences (Ed. by P.J. Raney), 285-324. Acoustic analysis of solid rocket motor cavities by a finite element method.

1.6 J.H. JAMES 1973 Admiralty Research Laboratory Report ARL/R/R4. Acoustic finite element analysis of axisymmetric fluid regions.

1.7 Y. KAGAWA and T. OMOTE 1976 Journal of Acoustical Society of America 60, 1003-1013. Finite element simulation of acoustic filters of arbitrary profile with circular cross section.

1.8 Y. KAGAWA, T. YAMABUCHI and A. MORI 1977 Journal of Sound and Vibration 53, 357-374. Finite element simulation of axisymmetric acoustic transmission system with sound absorbing wall.

1.9 M. PETYT and S.P. LIM 1980 International Journal for Numerical Methods in Engineering 13, 109-122. Finite element analysis of the noise inside a mechanically excited cylinder.

1.10 A. CRAGGS 1972 Journal of Sound and Vibration 23, 331-339. The use of simple three-dimensional acoustic finite elements for determining the natural modes and frequencies of complex shaped enclosures.

1.11 A. CRAGGS 1976 Journal of Sound and Vibration 48, 377-392. A finite element method for damped acoustic systems, an application to evaluate the performance of reactive mufflers.

1.12 M. PETYT, J. LEA and G. KOOPMANN 1976 Journal of Sound and Vibration, 45, 495-502. A finite element method for determining the acoustic modes of irregular shaped cavities.

1.13 T. SHUKU and K. ISHIHARA 1973 Journal of Sound and Vibration 29, 67-76. The analysis of the acoustic field in irregularly shaped rooms by the finite element method.

1.14 A. CRAGGS and G. STEAD 1976 Acustica 35, 89-98. Sound transmission between enclosures - a study using plate and acoustic finite elements.

1.15 A. CRAGGS 1971 Journal of Sound and Vibration 15, 509-528. The transient response of a coupled plate-acoustic system using plate and acoustic finite elements.

1.16 S.P. LIM and M. PETYT 1980 Recent Advances in Structural Dynamics (Ed. by M. Petyt) 447-455. Free vibrations of a cylinder partially filled with a liquid.

# 2. FINITE ELEMENT ANALYSIS OF IRREGULAR SHAPED CAVITIES WITH NON RIGID WALLS

## 2.1 Introduction

In practice, irregular shaped cavities will have part of the bounding surface covered in sound absorbing material and part of it will be flexible. Examples of this include the interior of passenger cars and civil aircraft. The methods described in the previous chapter will be extended to this situation in the following sections.

## 2.2 Equations of Motion and Boundary Conditions

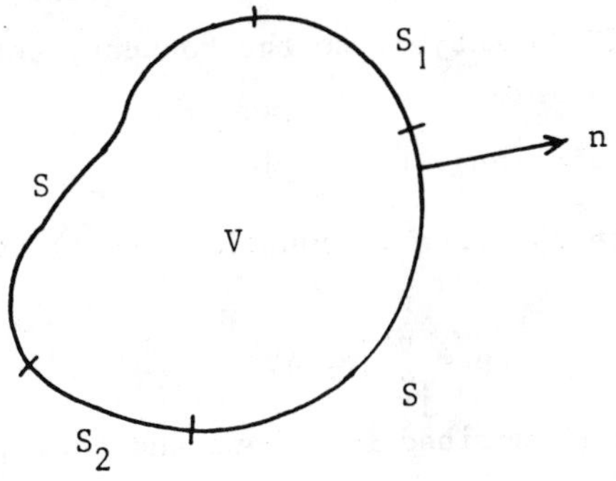


Fig. 2.1 Irregular Shaped Cavity with Non Rigid Walls

Consider a cavity of volume V enclosed by a non-rigid surface as shown in Figure 2.1. Part of the surface S is acoustically hard, the part $S_1$ is flexible and vibrating with a normal velocity $v_n$, and the part $S_2$ is covered with sound absorbing material, the specific acoustic impedance of which is $Z_s$. Within V the acoustic pressure, p, must satisfy the Helmholtz equation

$$\nabla^2 p + (\omega/c_o)^2 \, p = 0 \qquad (2.1)$$

where $\omega$ is the frequency of vibration and $c_o$ the speed of sound. Over the bounding surface the fluid particle velocity normal to the surface is equal to the normal velocity of the surface. This gives rise to the following boundary conditions:

$$\frac{\partial p}{\partial n} = 0 \qquad \text{over } S$$

$$\frac{\partial p}{\partial n} = - i\, \rho_a \omega\, v_n \qquad \text{over } S_1 \tag{2.2}$$

and

$$\frac{\partial p}{\partial n} = - i\, \rho a \omega \frac{p}{Z_s} \qquad \text{over } S_2$$

where $\rho_a$ is the density of the acoustic medium, and n is the outward normal to the surface.

The acoustic pressures within V, due to a velocity distribution $v_n$ over $S_1$, which varies harmonically with frequency $\omega$, are obtained by solving equation (2.1) subject to the boundary conditions (2.2).

## 2.3 Equivalent Variational Principle

An approximate solution to equations (2.1) and (2.2) is sought in the form

$$p = \sum_{j=1}^{n} \phi_j\, \alpha_j \tag{2.3}$$

where the $\phi_j$ are prescribed functions and the $\alpha_j$ are unknown parameters. Introducing (2.3) into the left hand sides of (2.1) and (2.2) will produce errors. The parameters $\alpha_j$ are determined by requiring that the weighted average of these errors is zero, that is

$$\int_V (\nabla^2 p + (\omega/c_o)^2 p)\, \delta p\, dV - \int_S \frac{\partial p}{\partial n}\, \delta p\, dS$$

$$- \int_{S_1} (\frac{\partial p}{\partial n} + i\rho_a \omega v_n)\, \delta p\, dS - \int_{S_2} (\frac{\partial p}{\partial n} + i\rho_a \omega \frac{p}{Z_s})\, \delta p\, dS = 0 \tag{2.4}$$

where

$$\delta p = \sum_{j=1}^{n} \phi_j\, \delta\alpha_j \tag{2.5}$$

Applying Green's theorem to the first term in (2.4) gives

$$- \int_V \nabla p . \nabla (\delta p) + \int_V (\omega/c_o)^2 \, p \, \delta p$$

$$- \int_{S_1} i \, \rho_a \omega v_n \, \delta p \, dS - \int_{S_2} i \, \rho_a \omega \frac{p}{Z_s} \, \delta p \, dS = 0 \qquad (2.6)$$

This equation can be written in the alternative form

$$\delta \left[ \frac{1}{2} \int_V \{(\nabla p)^2 - (\omega/c_o)^2 \, p^2\} \, dV + \int_{S_1} i\rho_a \omega \, p v_n \, dS \right.$$

$$\left. + \frac{1}{2} \int_{S_2} i \, \rho_a \omega \frac{p^2}{Z_s} \, dS \right] = 0 \qquad (2.7)$$

Equations (2.1) and (2.2) have now been replaced by the equivalent variational principle (2.7).

## 2.4 Solution by Finite Elements

An approximate solution to the variational principle (2.7) can be obtained using finite element techniques. The volume V is represented by an assemblage of three-dimensional finite elements. For example, the element described in Section 1.8. The pressure distribution within an element is approximated by an expression of the form

$$p = \lfloor N(x,y,z) \rfloor_e \, \{p\}_e \qquad (2.8)$$

This is used to evaluate the first two terms in equation (2.7) in the way described in the first chapter. Thus

$$\int_{V_e} (\nabla p)^2 \, dV = \{p\}_e^T [k]_e \{p\}_e$$

$$\int_{V_e} \frac{1}{c_o^2} \, p^2 dV = \{p\}_e^T [m]_e \{p\}_e \qquad (2.9)$$

where

$$[k]_e = \int_{V_e} [B]_e^T \; [B]_e \; dV$$

$$[m]_e = \int_{V_e} \frac{1}{c_o^2} \; \lfloor N \rfloor_e^T \; \lfloor N \rfloor_e \; dV \tag{2.10}$$

and

$$[B]_e = \begin{bmatrix} \frac{\partial}{\partial x} \\ \frac{\partial}{\partial y} \\ \frac{\partial}{\partial z} \end{bmatrix} \lfloor N \rfloor_e \tag{2.11}$$

If an element is in contact with either $S_1$ or $S_2$ it will contribute to one of the surface integrals in (2.7). Denote the variation of pressure over one surface of an element by $\lfloor N_a \rfloor_e$. This is obtained from the function $\lfloor N \rfloor_e$. For example, in the case of the 20 node isoparametric element described in Section 1.8, the variation of pressure over one of its surfaces will be given by the functions defined in Section 1.6.3 for an 8-node, two-dimensional element. Also, the normal velocity distribution over an element of $S_1$ (assumed identical in shape to one face of an acoustic element), is approximated by

$$v_n = \lfloor N_s \rfloor_e \{v\}_e \tag{2.12}$$

where $\{v\}_e$ is a column matrix of nodal velocity values.

Substituting the pressure and velocity variations into the third and fourth integrals in (2.7) gives

$$\int_{S_e} i\rho_a \, \omega p \, v_n \, dS = i\omega \, \{p\}_e^T \, [s]_e \, \{v\}_e \tag{2.13}$$

$$\int_{S_e} i\rho_a \, \omega \, \frac{p^2}{Z_s} \, dS = i\omega \, \{p\}_e^T \, [d]_e \, \{p\}_e \tag{2.14}$$

where

$$[s]_e = \int_{S_e} \rho_a \lfloor N_a \rfloor_e^T \lfloor N_s \rfloor_e \, dS \tag{2.15}$$

$$[d]_e = \int_{S_e} (\rho_a / Z_s) \lfloor N_a \rfloor_e^T \lfloor N_a \rfloor_e \, dS \tag{2.16}$$

Adding the contributions from each element and substituting into equation (2.7) gives the following equation

$$[\, \underset{\sim}{K}_a - \omega^2 \underset{\sim}{M}_a + i\omega \underset{\sim}{D}_a ]\{p\} = - i\omega \, [S]\{v\} \tag{2.17}$$

The matrices $[\underset{\sim}{K}_a]$ and $[\underset{\sim}{M}_a]$ are assembled in the way described in the first chapter, and the matrices $[D_a]$ and $[S]$ are formed from the element matrices $[d]_e$ and $[s]_e$ in the same way.

The specific acoustic impedance, $Z_s$, of the surface $S_2$ is complex and a function of frequency. This means that the matrix $[D_a]$ is also complex and a function of frequency. Writing

$$[D^i_a] = [D^r_a] + i \, [D^i_a] \tag{2.18}$$

equation (2.17) becomes

$$[\, \underset{\sim}{K}_a - \omega \underset{\sim}{D}^i_a - \omega^2 \underset{\sim}{M}_a + i\omega \underset{\sim}{D}^r_a \,] \{p\} = - i\omega \, [S]\{v\} \tag{2.19}$$

## 2.5 Transmission Loss of Acoustic Filters

Equation (2.19) has been used to analyse the performance of mufflers [2.1, 2.2] and acoustic filters [2.3, 2.4] both without and with sound absorbing walls. As an illustration of this the case without sound absorbing walls will be considered.

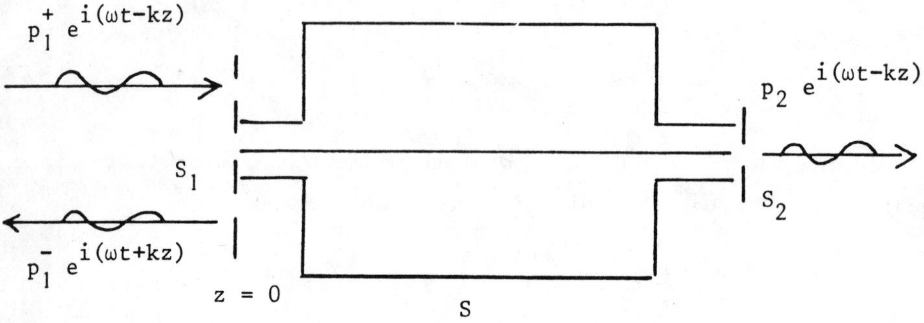


Fig. 2.2 Model for Transmission Loss Calculations

The transmission loss of acoustic filters can be calculated by considering the model shown in Figure 2.2. The system is excited by a plane wave moving from left to right. If the origin, z = 0, is taken to be at the entrance to the filter, the incident wave pressure is $p_1^+ e^{i\omega t}$ and the reflected wave pressure is $p_1^- e^{i\omega t}$. Thus, the total pressure at the input is $p_1 = (p_1^+ + p_1^-)e^{i\omega t}$ and the particle velocity $v_1 = (1/\rho_a c_o)(p_1^+ - p_1^-)e^{i\omega t}$. The 'prescribed' velocity over $S_1$ can, therefore, be expressed in terms of a given incident pressure and an unknown reflected pressure. It is assumed that a plane wave crosses the surface $S_2$ and so the specific acoustic impedance is $Z_s = \rho_a c_o$. This means that the matrix $[D_a]$ is real and so $[D_a^i] = 0$.

Equation (2.19) now becomes

$$\left[ \underset{\sim}{K}_a - \omega^2 \underset{\sim}{M}_a + i\,\omega\, \underset{\sim}{D}_a^r \right] \{p\} = - i\omega\, [S]\, \{v_1\} \tag{2.20}$$

The column matrix of pressures can be partitioned as follows

$$\{p\}^T = \lfloor \underset{\sim}{p}_1\ \underset{\sim}{p}_I\ \underset{\sim}{p}_2 \rfloor \tag{2.21}$$

where subscripts 1 and 2 refer to node points on surfaces $S_1$ and $S_2$ and I denotes all other node points. Matrices $[D^r_a]$ and $[S]$ take the following form:

$$[D^r_a] = \begin{bmatrix} 0 & 0 & 0 \\ 0 & 0 & 0 \\ 0 & 0 & \underset{\sim}{D}_{22} \end{bmatrix}, \quad [S] = \begin{bmatrix} \underset{\sim}{S}_{11} \\ 0 \\ 0 \end{bmatrix} \tag{2.22}$$

Equation (2.16) indicates that $[D_{22}]$ is a square symmetric matrix. Equation (2.15) indicates that $[S_{11}]$ will also be a square symmetric matrix, provided the variation of pressure and velocity are assumed to be the same.

Over $S_1$

$$\underset{\sim}{p}_1 = (\overset{+}{\underset{\sim}{p}_1} + \overset{-}{\underset{\sim}{p}_1}), \quad \underset{\sim}{v}_1 = \frac{1}{\rho_a c_o} (\overset{+}{\underset{\sim}{p}_1} - \overset{-}{\underset{\sim}{p}_1}) \tag{2.23}$$

Substituting (2.23) into (2.20) and rearranging so that unknown pressures are on the left and known pressures on the right gives:

$$[\underset{\sim}{K}_a - \omega^2 \underset{\sim}{M}_a + i\omega \underset{\sim}{D}_m] \{p_m\} = \{Q_m\} \tag{2.24}$$

where the modified column matrix of pressures is

$$\{p_m\}^T = \lfloor \overset{-}{\underset{\sim}{p}_1} \ \underset{\sim}{p}_I \ \underset{\sim}{p}_2 \rfloor \tag{2.25}$$

and the column matrix of source terms is

$$\{Q_m\} = \begin{bmatrix} - [\underset{\sim}{K}^1_a - \omega^2 \underset{\sim}{M}^1_a] - i(\omega/\rho_a c_o) [S_{11}] \\ 0 \\ 0 \end{bmatrix} \{\overset{+}{p_1}\} \tag{2.26}$$

The matrix $[\underset{\sim}{K}^1_a - \omega^2 \underset{\sim}{M}^1_a]$ represents the columns of the matrix $[\underset{\sim}{K}_a - \omega^2 \underset{\sim}{M}_a]$ which correspond to the input nodes.

The modified damping matrix is

$$[D_m] = \begin{bmatrix} (1/\rho_a\omega_o)S_{11} & 0 & 0 \\ 0 & 0 & 0 \\ 0 & 0 & D_{22} \end{bmatrix} \tag{2.27}$$

Equation (2.24) represents a set of linear equations with complex coefficients, which could be solved, for example, by means of Crout decomposition.

When the inlet and outlet of the system are small compared with the wavelength under consideration, the acoustic pressure can be assumed to be uniformly distributed over them. In this case the transmission loss is given by

$$TL = -20 \log_{10} \left| \frac{p_2}{p_1^+} \right| \tag{2.28}$$

2.6 Structural-Acoustic Interaction

If $S_1$ in Figure 2.1 represents the surface of a vibrating structure, then the velocity distribution will be given by the solution of the equations of motion of the structure. These can be formulated using Hamilton's Principle. This principle states that "among all the displacements which satisfy the geometrical boundary conditions and the prescribed conditions at $t = t_1$ and $t = t_2$, the actual solution renders the integral $\int_{t_1}^{t_2} (T - U + W)\, dt$ stationary". In this integral T denotes the kinetic energy of the system, U the strain energy and W the work done by the non-conservative forces. The principle can be stated briefly by the equation

$$\delta \int_{t_1}^{t_2} (T - U + W)\, dt = 0 \tag{2.29}$$

If the surface is a flat plate, then the energy expressions are given by

$$T = \frac{1}{2} \int_A \rho_s \{ h\, \dot{w}^2 + \frac{h^3}{12}\, \dot{\theta}_x^2 + \frac{h^3}{12}\, \dot{\theta}_y^2 \}\, dA \tag{2.30}$$

$$U = \frac{1}{2}\int_A \frac{h^3}{12}\{\chi\}^T[D^f]\{\chi\}\,dA + \int_A \kappa\{\gamma\}^T[D^s]\{\gamma\}\,dA \qquad (2.31)$$

$$\delta W = \int_A p_z\ \delta w\ dA \qquad (2.32)$$

where $\rho_s$ is the density of the material, h the thickness of the plate, w the lateral displacement, $\theta_x$ and $\theta_y$ the rotations of a normal about the x and y axes, and $p_z$ the lateral load per unit area. The 'strain' matrices are defined as

$$\{\chi\} = \{\ -\partial\theta_y/\partial x,\ \ \partial\theta_x/\partial y,\ (\partial\theta_x/\partial x - \partial\theta_y/\partial y) \qquad (2.33)$$

$$\{\gamma\} = \{\ (\theta_y + \partial w/\partial x),\ \ (-\theta_x + \partial w/\partial y)\ \} \qquad (2.34)$$

The elasticity matrices are given by

$$[D^f] = \frac{E}{(1-\nu^2)}\begin{bmatrix} 1 & \nu & 0 \\ \nu & 1 & 0 \\ 0 & 0 & \frac{1}{2}(1-\nu) \end{bmatrix} [D^s] = \frac{E}{2(1+\nu)}\begin{bmatrix} 1 & 0 \\ 0 & 1 \end{bmatrix} \qquad (2.35)$$

$\kappa$ is a constant which is introduced to account for the variation of the shear stresses and strains through the thickness. It is usual to take $\kappa$ to be either $\pi^2/12$ or 5/6.

Expressions (2.30) and (2.31) include rotary inertia and shear deformation effects and therefore apply to thick plates.

An approximate solution to Hamilton's Principle, (2.29), can be obtained using finite element techniques. The plate is represented by an assemblage of two-dimensional finite elements. For example, an eight-node isoparametric element (see Figure 1.10(a). In this case both the geometry and displacements are represented by the same functions, namely

$$x = \sum_{j=1}^{8} N_j\ (\xi,\eta)x_j\ ,\qquad y = \sum_{j=1}^{8} N_j\ (\xi,\eta)y_j \qquad (2.36)$$

$$w = \sum_{j=1}^{8} N_j(\xi,\eta) w_j, \qquad \theta_x = \sum_{j=1}^{8} N_j(\xi,\eta)\theta_{xj}, \quad \theta_y = \sum_{j=1}^{8} N_j(\xi,\eta)\,\theta_{yj} \tag{2.37}$$

where the functions $N_j(\xi,\eta)$ are defined in equations (1.53) to (1.55).

Substituting (2.37) into (2.30) and (2.31) and integrating over a single element gives

$$\begin{aligned} T_e &= \frac{1}{2}\{\dot{q}\}_e^T \; [m]_e \, \{\dot{q}\}_e \\ U_e &= \frac{1}{2}\{q\}_e^T \; [k]_e \, \{q\}_e \end{aligned} \tag{2.38}$$

where

$$[m]_e = \int_{A_e} [N]_e^T \begin{bmatrix} \rho_s h & & \\ & \rho_s h^3/12 & \\ & & \rho_s h^3/12 \end{bmatrix} [N]_e \, dA \tag{2.39}$$

$$[k]_e = \int_{A_e} \frac{h^3}{12} \, [B^f]_e^T \, [D^f] \, [B^f]_e \, dA + \int_{A_e} \kappa [B^s]_e^T \, [D^s] [B^s]_e \, dA$$

In these expressions

$$\{q\}_e = \{w_1 \; \theta_{x1} \; \theta_{y1} \; \dots \; w_8 \; \theta_{x8} \theta_{y8}\} \tag{2.40}$$

$$[N_e] = [[N_1] \; \dots\dots \; [N_8] \; ]$$

$$[B^f]_e = [[B_1^f] \; \dots \; [B_8^f] \; ] \; , \qquad [B^s]_e = [[B_1^s] \; \dots [B_8^2] \; ] \tag{2.41}$$

and

$$[N_j] = \begin{bmatrix} N_j(\xi,\eta) & 0 & 0 \\ 0 & N_j(\xi,\eta) & 0 \\ 0 & 0 & N_j(\xi,\eta) \end{bmatrix} \tag{2.42}$$

$$\left[B_j^f\right] = \begin{bmatrix} 0 & 0 & \partial N_j/\partial x \\ 0 & -\partial N_j/\partial y & 0 \\ 0 & -\partial N_j/\partial x & \partial N_j/\partial y \end{bmatrix} \qquad \left[B_j^s\right] = \begin{bmatrix} \partial N_j/\partial x & 0 & N_j \\ \partial N_j/\partial y & -N_j & 0 \end{bmatrix} \tag{2.43}$$

The integrals in (2.39) are transformed to $(\xi,\eta)$ coordinates by the technique described in Section 1.6.5.

Both the external loading, $p_z^e$, and the acoustic pressure, p, in the cavity contribute to the applied loading, $p_z$, in equation (2.32), that is

$$p_z = p_z^e + p \tag{2.44}$$

In this equation, both loadings are assumed to act in the z-direction, which coincides with the outward normal to the cavity.

Assuming that

$$w = \lfloor N_s \rfloor_e \{q\}_e \;, \qquad p = \lfloor N_a \rfloor_e \{p\} \tag{2.45}$$

as in Section 1.3, then equation (2.32) gives

$$\delta W_e = \{\delta w\}_e^T \{f_m\}_e + \{\delta w\}_e^T \frac{1}{\rho_a} [s]_e^T \{p\}_e \tag{2.46}$$

where

$$\{f_m\}_e = \int_{A_e} p_z^e \lfloor N_s \rfloor_e^T \, dA \tag{2.47}$$

and $[s]_e$ is defined by equation (2.15).

Adding the contribution from each element and substituting into equation (2.29) gives the following equation (assuming harmonic motion)

$$\left[ \underset{\sim}{K}_s - \omega^2 \underset{\sim}{M}_s + i\omega \underset{\sim}{D}_s \right] \{q\} - \frac{1}{\rho_a} [S]^T \{p\} = \{f_m\} \tag{2.48}$$

In this equation a damping matrix $[D_s]$ has been introduced. It is not usual to derive this matrix from a dissipation function for each element in the same manner as the strain and kinetic energies. Instead, simplified damping models for the complete structure are used.

Putting $\{v\} = i\,\omega\,\{q\}$ into equation (2.17) gives

$$\left[ \underset{\sim}{K}_a - \omega^2 \underset{\sim}{M}_a + i\omega \underset{\sim}{D}_a \right] \{p\} - \omega^2 \left[ S \right] \{q\} = 0 \qquad (2.49)$$

Equations (2.48) and (2.49) are coupled equations for the structural displacements and acoustic pressures. They have been used for interior noise studies of buildings [2.5], aircraft [2.6,2.7] and automobile interiors [2.8-2.11]. The equations can be solved either directly or by means of modal synthesis techniques [2.6,2.12,2.13]. These techniques have also been extended to study sound transmission between enclosures [2.14, 2.15].

## 2.7 References

2.1 A. CRAGGS 1976 Journal of Sound and Vibration 48, 377-392. A finite element method for damped acoustic systems: an application to evaluate the performance of reactive mufflers.

2.2 A. CRAGGS 1977 Journal of Sound and Vibration 54, 285-296. A finite element method for modelling dissipative mufflers with a locally reactive lining.

2.3 Y. KAGAWA and T. OMOTE 1976 Journal Acoustical Society America 60, 1003-1013. Finite element simulation of acoustic filters of arbitrary profile with circular cross-section.

2.4 Y. KAGAWA, T. YAMABUCHI and A. MORI 1977 Journal of Sound and Vibration 53, 357-374. Finite element simulation of axisymmetric acoustic transmission system with sound absorbing wall.

2.5 A. CRAGGS 1971 Journal of Sound and Vibration 15, 509-528. The transient response of a coupled plate-acoustic system using plate and acoustic finite elements.

2.6 M. PETYT and S.P. LIM 1980 International Journal for Numerical Methods in Engineering 13, 109-122. Finite element analysis of the noise inside a mechanically excited cylinder.

2.7 J.F. UNRUH 1980 AIAA Paper No. 80-1037 Structure-borne noise prediction for a single engine general aviation aircraft.

2.8 J.A. WOLF, Jr. and D.J. NEFSKE 1975 NASA Technical Memorandum X-3278 615-631 NASTRAN modelling and analysis of rigid and flexible walled acoustic cavities.

2.9 J.A. WOLF, Jr., D.J. NEFSKE and L.J. HOWELL 1976 Transactions of the Society of Automotive Engineers 85, 857-864. Structural-acoustic finite element analysis of the automobile passenger compartment.

2.10 D.J. NEFSKE and L.J. HOWELL 1978 Transactions of the Society of Automotive Engineers 87, 1726-1737. Automobile interior noise reduction using finite element methods.

2.11 D.J. NEFSKE, J.A. WOLF, Jr. and L.J. HOWELL 1982. Journal of Sound and Vibration 80, 247-266. Structural-acoustic finite element analysis of the automobile passenger compartment: a review of current practice.

2.12 J.A. WOLF, Jr. 1977 American Institute of Aeronautics and Astronautics Journal 15, 743-745. Modal synthesis for combined structural-acoustic systems.

2.13 J.F. UNRUH 1980 Journal of Aircraft 17, 434-441. Finite element sub-volume technique for structural-borne interior noise prediction.

2.14 A. CRAGGS 1973 Journal of Sound and Vibration 30, 343-357. An acoustic finite element approach for studying boundary flexibility and sound transmission between irregular enclosures.

2.15 A. CRAGGS and G. STEAD 1976 Acustica 35, 89-98. Sound transmission between enclosures - a study using plate and acoustic finite elements.

# 3. FINITE ELEMENT METHODS FOR DUCT ACOUSTICS

## 3.1 Introduction

The method of determining the propagation of sound in a duct, due to a given source distribution in the form of an input pressure or velocity, depends upon whether the duct is uniform or non-uniform and whether there is no mean flow or a mean flow present. All of these cases are considered in the following sections. Both two-dimensional and axisymmetric ducts have been analysed. The methods will be illustrated by considering two-dimensional ducts. Axisymmetric ducts can be reduced to a two-dimensional analysis by the technique described in the first chapter.

## 3.2 Ducts without mean flow

In previous chapters various acoustical problems were solved by obtaining solutions to the Helmholtz equation satisfying the appropriate boundary conditions. This approach can also be used for ducts without mean flow. A second approach, using the linearized equations of momentum and continuity for the fluid, will also be presented as an introduction to one of the methods used for ducts with mean flow. Methods for both uniform and non-uniform ducts are presented.

### 3.2.1 Uniform ducts

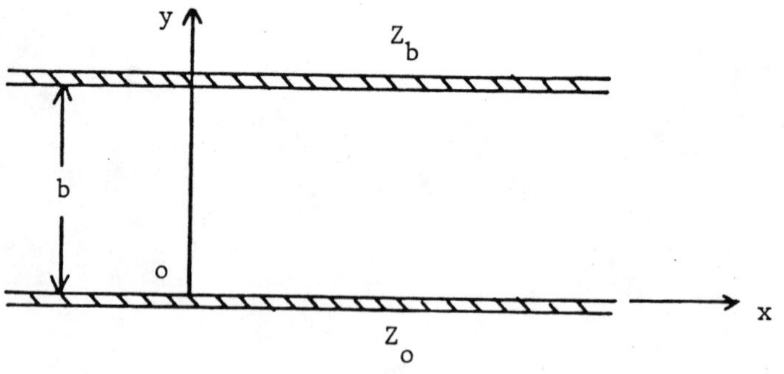


Fig. 3.1 Infinite, two-dimensional, uniform duct

Consider an infinite, two-dimensional, uniform duct of width b as shown in Figure 3.1. The upper and lower walls of the duct are acoustically treated with materials having specific impedances $Z_b$ and $Z_o$ respectively. The acoustic pressure, p, within the duct must satisfy the Helmholtz equation:

$$\nabla^2 p + (\omega/c_o)^2 p = 0 \tag{3.1}$$

where $\omega$ is the frequency and $c_o$ the speed of sound. Over the duct walls the fluid particle velocity normal to the surface is equal to the normal velocity of the surface. This gives rise to the boundary conditions:

$$\frac{\partial p}{\partial y} = i \rho_a \omega \frac{p}{Z_o} \quad \text{over } y = 0$$

$$\frac{\partial p}{\partial y} = -i \rho_a \omega \frac{p}{Z_b} \quad \text{over } y = 0 \tag{3.2}$$

where $\rho_a$ is the density of the acoustic medium.

The eigenfrequencies and eigenfunctions of the duct are obtained by solving equation (3.1) subject to the boundary conditions (3.2). The propagation of sound in the duct can be expressed in terms of these eigenfunctions.

Since the duct is uniform, the solution of (3.1) takes the form

$$p(x,y) = p(y)e^{-ik_x x} \tag{3.3}$$

where $k_x$ is a complex propagation constant in the x-direction.

Substituting (3.3) into (3.1) gives

$$\frac{d^2p}{dy^2} + k_y^2 \; p = 0 \tag{3.4}$$

where

$$k_y^2 = (\omega/c_o)^2 - k_x^2 \tag{3.5}$$

In Section 2.3 it is shown that a differential equation of the form (3.4) subject to the boundary conditions (3.2) can be replaced by an equivalent variational principle, namely:

$$\delta\left[\frac{1}{2}\int_{0}^{b}\{(\frac{dp}{dy})^2 - k_y^2\, p^2\}d_y + \frac{1}{2}\, i\rho_1\omega\, \{\, \frac{1}{Z_0}\,[p(0)]^2 + \frac{1}{Z_b}\,[p(b)]^2\}\right] = 0 \tag{3.6}$$

An approximate solution to this variational principle can be obtained using finite element techniques. The cross-section of the duct is represented by an assemblage of one-dimensional elements. For example, reference [3.1] uses two elements: a linear one with p as the only degree of freedom at two nodes, and a cubic one with p and dp/dy as degrees of freedom at the two nodes. The pressure distribution within an element is approximated by an expression of the form

$$p = \lfloor N(y) \rfloor_e \{p\}_e \tag{3.7}$$

This is used to evaluate the first two terms in equation (3.6) in the way described in previous chapters. Thus

$$\begin{aligned} \int_{-a}^{+a} (\frac{dp}{dy})^2\, dy &= \{p\}_e^T\, [k]_e\, \{p\}_e \\ \int_{-a}^{+a} p^2\, dy &= \{p\}_e^T\, [m]_e\, \{p\}_e \end{aligned} \tag{3.8}$$

where 2a is the length of the element and

$$\begin{aligned} [k]_e &= \int_{-a}^{+a} \lfloor dN/dy \rfloor_e^T\, \lfloor dN/dy \rfloor_e\, dy \\ [m]_e &= \int_{-a}^{+a} \lfloor N \rfloor_e^T\, \lfloor N \rfloor_e\, dy \end{aligned} \tag{3.9}$$

Adding the contributions from each element and substituting into equation (3.6) gives the following equation

$$[(\underset{\sim}{K}_a + i\,\omega\,\underset{\sim}{D}_a) - k_y^2\,\underset{\sim}{M}_a]\,\{p\} = 0 \tag{3.10}$$

The matrices $[K_a]$ and $[M_a]$ are assembled in the way described in previous chapters, and the matrix $[D_a]$ is formed from the third and fourth terms in (3.6).

If the cross-section of the duct is divided into four linear elements then $[K_a]$ is given by equation (1.20) and $[M_a]$ will be the same as equation (1.21) without the factor $(1/c_o^2)$, and $a = b/8$. Neither matrix requires the factor A in the present application. The matrix $[D_a]$ is defined as

$$[D_a] = \begin{bmatrix} \rho_a/Z_c & 0 & 0 & 0 & 0 \\ 0 & 0 & 0 & 0 & 0 \\ 0 & 0 & 0 & 0 & 0 \\ 0 & 0 & 0 & 0 & 0 \\ 0 & 0 & 0 & 0 & \rho_a/Z_b \end{bmatrix} \tag{3.11}$$

The eigenvalues, $k_y^2$, and corresponding eigenvectors of equation (3.10) are determined for a given frequency, $\omega$. Because of the presence of the term $i\omega[D_a]$ in (3.10), its eigenvalues will be complex. The axial wavenumber, $k_x$, is from (3.5) given by

$$k_x^2 = (\omega/c_o)^2 - k_y^2 \tag{3.12}$$

In general this will be complex also. The real part of $k_x$ represents the phase distribution of the mode, and the imaginary part describes the attenuation of the mode amplitude. Wave propagation is possible as long as the wavenumber has a real part.

An alternative procedure, which has no real advantage for the no mean flow case, uses the linearized equations of momentum and continuity for the fluid. It is a method which is used when a mean flow is present. Without a mean flow the equations are:

$$\begin{bmatrix} ik & 0 & \partial()/\partial x \\ 0 & ik & \partial()/\partial y \\ \partial()/\partial x & \partial()/\partial y & ik \end{bmatrix} \begin{bmatrix} u^* \\ v^* \\ p^* \end{bmatrix} = 0 \tag{3.13}$$

where $u^* = u/c_o$, $v^* = v/c_o$, $p^* = p/\rho_a c_o^2$, $k = \omega/c_o$. $u$, $v$ being the components of the perturbation velocity in the x,y directions. The boundary conditions (3.2) become, in terms of non-dimensional quantities

$$\begin{aligned} v^* &= -\,p^*/\bar{Z}_o \qquad \text{over } y = 0 \\ v^* &= \;\; p^*/\bar{Z}_b \qquad \text{over } y = b \end{aligned} \tag{3.14}$$

where $\bar{Z} = Z/\rho_a c_o$ .

Since the duct is uniform the solution of (3.13) takes the form

$$\lfloor u^* \; v^* \; p^* \rfloor = \lfloor \bar{u} \; \bar{v} \; \bar{p} \rfloor \; e^{-ik\mu x} \tag{3.15}$$

where $\mu$ is a non-dimensional axial wavenumber. Substituting (3.15) into (3.13) gives

$$\begin{bmatrix} ik & 0 & -ik\mu \\ 0 & ik & d()/dy \\ -ik\mu & d()/dy & ik \end{bmatrix} \begin{bmatrix} \bar{u} \\ \bar{v} \\ \bar{p} \end{bmatrix} = 0 \tag{3.16}$$

and (3.14) become

$$\bar{v} = -\,\bar{p}/\bar{Z}_o, \quad \bar{v} = \bar{p}/\bar{Z}_b \quad \text{over } y = 0,\ b \text{ respectively} \tag{3.17}$$

An approximate solution to equations (3.16) and (3.17) can be obtained using the method of weighted residuals. Equating the weighted average of the errors to zero gives

$$\int_0^b \left[ \delta\bar{u}\,(ik\bar{u} - ik\mu\bar{p}) + \delta\bar{v}\,(ik\bar{v} + d\bar{p}/dy) + \delta\bar{p}\,(-ik\mu\bar{u} + d\bar{v}/dy + ik\bar{p})\right] dy \tag{3.18}$$

$$+ \left[\bar{v}(0) + \bar{p}(0)/\bar{Z}_o\right] \delta\bar{p}(0) - \left[\bar{v}(b) - \bar{p}(b)/\bar{Z}_b\right]\delta\bar{p}(b) = 0$$

Now

$$\int_0^b \delta\bar{p}\, d\bar{v}/dy\, dy = \bar{v}(b)\, \delta p(b) - \bar{v}(0)\, \delta p(0) - \int_0^b \delta\bar{p}_y\, \bar{v}\, dy \tag{3.19}$$

Substituting (3.19) into (3.18) gives

$$\int_0^b \left[ \delta\bar{u}(ik\bar{u} - ik\mu\bar{p}) + \delta\bar{v}(ik\bar{v} + d\bar{p}/dy) + \delta\bar{p}\,(-ik\mu\bar{u} + ik\bar{p}) - \delta\bar{p}_y\bar{v}\right] dy$$

$$+ \delta\bar{p}(0)\, \bar{p}(0)/\bar{Z}_o + \delta\bar{p}(b)\, \bar{p}(b)/\bar{Z}_b = 0 \tag{3.20}$$

The cross-section of the duct is now represented by an assemblage of one-dimensional elements. The distribution of velocity and pressure within an element is approximated by

$$\bar{u} = \lfloor N(y)\rfloor_e \{\bar{u}\}_e\,, \quad \bar{v} = \lfloor N(y)\rfloor_e \{\bar{v}\}_e, \quad \bar{p} = \lfloor N(y)\rfloor_e \{\bar{p}\}_e \tag{3.21}$$

These expressions are used to evaluate the integral in (3.20). This gives

$$\int_{-a}^{+a} [\ldots\ldots] dy = \{q\}_e^T [k]_e \{q\}_e - \mu\{\delta q\}_e^T [m]_e \{q\}_e \tag{3.22}$$

where 2a is the length of the element,

$$\{q\}_e^T = \lfloor \{\bar{u}\}_e^T \ \{\bar{v}\}_e^T \ \{\bar{p}\}_e^T \rfloor \tag{3.23}$$

$$[k]_e = \int_{-a}^{+a} \begin{bmatrix} ik\lfloor N\rfloor_e^T \lfloor N\rfloor_e & 0 & 0 \\ 0 & ik\lfloor N\rfloor_e^T \lfloor N\rfloor_e & \lfloor N\rfloor_e^T \lfloor dN/dy\rfloor_e \\ 0 & -\lfloor dN/dy\rfloor_e^T \lfloor N\rfloor_e & ik\lfloor N\rfloor_e^T \lfloor N\rfloor_e \end{bmatrix} \qquad (3.24)$$

$$|m|_e = \int_{-a}^{+a} \begin{bmatrix} 0 & 0 & ik\lfloor N\rfloor_e^T \lfloor N\rfloor_e \\ 0 & 0 & 0 \\ ik\lfloor N\rfloor_e^T \lfloor N\rfloor_e & 0 & 0 \end{bmatrix} \qquad (3.25)$$

Adding the contributions from each element gives the following equation for arbitrary variations $\{\delta q\}$ :

$$[\underset{\sim}{K}_d - \mu \underset{\sim}{M}_d] \quad \{q\} = 0 \qquad (3.26)$$

The matrices $[K_d]$ and $[M_d]$ are assembled in the usual way. However, $[K_d]$ will also have additional contributions $1/\bar{Z}_o$ and $1/\bar{Z}_b$ added to the diagonal terms which correspond to the pressures at $y = 0$ and $b$ respectively. These arise from the last two terms in (3.20).

Equation (3.26) is a linear eigenvalue problem for the axial wavenumber $\mu$. In general this will be complex because of the presence of complex terms in $[K_d]$ and $[M_d]$. Because these matrices are also functions of $k = \omega/c_o$, then again the axial wavenumbers are determined for a given frequency.

3.2.2 Non-uniform ducts

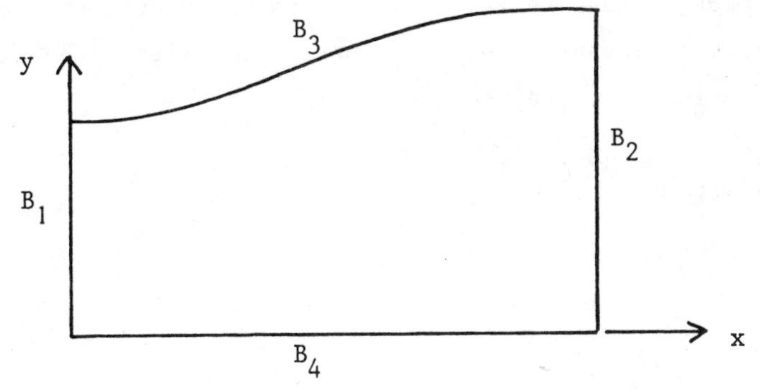


Fig. 3.2 Two-dimensional, non-uniform duct

Figure 3.2 shows a two-dimensional duct of variable corss-section and wall impedance. the walls of the duct are the boundaries $B_3$ and $B_4$. Because of the varying geometry, it is necessary to analyse a finite length of the duct which is terminated by the boundaries $B_1$ and $B_2$. It is usual to specify the pressure distribution over $B_1$ and the specific acoustic impedance over $B_2$. In many cases this is taken to be $\rho_a c_o$, the value for plane waves. Reference [3.2] considers the case when two semi-infinite, uniform ducts are joined by a transition section of variable cross-section. In this case the pressures in the uniform sections are represented by infinite sums of known eigen functions and equality of momentum and continuity across $B_1$ and $B_2$ is required.

Both Helmholtz equation and momentum/continuity equation formulations are possible. The Helmholtz equation formulation is identical to the analysis of acoustical filters presented in Section 2.5. Examples of the use of this approach can be found in references [3.2-3.4]. In these references 4, 8 and 12-node isoparametric elements are used.

The momentum/continuity equation formulation is used in references [3.2, 3.5] in which 8 and 4-node isoparametric elements are used respectively. The development of the equations parallels that given for uniform ducts, the main difference being that the assumption (3.15) cannot be made. The integral in (3.18) is taken over the area bounded by $B_1$-$B_4$ in Figure 3.2 and also along these boundaries. The expression is simplified using the divergence theorem.

## 3.3 Ducts with mean flow

### 3.3.1 Uniform ducts

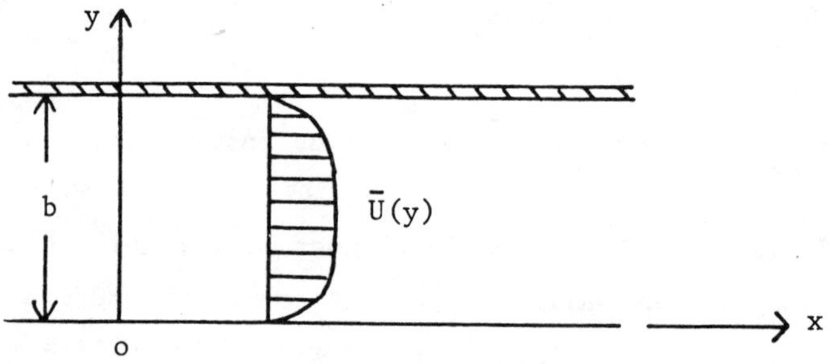


Fig. 3.3 Two-dimensional, uniform duct with shear flow

Figure 3.3 shows an infinite, two-dimensional, uniform duct which contains a fluid moving with a steady state axial velocity $\bar{U}(y)$ which is a function of the transverse coordinate y only. In this case, the linearized equations of momentum and continuity (3.13) become

$$\begin{bmatrix} ik + M\,\partial()/\partial x & dM/dy & \partial()/\partial x \\ 0 & (ik + M\,\partial()/\partial x) & \partial()/\partial y \\ \partial()/\partial x & \partial()/\partial y & (ik + M\,\partial()/\partial x) \end{bmatrix} \begin{bmatrix} u^* \\ v^* \\ p^* \end{bmatrix} = 0 \qquad (3.27)$$

where $M = \bar{U}/c_0$. The duct, in this case, is assumed to have a rigid wall at y = 0 and an acoustically treated wall at y = b. The specific acoustic impedance of the material is Z. The boundary conditions are therefore

$$v^* = 0 \text{ on } y = 0, \quad v^* = p^*/\bar{Z} - (i/k)M\,\partial(p^*/\bar{Z})\,/\,\partial x \text{ on } y = b \qquad (3.28)$$

Again the solution takes the form (3.15). Substituting into (3.27) gives

$$\begin{bmatrix} ik(1-\mu M) & dM/dy & -ik\mu \\ 0 & ik(1-\mu M) & d()/dy \\ -ik\mu & d()/dy & ik(1-\mu M) \end{bmatrix} \begin{bmatrix} \bar{u} \\ \bar{v} \\ \bar{p} \end{bmatrix} = 0 \quad (3.29)$$

and (3.28) become

$$\bar{v} = 0 \text{ on } y = 0, \quad \bar{v} = (1-\mu M)\bar{p}/\bar{Z} \quad (3.30)$$

Equations (3.29) together with the boundary conditions (3.30) can be solved using the method of weighted residuals as described in Section 3.2.1. Further details can be found in reference [3.6]. This reference uses a quadratic element with three nodes having three degrees of freedom at each node. A comparison of results with those produced by other techniques indicates the presence of a number of spurious eigenvalues and eigenvectors. Reference [3.7] demonstrates that these can be eliminated by using cubic elements with two nodes having six degrees of freedom at each node.

An alternative procedure is to eliminate $\bar{u}$ and $\bar{v}$ from equations (3.29). This gives

$$\frac{d^2\bar{p}}{dy^2} + \frac{2\mu dM/dy}{(1-\mu M)}\frac{d\bar{p}}{dy} + k^2\left[(1-\mu M)^2 - \mu^2\right]\bar{p} = 0 \quad (3.31)$$

This equation reduces to equation (3.4) when there is no mean flow. The boundary conditions for the situation illustrated in Figure 3.3 are

$$\frac{d\bar{p}}{dy} = 0 \text{ on } y = 0, \quad \frac{d\bar{p}}{dy} = -ik(1-\mu M)\frac{\bar{p}}{\bar{Z}} \text{ on } y = b \quad (3.32)$$

An approximate solution to equations (3.16) and (3.17) can be obtained using the method of weighted residuals in manner similar to that presented in Section 3.2.1. Equating the weighted average of the errors to zero gives

$$\int_0^b \left[ \delta\bar{p}(1-\mu M) \frac{d^2\bar{p}}{dy^2} + 2\mu\ \delta\bar{p} \frac{dM}{dy} \frac{d\bar{p}}{dy} + k^2 (1-\mu M) \{(1-\mu M)^2 - \mu^2\}\delta\bar{p}\ \bar{p} \right]dy$$

$$+ \delta\bar{p}(0) \frac{d\bar{p}(0)}{dy} - \delta\bar{p}(b) \left[ \frac{d\bar{p}(0)}{dy} + \frac{ik\bar{p}(b)}{\bar{Z}} \right] = 0 \qquad (3.33)$$

In this and the following equations it is assumed that $M(0) = 0 = M(b)$, for simplicity. Now

$$\int_0^b \delta\bar{p}(1-\mu M) \frac{d^2\bar{p}}{dy^2} dy = \left[\delta\bar{p}(1-\mu M) \frac{d\bar{p}}{dy}\right]_0^b$$

$$- \int_0^b \left[\delta\left(\frac{d\bar{p}}{dy}\right) (1-\mu M) \frac{d\bar{p}}{dy} - \delta\bar{p}\,\mu \frac{dM}{dy} \frac{d\bar{p}}{dy}\right] dy \qquad (3.34)$$

Substituting (3.34) into (3.33) and rearranging gives

$$\int_0^b \left[k^2\ \delta\bar{p}\ \bar{p} - \delta\left(\frac{d\bar{p}}{dy}\right) \frac{d\bar{p}}{dy}\right] dy - ik\ \delta\bar{p}(b)\ \bar{p}(b)/\bar{Z}$$

$$+ \int_0^b \left[\delta\left(\frac{d\bar{p}}{dy}\right) M \frac{d\bar{p}}{dy} + 3\,\delta\bar{p} \frac{dM}{dy} \frac{d\bar{p}}{dy} - 3\ k^2\ M\ \delta\bar{p}\ \bar{p}\right] dy\ \mu$$

$$+ \int_0^b k^2(3M^2-1)\ \delta\bar{p}\ \bar{p}\ dy\ \mu^2 + \int_0^b k^2(M-M^3)\ \delta\bar{p}\ \bar{p}\ dy\ \mu^3 = 0 \qquad (3.35)$$

Representing the cross-section of the duct by an assemblage of one-dimensional elements assuming

$$\bar{p} = \lfloor N(y) \rfloor_e \{\bar{p}\}_e \qquad (3.36)$$

within each element gives

$$\int_{-a}^{+a} [\ldots\ldots]\ dy = \{\delta\bar{p}\}_e^T \left[\underset{\sim}{a}_e + \underset{\sim}{b}_e\mu + \underset{\sim}{c}_e\mu^2 + \underset{\sim}{d}_e\mu^3\right] \{p\}_e \qquad (3.37)$$

where

$$[a]_e = \int_{-a}^{+a} \left[k^2 \lfloor N \rfloor_e^T \lfloor N \rfloor_e - \lfloor dN/dy \rfloor_e^T \lfloor dN/dy \rfloor_e\right] dy \qquad (3.38)$$

$$[b]_e = \int_{-a}^{+a} [M \lfloor dN/dy \rfloor_e^T \lfloor dN/dy \rfloor_e + 3\ dM/dy \lfloor N \rfloor_e^T \lfloor dN/dy \rfloor_e$$

$$- 3k^2 M \lfloor N \rfloor_e^T \lfloor N \rfloor_e]\ dy \qquad (3.39)$$

$$[c]_e = \int_{-a}^{+a} k^2\ (3M^2-1)\ \lfloor N \rfloor_e^T \lfloor N \rfloor_e\ dy \qquad (3.40)$$

$$[d]_e = \int_{-a}^{+a} k^2\ (M-M^3)\ \lfloor N \rfloor^T \lfloor N \rfloor_e\ dy \qquad (3.41)$$

Adding the contributions from each element gives the following equation for arbitrary values of $\{\delta\bar{p}\}$

$$[\ \underset{\sim}{A} + \underset{\sim}{B}\ \mu + \underset{\sim}{C}\ \mu^2 + \underset{\sim}{D}\ \mu^3\ ]\ \{\bar{p}\} = 0 \qquad (3.42)$$

The matrices $[A]$, $[B]$, $[C]$ and $[D]$ are assembled in the usual way. However, $[A]$ will also have an additional contribution $-\ ik/\bar{Z}$ added to the diagonal term which corresponds to the pressure at $y = b$.

Equation (3.42) is a cubic eigenproblem in $\mu$ which can be transformed to a linear eigenproblem and then solved by standard techniques. Examples of the use of this technique can be found in references [3.8, 3.9]. Reference [3.8] uses three-node quadratic elements whilst reference [3.9] uses two-node cubic elements. Reference [3.9] also considers the effect of a temperature variation across the duct.

### 3.3.2 Non-uniform ducts

Consider the non-uniform duct shown in Figure 3.2 in which a mean flow is present. In the following discussion it will be assumed that the velocity distribution of the mean flow is known. Again both momentum/continuity equation and pressure equation formulations are possible.

References [3.5, 3.10, 3.11] use the momentum/continuity equation formulation. Four and 8-node isoparametric elements are used respectively.

The formulation follows that of uniform ducts in Section 3.3.1, the main difference being that there are additional terms in equation (3.27) due to the fact that the mean flow velocity has two components now. In addition, assumption (3.15) cannot be made. The comments made in Section 3.2.2 regarding the integration apply.

A pressure formulation is used in reference [3.3]. The duct is idealised using 12-node isoparametric elements. However, the analysis has been simplified by assuming that only the axial component of the mean flow velocity is significant.

References [3.12-3.14] use a velocity potential formulation. The duct is idealised by means of triangular elements with three nodes having one degree of freedom, the velocity potential, at each node. The formulation is very similar to a pressure formulation.

## 3.4 References


3.1 W. WATSON and D.L. LANSING 1976 NASA TN D-8186. A comparison of matrix methods for calculating eigenvalues in acoustically lined ducts.

3.2 R.J. ASTLEY and W. EVERSMAN 1978 Journal of Sound and Vibration 57, 367-388. A finite element method for the transmission in non-uniform ducts without flow: comparison with the method of weighted residuals.

3.3 S-F.LING 1976 Purdue University, Ph.D. Thesis. A finite element method for duct acoustic problems.

3.4 L. CEDERFELDT 1979 Swedish Council for Building Research. Document D4: 1979. On the use of finite element method on some acoustical problems.

3.5 A.L. ABRAHAMSON 1977 American Institute of Aeronautics and Astronautics Paper 77-1301. A finite element algorithm for sound propagation in axisymmetric ducts containing compressible mean flow.

3.6 R.J. ASTLEY and W. EVERSMAN 1979 Journal of Sound and Vibration 65, 61-74. A finite element formulation of the eigenvalue problem in lined ducts with flow.

3.7 R.J. ASTLEY and W. EVERSMAN 1980 Journal of Sound and Vibration 69, 13-25. The finite element duct eigenvalue problem: an improved formulation using Hermitian elements and no flow condensation.

3.8 M.S.Y. FAHMY 1979 University of Southampton, ISVR Technical Report No. 105. A finite element formulation of the eigenvalue problem of sound propagation in uniform ducts with shear flow.

3.9 S.B. DONG and C.Y. LIU 1979 Journal of Acoustical Society of America 66, 548-555. A finite element analysis of sound propagation in a non uniform moving medium.

3.10 R.J. ASTLEY and W. EVERSMAN 1981 Journal of Sound and Vibration 74, 103-121. Acoustic transmission in non-uniform ducts with mean flow, Part II: The finite element method.

3.11 R.J. ASTLEY and W. EVERSMAN 1981 Journal of Sound and Vibration 76, 595-601. A note on the utility of a wave envelope approach in finite element duct transmission studies.

3.12 R.K. SIGMAN, R.K. MAJJIGI and B.T. ZINN 1978 American Institute of Aeronautics and Astronautics Journal 16, 1139-1145. Determination of turbo fan inlet acoustics using finite elements.

3.13 R.K. SIGMAN and B.T. ZINN 1980 18th Aerospace Sciences Meeting, Paper No. AIAA-80-0085. Theoretical determination of nozzle admittances using a finite element approach.

3.14 K.J. BAUMEISTER and R.K. MAJJIGI 1980 American Institute of Aeronautics and Astronautics Journal 18, 509-514. Application of velocity potential function to acoustic duct propagation using finite elements.

# WAVE PROPAGATION ABOVE LAYERED MEDIA

Dominique HABAULT
Laboratoire de Mécanique et d'Acoustique
31, chemin Joseph Aiguier, B.P. 71,
13277 Marseille cedex 9, France.

## 1. - INTRODUCTION

This lecture is devoted to the study of the diffraction of a spherical wave by an absorbing plane.

Three methods are proposed to obtain an exact representation of the emitted pressure. They are very general and can be applied to many kinds of absorbing media (homogeneous, stratified,...). They lead to three different but equivalent representations.

Of course, this diffraction problem is of great interest for engineering purposes. Indeed, the prediction of the ground effect is one of the main problems of the outdoor sound propagation. The ground effect must be taken into account as well as the diffraction by obstacles (buildings, barriers,...) and the effects of the atmospheric conditions (gradient of temperature, turbulence,...).

Here we consider the diffraction, by an homogeneous plane ground, of a spherical wave emitted by a harmonic, point source in a three-dimensional space. We assume that the air behaves like a homogeneous fluid at rest, i.e. we consider idealized atmospheric conditions.
This diffraction problem has been extensively studied /1 - 16/. The first step is the choice of a ground model. In most of the articles dedicated to ground effects, the ground is assumed to be locally reacting, i.e. it is only characterized by a specific normal impedance. This model seems to be often satisfying for grassy surfaces,... The other models studied are the layer of homogeneous, isotropic, porous medium of infinite or finite thickness. Thus, the ground is characterized by a complex sound speed, a complex density and the thickness of the layer. A fourth model seems to be of great interest for the ground effect : a layer of porous medium with a porosity decreasing with the depth.

The examples proposed in this lecture correspond to the first and third models ("local reaction", finite thickness layer of homogeneous medium).

Most of the authors have started with the expression of the two dimensional Fourier transform of the acoustic pressure (or the velocity potential). Then, using saddle point and steepest descent methods, they have obtained exact representations of the sound field (including a Laplace type integral and a surface wave term) or more frequently, approximations of the acoustic far field. These approximations are given by the first terms of asymptotic series or by using the error function.

The methods presented here lead to exact expressions of the acoustic field. They can also be applied for other diffraction problems (diffraction by membranes, plates,...) /17-18/ and for electromagnetic cases. Finally, let us note that similar solutions can be obtained for two-dimensional cases.

The first method, described in chapter 2, is the most general. The main point is to decompose the bidimensional Fourier transform of the acoustic pressure to make appear Fourier transforms of well known functions or functions easy to evaluate by a residue method. Thus, this leads to an exact expression of the pressure using layer potentials.

Definitions and properties of layer potentials can be found in /19/.

The pressure is expressed as the sum of the incident field, the reflected field (from the image source) and the field emitted by layers of monopoles and dipoles located in the plane passing through the image source and parallel to the surface. This kind of representation is easy to obtain for a lot of diffraction problems and has an easy physical interpretation. But it is not very convenient for numerical purposes.

The second method, described in chapter 3, consists of representing the acoustic field as the sum of the incident field, the reflected field, a surface wave term and a line integral. This integral is interpreted as the radiation of a line source located on a vertical axis passing through the primary source and its image. Many papers have been devoted to the existence of a "surface wave". Many authors have argued that they have measured it. But in the light of chapters 2 and 3, it is clear that the surface wave term is mainly a mathematical tool; it appears or not, depending on the choosen representation. So, the interest of this chapter is to show how a surface wave term can appear. The expression obtained for the pressure has an easy physical interpretation. Furthermore, it can be computed easily.

The aim of the last method, presented in chapter 4, is to obtain another representation of the acoustic field which can be computed by very rapid and simple techniques, on a pocket computer. The pressure is expressed as the sum of the incident and reflected field, a surface wave term and a Laplace type integral. Because of its exponentially decreasing term, this integral can be evaluated more rapidly than the oscillating integral obtained in chapter 3.

## 2. - REPRESENTATION OF THE PRESSURE, USING LAYER POTENTIALS

Let $R^3(0,x,y,z)$ be the three-dimensional space. The aim of the chapter is to obtain an exact representation of the sound field emitted by a spherical, harmonic ($e^{-i.\omega t}$) point source, over an absorbing plane.

The method presented here is quite general. It has been applied to problems of diffraction by plates, membranes and several ground models.

In section 2.1, we give the details of the method for the "local reaction" case. The section 2.2 corresponds to the model of a finite depth layer of porous medium. These two cases were studied in /20-21/.

### 2.1. - "Local reaction" case

In this paragraph, the absorbing plane is assumed to be completely described by a complex, constant parameter : the normal specific impedance $\zeta$. This model is the simplest one, it is called the "locally reacting surface" model.

Let $(z > 0)$ be the half-space occupied by a fluid characterized by a density $\rho_1$ and a sound speed $c_1$ . p is the pressure emitted by a point spherical source located at $S = (0, 0, s > 0)$. The plane $(z = 0)$ represents the surface of the ground. Then, p must satisfy the system :

(2-1) $$(\Delta + k^2)\, p(x, y, z) = \delta(x) \otimes \delta(y) \otimes \delta(z - s) \qquad \text{for } z > 0$$

(2-2) $$\left(\frac{\partial}{\partial n} + \frac{ik}{\zeta}\right) p(x, y, z) = 0 \qquad \text{for } z = 0$$

(2-3) limit absorption principle.

$\delta$ is the Dirac measure; $k = \omega/c_1$ is the wave number; $\vec{n}$ is the vector normal to the plane $(z = 0)$, exterior to $(z < 0)$ . $\frac{\partial}{\partial n} = \nabla . \vec{n}$ (see fig. 1).

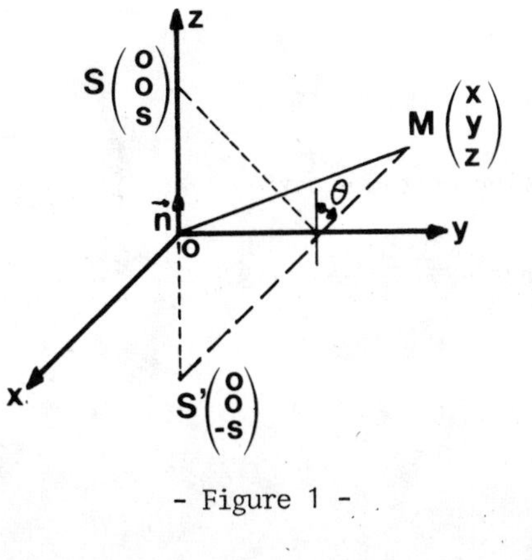


- Figure 1 -

The limit absorption principle is equivalent to the Sommerfeld condition. It consists of replacing $\omega$ by $(\omega + i\varepsilon)$, $\varepsilon > 0$. Then, the new system of equations obtained has a unique bounded solution and the solution of the system (2-1 to 3) is given by $p = \lim_{\substack{\varepsilon \to 0 \\ \varepsilon > 0}} p_\varepsilon$

This limit absorption principle is described by C. Wilcox /22/. It is shown that the limit p is unique.

To obtain $p_\varepsilon$ , use is made of the two dimensional Fourier transform with respect to the x and y axis. Because of the symmetry of the problem it can be defined by:

(2-4) $$\mathcal{F} f(\rho, z) = \hat{f}(\xi, z) = 2\pi \int_0^{+\infty} f(\rho, z)\, J_0(\xi\rho)\, \rho\, d\rho$$

and

(2-5) $$\mathcal{F}^{-1}(\hat{f}(\xi, z)) = f(\rho, z) = \frac{1}{4\pi} \int_{-\infty}^{+\infty} \hat{f}(\xi, z)\, H_0^{(1)}(\xi\rho)\, \xi\, d\xi$$

where $\rho = \sqrt{x^2+y^2}$ is the radial coordinate,

$J_0$ is the Bessel function of zero order,

$H_0^{(1)}$ is the Hankel function of zero order . $H_0^{(1)}(-z) = H_0^{(1)}(z\, e^{i\pi})$

Then, the system of equations becomes :

(2-6) $$\left(\frac{\partial^2}{\partial z^2} + \bar{k}^2 - \xi^2\right) \hat{p}_\varepsilon(\xi, z) = \delta(z - s) \qquad \text{for } z > 0$$

(2-7) $$\left(\frac{\partial}{\partial z} + i\frac{\bar{k}}{\zeta}\right) \hat{p}_\varepsilon(\xi, z) = 0 \qquad \text{for } z = 0$$

where $\bar{k} = (\omega + i\varepsilon)/c_1$.

A general solution of the equation (2-6) is given by :

(2-8) $$\hat{p}_\varepsilon(\xi, z) = \frac{e^{i K|z-s|}}{2 i K} + \hat{A}(\xi)\, \frac{e^{i K(z+s)}}{2 i K} \qquad \text{for } z \geqslant 0$$

where $K^2 = \bar{k}^2 - \xi^2$ ; $\mathrm{Im}\, K > 0$.

$\hat{A}(\xi)$ is the plane wave reflection coefficient. It is determined by the equation (2-7):

(2-9) $$\hat{A}(\xi) = \frac{K - \bar{k}/\zeta}{K + \bar{k}/\zeta}$$

The next step is to decompose $\hat{p}_\varepsilon$ . The following method is used: let multiply the numerator and the denominator of $\hat{A}(\xi)$ by the expression $(K - \bar{k}/\zeta)$, to remove the root K from the denominator.

$$\hat{A}(\xi) = \frac{K^2 + \bar{k}^2/\zeta^2 + 2iK \, . \, i\bar{k}/\zeta}{K^2 - \bar{k}^2/\zeta^2}$$

Let $\hat{B}(\xi^2)$, $B_\infty$ and $\hat{\mu}_1(\xi^2)$ be defined by :

$$\hat{B}(\xi^2) = \frac{K^2 + \bar{k}^2/\zeta^2}{K^2 - \bar{k}^2/\zeta^2} \quad ; \quad B_\infty = \lim_{|\xi| \to \infty} \hat{B}(\xi^2) = 1$$

$$\hat{\mu}_1(\xi^2) = \hat{B}(\xi^2) - B_\infty = \frac{2\bar{k}^2/\zeta^2}{K^2 - \bar{k}^2/\zeta^2}$$

$$\hat{\mu}_2(\xi^2) = \frac{2i\bar{k}/\zeta}{K^2 - \bar{k}^2/\zeta^2}$$

Such a decomposition of the term $\hat{A}(\xi) . e^{iK|z+s|}/2iK$ points out three different waves emitted by the image source $S' = (0, 0, -s)$.

(2-10) $$\hat{A}(\xi)\frac{e^{iK|z+s|}}{2iK} = B_\infty \frac{e^{iK|z+s|}}{2iK} + \hat{\mu}_1(\xi^2)\frac{e^{iK|z+s|}}{2iK} + \hat{\mu}_2(\xi^2)\, iK \frac{e^{iK|z+s|}}{2iK}$$

Because the inverse Fourier transform of $e^{iK|z+s|}/2iK$ is

$-e^{i\bar{k}r(S',X)}/4\Pi r(S', X)$ , it is easily seen that :

- the first term of (2-10) is the Fourier transform of a reflected wave emitted by S', with a constant amplitude,

- the second term is a plane wave with an amplitude depending on $\xi^2$. So, its inverse Fourier transform can be interpreted as the expression of the radiation of sources (monopoles) situated in the plane (z = -s).

- the third term is analogous to the z-derivative of the second one. It also can be interpreted as the expression of the radiation of dipoles situated in the plane (z = -s).

Now, it is easy to obtain p because $\hat{\mu}_1$ and $\hat{\mu}_2$ are the Fourier transforms of well-known functions :

$$\hat{\mu}_1(\xi^2) = \frac{2\bar{k}^2/\zeta^2}{-\xi^2+\bar{\alpha}^2} \qquad \text{where} \qquad \bar{\alpha}^2 = \bar{k}^2(1-1/\zeta^2)$$

$$\Rightarrow \mu_1(\rho) = \frac{2\bar{k}^2}{\zeta^2}\,\mathscr{F}^{-1}\left[\frac{1}{-\xi^2+\bar{\alpha}^2}\right] = \frac{2\bar{k}^2}{\zeta^2}\left(-\frac{i}{4}\right) H_0^{(1)}(\bar{\alpha}\rho)$$

$$\hat{\mu}_2(\xi^2) = \frac{2i\bar{k}/\zeta}{-\xi^2+\bar{\alpha}^2} \quad \Rightarrow \mu_2(\rho) = \frac{2i\bar{k}}{\zeta}\left(-\frac{i}{4}\right) H_0^{(1)}(\bar{\alpha}\rho)$$

and :

$$\mathscr{F}^{-1}\left[\hat{\mu}_1(\xi^2)\,\frac{e^{iK|z+s|}}{2iK}\right] = -\left(\mu_1(\rho)\otimes\delta_{z=-s} * \frac{e^{i\bar{k}r}}{4\Pi r}\right)(M) \tag{2-11}$$

$$= -\int_{z'=-s} \mu_1(\rho(P))\,\frac{e^{i\bar{k}r(M,P)}}{4\Pi r(M,P)}\,d\sigma(P)$$

where M = (x,y,z) ; P = (x',y', -s) is a point of the plane (z = -s) ;

$\rho(P) = \sqrt{x'^2+y'^2}$ is the radial coordinate of P ; $d\sigma(P) = dx'dy'$.

This integral is a simple layer potential, of density $\mu_1$. Definitions and properties of layer potentials can be found in /19/.

Likewise,

$$\mathscr{F}^{-1}\left[\hat{\mu}_2(\xi^2)\,iK\,\frac{e^{iK|z+s|}}{2iK}\right] = -\frac{\partial}{\partial z}\int_{z'=-s} \mu_2(\rho(P))\,\frac{e^{i\bar{k}r(M,P)}}{4\Pi r(M,P)}\,d\sigma(P)$$

Finally, the pressure p(M) is given by $\lim_{\varepsilon \to 0} p_\varepsilon$ and can be represented by :

(2-12)
$$p(M) = -\frac{e^{ikr(S,M)}}{4\Pi r(S,M)} - \frac{e^{ikr(S',M)}}{4\Pi r(S',M)}$$

$$+ \frac{ik^2}{2\zeta^2} \int_{z'=-s} H_0^{(1)}(\alpha\rho(P)) \frac{e^{ikr(M,P)}}{4\Pi r(M,P)} d\sigma(P)$$

$$- \frac{k}{2\zeta} \frac{\partial}{\partial z} \int_{z'=-s} H_0^{(1)}(\alpha\rho(P)) \frac{e^{ikr(M,P)}}{4\Pi r(M,P)} d\sigma(P)$$

where $\alpha^2 = k^2(1 - 1/\zeta^2)$ , $\operatorname{Im} \alpha > 0$

The layer potential and its derivative are indefinitely differentiable, if M is out of the plane (z' = -s). The solution p is indefinitely differentiable everywhere in the half-space (z > 0) except in S.

2.2. - Finite depth layer of porous medium

For this model, the decomposition of the reflexion coefficient $\hat{A}(\xi)$ to obtain Fourier transform of well-known functions is not obvious. So, use is made of integration in the complex $\xi$ - plane.

The half-space (z > 0) is assumed to be occupied by a fluid (density $\rho_1$ and sound speed $c_1$). The layer of homogeneous, isotropic porous medium corresponds to (0 < z < -h) and is characterized by the complex density $\rho_2$ and the complex sound speed $c_2$. The plane (z = -h) is assumed to be a perfectly reflecting plane (Neumann condition). Let $p_1$ and $p_2$ be the pressure in (z ⩾ 0) and (0 ⩽ z ⩽ -h) respectively. They satisfy the following system :

(2-13) $$(\Delta + k_1^2)\, p_1(x, y, z) = \delta(x) \otimes \delta(y) \otimes \delta(z - s) \quad \text{for } z > 0$$

(2-14) $$(\Delta + k_2^2)\, p_2(x, y, z) = 0 \quad \text{for } 0 < z \leqslant -h$$

(2-15) $$p_1(x, y, z) = p_2(x, y, z) \quad \text{for } z = 0$$

(2-16) $$\frac{1}{\rho_1} \frac{\partial}{\partial z} p_1(x, y, z) = \frac{1}{\rho_2} \frac{\partial}{\partial z} p_2(x, y, z) \quad \text{for } z = 0$$

(2-17) $$\frac{\partial}{\partial z} p_2(x, y, z) = 0 \qquad \text{for } z = -h$$

(2-18) limit absorption principle for $p_1$ and $p_2$.

where $k_i = \omega / c_i$.

Again, we use the two dimensional Fourier transform. The transforms of equ.(2-13 and 14) imply that :

$$\hat{p}_1(\xi, z) = \frac{e^{iK_1|z-s|}}{2iK_1} + \hat{A}(\xi)\frac{e^{iK_1(z+s)}}{2iK_1} \qquad \text{for } z \geqslant 0$$

$$\hat{p}_2(\xi, z) = \hat{A}_1(\xi)\frac{e^{iK_2 z}}{2iK_2} + \hat{A}_2(\xi)\frac{e^{-iK_2 z}}{2iK_2} \qquad \text{for } 0 \leqslant z \leqslant -h$$

where : $K_i^2 = k_i^2 - \xi^2$ ; $\operatorname{Im} K_i > 0$.

The conditions (2-16 and 17) lead to :

$$\hat{A}(\xi) = \frac{K_1 \rho_2 \cos K_2 h + iK_2 \rho_1 \sin K_2 h}{K_1 \rho_2 \cos K_2 h - iK_2 \rho_1 \sin K_2 h}$$

$$= \frac{(\rho_2^2 K_1^2 \cos^2 K_2 h - \rho_1^2 K_2^2 \sin^2 K_2 h) + iK_1(2\rho_1\rho_2 K_2 \sin K_2 h \,.\, \cos K_2 h)}{D(\xi^2)}$$

with $D(\xi^2) = \rho_2^2 K_1^2 \cos^2 K_2 h + \rho_1^2 K_2^2 \sin^2 K_2 h$

Using the same notations than in 2.1, one can write :

$$\hat{B}(\xi) = \frac{\rho_2^2 K_1^2 \cos^2 K_2 h - \rho_1^2 K_2^2 \sin^2 K_2 h}{D(\xi^2)} = \frac{h_1(\xi^2)}{D(\xi^2)} ;$$

$$B_\infty = \frac{\rho_2^2 + \rho_1^2}{\rho_2^2 - \rho_1^2} ; \qquad \hat{\mu}_1(\xi^2) = \frac{h_1(\xi^2)}{D(\xi^2)} - \frac{\rho_2^2 + \rho_1^2}{\rho_2^2 - \rho_1^2} ;$$

$$\hat{\mu}_2(\xi^2) = \frac{2\rho_1\rho_2 K_2 \sin K_2 h \,.\, \cos K_2 h}{D(\xi^2)} = \frac{h_2(\xi^2)}{D(\xi^2)}$$

Because the functions $h_1$ and $h_2$ are even in $K_2$, no branch integral appears after an integration in the complex plane.

Then, on integrating in the upper half-plane ($\text{Im}\ \xi > 0$) by the residue method, it is easy to find :

$$\mu_i(\rho) = \frac{i}{4} \sum_n \frac{h_i(\lambda_n^2)}{D'(\lambda_n^2)} H_0^{(1)}(\lambda_n \rho)$$

where $\lambda_n^2$ are the zeros of $D(\xi^2)$ such that $\text{Im}(\lambda_n) > 0$, and

$$D'(\lambda_n^2) = \frac{d}{d\xi^2} D(\xi^2) \Big|_{\xi^2 = \lambda_n^2}$$

Finally the pressure emitted in the half space ($z \geqslant 0$) over a finite depth layer of porous medium can be expressed as :

(2-19)
$$p_1(M) = -\frac{e^{ikR(S,M)}}{4\Pi R(S,M)} - \frac{\rho_2^2 + \rho_1^2}{\rho_2^2 - \rho_1^2} \frac{e^{ikR(S',M)}}{4\Pi R(S',M)}$$

$$-\frac{i}{4} \int_{z'=-s} \sum_n \frac{h_1(\lambda_n^2)}{D'(\lambda_n^2)} H_0^{(1)}(\lambda_n \rho(P)) \frac{e^{ikr(M,P)}}{4\Pi r(M,P)} d\sigma(P)$$

$$-\frac{i}{4} \frac{\partial}{\partial z} \int_{z'=-s} \sum_n \frac{h_2(\lambda_n^2)}{D'(\lambda_n^2)} H_0^{(1)}(\lambda_n \rho(P)) \frac{e^{ikr(M,P)}}{4\Pi r(M,P)} d\sigma(P)$$

## 2.3. - Conclusion

The method proposed here to obtain an exact expression of the sound field is very general. It can be summarized as follows :

- use of the two-dimensional Fourier transform ,
- calculation of the plane wave reflection coefficient $\hat{A}$, using the boundary conditions,
- transformation and decomposition of $\hat{A}$ to obtain Fourier transforms of well-known functions or to avoid branch integrals if complex integration is used,
- representation of the solution as a sum of the incident wave, a reflected wave and layer potentials.

## 3. - REPRESENTATION OF THE PRESSURE, USING SURFACE WAVES

Since problems of diffraction by a plane are studied, many authors have proposed solutions involving a "surface wave" term. This term has the general form:

$$OS(M) = H_0^{(1)}(\alpha \sqrt{x^2 + y^2})\, e^{i\beta z} , \quad \text{if } M = (x, y, z > 0)$$

where $\alpha$ and $\beta$ are complex coefficients ; Im $\alpha$ and Im $\beta$ positive ; OS represents a wave which exponentially decays when the height z increases.

In this chapter, it is shown how the surface wave term appears in the expression of the sound field. In section 3.1, it is shown that a layer potential contains a surface wave term. In "good" cases, it can be represented by the sum of a surface wave term and a line integral. In section 3.2, the "local reaction" model is studied as an example.

So, this chapter gives another decomposition and another interpretation of the sound field. Furthermore, this new representation lead to easier numerical computations.

The results proposed here can be found in ref. /23/.

### 3.1. - Layer potential and surface wave

As shown in chapter 2, a simple layer potential of density $\mu$ is expressed as :

(3-1) $$\varphi(M) = - \int_{\Sigma} \mu(P) \frac{e^{ikr(M,P)}}{4\Pi r(M,P)}\, d\sigma(P)$$

where $\Sigma$ is the plane (z' = -s), parallel to the surface and passing through the image source.

Because of the symetry of the problems studied, the density $\mu$ depends on the radial coordinate of P, only.

$\varphi(M)$ can be represented by the Fourier integral :

(3-2) $$\varphi(M) = \frac{1}{4\Pi} \int_{-\infty}^{+\infty} \hat{\mu}(\xi^2) \frac{e^{iK|z+s|}}{2iK} H_0^{(1)}(\xi\rho)\, \xi\, d\xi$$

where $\hat{\mu}$ denotes the two dimensional Fourier transform of $\mu$.

Let us write $\hat{\mu}(\xi^2) = N(\xi^2)/D(\xi^2)$

In equation (3-2), the integral can be evaluated by integration in the upper half plane (Im $\xi > 0$). It leads to :

(3-3) $$\varphi(M) = \frac{i}{4} \sum_j \frac{N(\alpha_j^2)}{D'(\alpha_j^2)} H_0^{(1)}(\alpha_j \rho) \frac{e^{iK(\alpha_j^2)(z+s)}}{2iK(\alpha_j^2)} + \psi(M) ,$$

where $\alpha_j^2$ are the roots of $D(\xi^2)$ ; for simplicity, they are assumed to be simple poles .

$$D'(\alpha_j^2) = \frac{d}{d\xi^2} D(\xi^2) \Big|_{\xi^2 = \alpha_j^2}$$

$$K^2(\alpha_j^2) = k^2 - \alpha_j^2 , \quad \operatorname{Im} K(\alpha_j^2) > 0 , \quad \operatorname{Im}(\alpha_j) > 0$$

Such a representation makes appear a surface wave term. Now, $\psi$ (M) can be evaluated, using the three dimensional Fourier transform :

$$\tilde{\psi}(\xi, \eta) = \int_{-\infty}^{+\infty} \int_{-\infty}^{+\infty} \int_{-\infty}^{+\infty} e^{-i(x\xi_1 + y\xi_2 + z\eta)} \psi(x, y, z) \, dx \, dy \, dz$$

From (3-1), the Fourier transform of the layer potential is readily obtained :

(3-4) $$\tilde{\varphi}(\xi, \eta) = \hat{\mu}(\xi^2) \frac{e^{i\eta s}}{k^2 - \xi^2 - \eta^2}$$

and the equality (3-3) implies :

(3-5) $$\tilde{\varphi}(\xi, \eta) = - \sum_j \frac{N(\alpha_j^2)}{D'(\alpha_j^2)} \frac{1}{\alpha_j^2 - \xi^2} \frac{e^{i\eta s}}{k^2 - \alpha_j^2 - \eta^2} + \tilde{\psi}(\xi, \eta)$$

From equations (3-4 and 5), $\tilde{\psi}$ (M) is obtained and decomposed as follows to facilitate the inverse Fourier transform :

(3-6) $$\tilde{\psi}(\xi, \eta) = \frac{e^{i\eta s}}{k^2 - \xi^2 - \eta^2} (\hat{\nu}(\xi^2) + \hat{\chi}(\eta))$$

with $$\hat{\nu}(\xi^2) = \frac{N(\xi^2)}{D(\xi^2)} + \sum_j \frac{N(\alpha_j^2)}{D'(\alpha_j^2)} \frac{1}{\alpha_j^2 - \xi^2}$$

and
$$\hat{\chi}(\eta) = \sum_j \frac{N(\alpha_j^2)}{D'(\alpha_j^2)} \frac{1}{k^2 - \alpha_j^2 - \eta^2}$$

It is obvious that :

$$\chi(|z|) = \sum_j \frac{N(\alpha_j^2)}{D'(\alpha_j^2)} \frac{e^{iK(\alpha_j^2)|z|}}{2iK(\alpha_j^2)}$$

Furthermore, if $\hat{\nu}(\xi^2)$ is a meromorphic function (that is, analytical except at isolated points), its inverse Fourier transform $\nu(\rho)$ is zero.
Finally, the layer potential can be represented by :

(3-7) $$\varphi(M) = \frac{i}{4} \sum_j \frac{N(\alpha_j^2)}{D'(\alpha_j^2)} H_0^{(1)}(\alpha_j \rho) \frac{e^{iK(\alpha_j^2)|z|}}{2iK(\alpha_j^2)}$$

$$- \int_{-\infty}^{+\infty} \chi(|z' + s|) \frac{e^{ikr'}}{4\Pi r'} dz' - \int_\Sigma \nu(\rho'(P)) \frac{e^{ikr(M,P)}}{4\Pi r(M,P)} d\sigma(P)$$

with : $\rho^2 = x^2 + y^2$ ; $r'^2 = \rho^2 + (z - z')^2$

The first integral can be interpreted as the radiation of monopoles located along the z-axis. This equality shows that a surface wave term is included in a layer potential.

From a numerical viewpoint, this representation is only interesting if $\nu(\rho) = 0$.

The z-derivative of a simple layer potential is readily obtained, on deriving the expression (3-7).

### 3.2. - Expression of the pressure for a locally reacting ground

It has been shown (eq. 2-12), that the layer potentials representing the pressure for the "impedance" case, have a density equal to $H_0^{(1)}(\alpha\rho)$. In this case, the only pole of $\hat{\mu}$ is $\alpha = k\sqrt{1 - 1/\zeta^2}$ and the function $\nu(\rho)$ is zero.
Then :

(3-8)

$$p(M) = -\frac{e^{ikR(S,M)}}{4\Pi R(S,M)} - \frac{e^{ikR(S',M)}}{4\Pi R(S',M)}$$

$$+ \frac{k}{4\zeta}(1 + \operatorname{sgn}\hat{\zeta})\, H_0^{(1)}(\alpha\rho)\, e^{-i\frac{k}{\zeta}|z+s|\operatorname{sgn}\hat{\zeta}}$$

$$+ \frac{ik}{\zeta}(1 + \operatorname{sgn}\zeta) \int_{-s}^{+\infty} e^{-\frac{ik}{\zeta}|z'+s|\operatorname{sgn}\zeta} \frac{e^{ikr}}{4\Pi r}\, dz'$$

$$+ \frac{ik}{\zeta}(\operatorname{sgn}\hat{\zeta} - 1) \int_{-\infty}^{-s} e^{\frac{ik}{\zeta}|z'+s|\operatorname{sgn}\hat{\zeta}} \frac{e^{ikr}}{4\Pi r}\, dz'$$

where the impedance $\zeta = \dot{\zeta} + i\hat{\zeta}$

$$r^2 = x^2 + y^2 + (z - z')^2$$

The pressure can be interpreted as the sum of the incident wave, the perfectly reflected wave, a surface wave term and the radiation of monopole sources situated on the z-axis.

Let us remark that the surface wave term must not be taken into account if the imaginary part of the impedance is negative. Furthermore, depending on the $\hat{\zeta}$ sign, only one integral must be calculated.

Obviously, this expression of p is indefinitely differentiable in the half-space ($z > 0$). Indeed, the first line integral has the same singularity than $H_0^{(1)}(\alpha\rho)$ with the opposite sign.

The computation of the integrals does not present any difficulty. A simple Simpson's rule can be used.

3.3. - Conclusion

It has been shown that it is always possible to represent the pressure using a surface wave term.

When the layer potential can be written as the sum of a surface wave and a line integral, the corresponding expression of the sound field can be computed by very efficient and rapid methods.

## 4. - REPRESENTATION OF THE PRESSURE USING LAPLACE TYPE INTEGRALS

The aim of this chapter is to obtain a third representation of the sound field which can be calculated on a pocket computer, for engineering purposes.

For it, use is made of Laplace type integrals. This kind of expression was proposed by Weyl /24/ in 1919, for the solution of an electromagnetic problem. More recently, S.I. Thomasson /11/ has obtained a similar result to express the pressure emitted over a "locally reacting plane".

The results presented here can be found in /25/ .

In section 4.1., a general method is described to express the radiation of a source as a Laplace integral. In section 4.2., this method is used for the representation of a layer potential of density $H_0^{(1)}$.

For such a density, the process leads to a Laplace type integral which can be analytically calculated. Fortunately, the solution of quite a number of diffraction problems /17, 20, 21/ can be obtained by using only layer potentials of density $H_0^{(1)}$ . Then, for these problems, the pressure can be expressed as a sum of surface wave terms and Laplace integrals. In section 4.3., the corresponding formula for the sound field over a "locally reacting surface" is proposed.

### 4.1. - Radiation of a source

First, we consider a harmonic source $(e^{-i\omega t})$ .Let M be a point of the $R^3$- space and f(M) denote the distribution representing the amplitude of the source. Then, its radiation is given by :

(4-1) $$\varphi(M) = \left( - \frac{e^{ikr}}{4\Pi r} \underset{(3)}{*} f \right) (M)$$

For any point M, a cartesian reference frame $(X_1, Y_1, Z_1)$ can be chosen such that the coordinates of M are $(0,0, z_1 > 0)$.

Let us define the three dimensional Fourier transform of f :

$$\mathscr{F}(f) = \tilde{f}(\xi_1, \eta_1, \zeta_1) = \int_{-\infty}^{+\infty}\int_{-\infty}^{+\infty}\int_{-\infty}^{+\infty} f(x_1, y_1, z_1)\, e^{-i(x_1\xi_1 + y_1\eta_1 + z_1\zeta_1)}\, dx_1\, dy_1\, dz_1$$

then

$$\mathscr{F}\left(\frac{e^{ikr}}{4\Pi r}\right) = [\, k^2 - (\xi_1^2 + \eta_1^2 + \zeta_1^2)]^{-1}$$

and :

(4-2) $$\varphi(M) = \frac{1}{(2\Pi)^3} \int_{-\infty}^{+\infty}\int_{-\infty}^{+\infty}\int_{-\infty}^{+\infty} \frac{e^{iz_1\zeta_1}}{k^2 - (\xi_1^2 + \eta_1^2 + \zeta_1^2)}\, \tilde{f}(\xi_1, \eta_1, \zeta_1)\, d\xi_1\, d\eta_1\, d\zeta_1$$

This formula is convenient because it is often easier to know $\tilde{\varphi}$ than $\varphi$ .

If $\tilde{f}$ is an analytic function of $\zeta_1$ , using the residue integration method, one obtains :

$$\varphi(M) = \frac{1}{(2\Pi)^2} \int_{-\infty}^{+\infty} \int_{-\infty}^{+\infty} \frac{e^{ig_1|z_1|}}{2ig_1} \tilde{h}(\xi_1, \eta_1)\, d\xi_1\, d\eta_1$$

where :

$$g_1^2 = k^2 - (\xi_1^2 + \eta_1^2)\ , \quad \operatorname{Im} g_1 > 0$$

$$\tilde{h}(\xi_1, \eta_1) = \tilde{f}(\xi_1, \eta_1, g_1)$$

Let $(u_1 , v_1 )$ be the polar coordinates such that :

(4-3)
$$\begin{cases} \xi_1 = u_1 \cos v_1 \\ \eta_1 = u_1 \sin v_1 \end{cases}$$

$\varphi$ is given by :

$$\varphi(M) = \frac{1}{(2\Pi)^2} \int_0^{+\infty} u_1 \frac{e^{ig_1 z_1}}{2ig_1} \int_0^{2\Pi} \tilde{h}(u_1 \cos v_1, u_1 \sin v_1)\, dv_1\, du_1$$

where $g_1^2 = k^2 - u_1^2$

Let h be defined by :

$$h(u_1) = \frac{1}{2\Pi} \int_0^{2\Pi} \tilde{h}(u_1 \cos v_1, u_1 \sin v_1)\, dv_1$$

Then, $u_1\, du_1 = -g_1\, dg_1$ implies :

(4-4)
$$\varphi(M) = -\frac{1}{2\Pi} \int_L \frac{e^{ig_1 z_1}}{2i} h(\sqrt{k^2 - g_1^2})\, dg_1$$

where L is the contour shown in figure 1

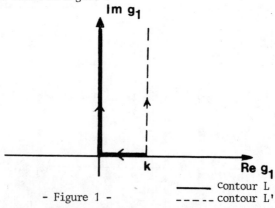

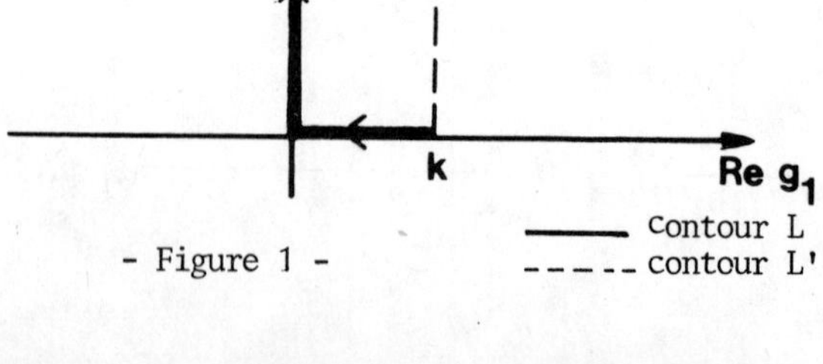


- Figure 1 -

And, finally, under some conditions of regularity for the function f (that is, f is such that no residues nor branch integrals appear), L is replaced by L' (the steepest descent contour).

Thus :

(4-5)
$$\varphi(M) = -\frac{e^{ikz_1}}{4\Pi}\int_0^{+\infty} h(\sqrt{t^2 - 2itk})\, e^{-z_1 t}\, dt$$

This is a very convenient representation for numerical calculations. The use of Laguerre polynomials (for example) provides a very rapid and efficient computation scheme.

Remark : Generally, f is not an analytic function. Then, $\varphi$ (M) is obtained as a sum of a Laplace integral, some residues and/or some branch integrals. Then, the problem is the calculation of the branch integrals.

### 4.2. - Simple layer potential radiation

In chapter 2, it has been shown that, for quite a number of diffraction problems, the solution can be expressed using simple layer potentials and derivatives of simple layer potentials. For several kinds of problems, the density of these potentials is the Hankel function $H_0^{(1)}$.

The aim of this section is to show how such a layer potential can be represented by the sum of a surface wave term and a Laplace integral. Then, the same kind of formula is given for the z-derivative of a simple layer potential.

Let us define $\varphi$ by :

(4-6)
$$\varphi(M) = -\frac{e^{ikr}}{4\Pi r} * \left(-\frac{i}{4} H_0^{(1)}(\alpha\rho) \otimes \delta_{z=0}\right)(M)$$

$$= \frac{i}{4}\int_{z=0} \frac{e^{ikr(M,P)}}{4\Pi r(M,P)} H_0^{(1)}(\alpha\rho(P))\, d\sigma(P)$$

$\alpha$ is a complex number, with Im $\alpha \geqslant 0$ and using the same notations than in (2-11).

a/ - New coordinates system :

Like in 4.1., the calculations are simplified if one uses the cartesian reference frame $(X_1, Y_1, Z_1)$ such that M = $(0, 0, z_1 > 0)$.

To obtain this new system, let us consider the spherical coordinates of M (see figure 2) :

$$z_1 = (x^2 + y^2 + z^2)^{1/2} \; ; \quad \vartheta = \text{Arc cot} \frac{z}{(x^2 + y^2)^{1/2}} \; ; \; \phi$$

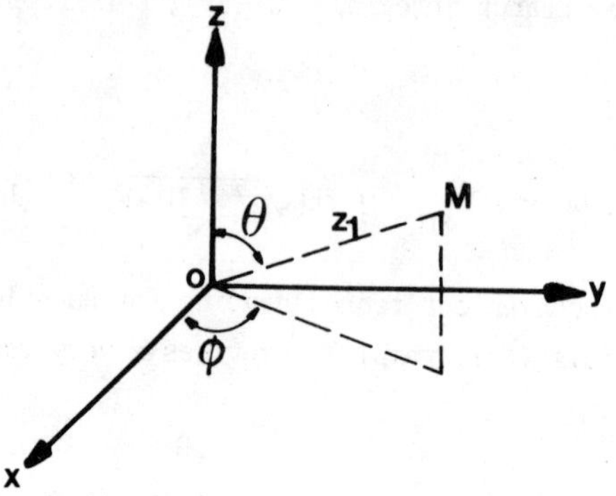


- Figure 2 -

Because of the symmetry of the problem, the expression of $\varphi$ does not depend on the angle $\phi$. So, let us assume that $\phi = \Pi/4$. Then, the new system $(x_1, y_1, z_1)$ is defined by :

$$x_1 = x\,\frac{\cos\vartheta + 1}{2} + y\,\frac{\cos\vartheta - 1}{2} - z\,\frac{\sqrt{2}}{2}\sin\vartheta$$

$$y_1 = x\,\frac{\cos\vartheta - 1}{2} + y\,\frac{\cos\vartheta + 1}{2} - z\,\frac{\sqrt{2}}{2}\sin\vartheta$$

$$z_1 = x\,\frac{\sqrt{2}}{2}\sin\vartheta + y\,\frac{\sqrt{2}}{2}\sin\vartheta + z\cos\vartheta$$

And, if $\hat{f}(\xi,\eta,\zeta)$ and $\tilde{f}(\xi_1,\eta_1,\zeta_1)$ are the respective Fourier transforms of the functions $f(x, y, z)$ and $f(x_1, y_1, z_1)$, they are related by :

(4-7) $$\tilde{f}(\xi_1,\eta_1,\zeta_1) = \hat{f}\Big[\xi_1\,\frac{\cos\vartheta + 1}{2} + \eta_1\,\frac{\cos\vartheta - 1}{2} + \zeta_1\,\frac{\sqrt{2}}{2}\sin\vartheta\,,$$

$$\xi_1\,\frac{\cos\vartheta - 1}{2} + \eta_1\,\frac{\cos\vartheta + 1}{2} + \zeta_1\,\frac{\sqrt{2}}{2}\sin\vartheta\,,\; -\xi_1\,\frac{\sqrt{2}}{2}\sin\vartheta - \eta_1\,\frac{\sqrt{2}}{2}\sin\vartheta + \zeta_1\cos\vartheta\Big]$$

The layer potential $\varphi$ (M) can be expressed by :

$$\varphi(M) = \frac{1}{(2\Pi)^3} \int_{-\infty}^{+\infty} d\xi_1 \int_{-\infty}^{+\infty} d\eta_1 \int_{-\infty}^{+\infty} e^{i\zeta_1 z_1} \; \tilde{\varphi}(\xi_1, \eta_1, \zeta_1) \, d\zeta_1$$

Use of the cylindrical coordinates $(u_1, v_1, \zeta_1)$ (see equation 4-3) leads to :

$$\varphi(M) = \frac{1}{(2\Pi)^3} \int_0^{2\Pi} dv_1 \int_0^{+\infty} u_1 \, du_1 \int_{-\infty}^{+\infty} \tilde{\varphi}(u_1, v_1, \zeta_1) \, e^{i\zeta_1 z_1} \, d\zeta_1 \tag{4-8}$$

b/ - <u>Decomposition of the simple layer potential</u>

The Fourier transform of the potential $\varphi$ is :

$$\hat{\varphi}(\xi, \eta, \zeta) = (k^2 - (u^2 + \zeta^2))^{-1} (\alpha^2 - u^2)^{-1}$$

with $u^2 = \xi^2 + \eta^2$

It can be decomposed to point out the Fourier transforms of well-known functions :

$$\hat{\varphi}(\xi, \eta, \zeta) = [(\beta^2 - \zeta^2)(\alpha^2 - u^2)]^{-1} - [(\beta^2 - \zeta^2)(k^2 - u^2 - \zeta^2)]^{-1}$$

with $\beta^2 = k^2 - \alpha^2$ ; $\operatorname{Im} \beta > 0$.

Indeed, the inverse Fourier transform of the first term is readily obtained :

$$\widehat{OS} = [(\beta^2 - \zeta^2)(\alpha^2 - u^2)]^{-1} = \mathscr{F}\left[ -\frac{i}{4} H_0^{(1)}(\alpha\rho) \frac{e^{i\beta|z|}}{2 i\beta} \right]$$

This is, again, a "surface wave" term .

Finally, we use the decomposition :

$$\left\{ \begin{aligned} \hat{\varphi} &= \widehat{OS} + \hat{R}_1 + \hat{R}_2 \\ \hat{R}_1 &= -\frac{1}{2\beta(\beta + \zeta)(k^2 - u^2 - \zeta^2)} \\ \hat{R}_2 &= -\frac{1}{2\beta(\beta - \zeta)(k^2 - u^2 - \zeta^2)} \end{aligned} \right. \tag{4-9}$$

Then,

$$
(4\text{-}10)\quad \begin{cases}
\varphi = OS + R_1 + R_2 \\[2ex]
OS = -\dfrac{i}{4} H_0^{(1)}(\alpha\rho)\, \dfrac{e^{i\beta|z|}}{2i\beta} \\[2ex]
R_1 = \dfrac{1}{2i\beta}\, \dfrac{e^{ikr}}{4\Pi r} * \left[ Y(-z)\, e^{-i\beta z} \otimes \delta_{x=0} \otimes \delta_{y=0} \right] \\[2ex]
R_2 = \dfrac{1}{2i\beta}\, \dfrac{e^{ikr}}{4\Pi r} \cdot \left[ Y(z)\, e^{i\beta z} \otimes \delta_{x=0} \otimes \delta_{y=0} \right]
\end{cases}
$$

where Y is the Heaviside function :

$$
Y(z) \begin{vmatrix} = 0 \text{ if } z < 0 \\ = 1 \text{ if } z > 0 \end{vmatrix}
$$

Because our interest is to evaluate the sound field at a point $M = (x, y, z \geqslant 0)$,

$OS = -\frac{i}{4} H_0^{(1)}(\alpha\rho)\, e^{i\beta z}/2i\beta$ ; it is not defined on the semi-axis $(x = 0, y = 0, z > 0)$. But $R_2$ has the same singularity (with the opposite sign) and then, $OS + R_2$ is analytical in the $(z > 0)$ half-space.

c/ - Calculation of the component $R_1$

Using the equalities (4-7 and 8), $R_1$ is also given by :

$$
(4\text{-}11)\quad R_1 = \frac{1}{(2\Pi)^3} \int_0^{2\Pi} dv_1 \int_0^{+\infty} u_1\, du_1 \int_{-\infty}^{+\infty} \frac{-1}{2\beta}\, \frac{e^{i\zeta_1 z_1}}{\beta - u_1 \cos(v_1 - \Pi/4)\sin\vartheta + \zeta_1 \cos\vartheta}\, \frac{d\zeta_1}{k^2 - u_1^2 - \zeta_1^2}
$$

The result of the $\zeta_1$ – integration is readily obtained by the residue method, using the contour $C$ of figure 3.

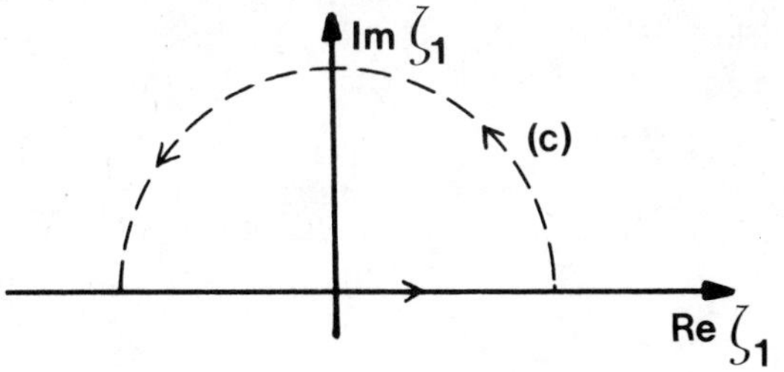


- Figure 3 -

Because the angle of incidence $\vartheta$ satisfies : $0 \leqslant \vartheta \leqslant \Pi/2$, there is only one pole in the upper half-plane (Im $\zeta_1 > 0$) and then :

$$R_1 = -\frac{1}{2\beta}\frac{1}{(2\Pi)^2}\int_0^{2\Pi} dv_1 \int_0^{+\infty} \frac{e^{i\sqrt{k^2-u_1^2}\,z_1}}{2i\sqrt{k^2-u_1^2}} \frac{u_1\, du_1}{\beta - u_1 \cos(v_1 - \Pi/4)\sin\vartheta + \sqrt{k^2-u_1^2}\cos\vartheta}$$

Again, the residue method makes the $v_1$ – integration easy.

Let $w_1$ be defined as $\exp(i(v_1 - \frac{\Pi}{4}))$. The contour is the circle of radius equal to one, in the $w_1$ – plane.

There is one and only one pole inside this circle :

$$w_1 = \frac{1}{u_1 \sin\vartheta}\{\beta + \sqrt{k^2-u_1^2}\cos\vartheta + \varepsilon_1\sqrt{(\beta+\sqrt{k^2-u_1^2}\cos\vartheta)^2 - u_1^2\sin^2\vartheta}$$

where $\varepsilon_1 = \pm 1$ depending on the condition $|w_1| < 1$.

Now, $R_1$ is given by :

$$R_1 = \frac{+1}{4\Pi\beta}\int_0^{+\infty}\frac{e^{i\sqrt{k^2-u_1^2}\,z_1}}{2i\sqrt{k^2-u_1^2}} \frac{u_1\, du_1}{\varepsilon_1\sqrt{(\beta+\sqrt{k^2-u^2}\cos\vartheta)^2 - u_1^2\sin^2\vartheta}}$$

or

$$R_1 = \frac{1}{8i\Pi\beta}\int_L \frac{e^{ik\gamma_1 z_1}\, d\gamma_1}{\varepsilon_1\sqrt{V^+(\gamma_1)}} \qquad \text{if} \quad 0 \leqslant \vartheta \leqslant \Pi/2 \tag{4-12}$$

where L is shown in figure 1,

and

$$V^{\pm}(\gamma_1) = (\frac{\beta}{k} \pm \gamma_1\cos\vartheta)^2 - (1-\gamma^2)\sin^2\vartheta$$

d/ - Calculation of the component $R_2$

Through the same way, $R_2$ is obtained as :

(4-13) $$R_2 = \frac{1}{8 i \Pi \beta} \int \frac{e^{ik\gamma_1 z_1} \, d\gamma_1}{\varepsilon_2 \sqrt{V^-(\gamma_1)}} \qquad \text{if} \quad \Pi/2 \leqslant \vartheta \leqslant \Pi$$

where $\varepsilon_2 = \pm 1$ to obtain $|w_2| < 1$

and

$$w_2 = \frac{1}{u_1 \sin\vartheta} \{ -\beta + \sqrt{k^2 - u_1^2} \cos\vartheta + \varepsilon_2 \sqrt{(\beta - \sqrt{k^2 - u_1^2} \cos\vartheta)^2 - u_1^2 \sin^2\vartheta} \}$$

But, let us recall that our aim is to obtain an expression of $R_2$ for $0 \leqslant \vartheta \leqslant \Pi/2$. Nevertheless, an analytical continuation of the integral (4-13) can be found, taking a branch integral into account.

Indeed, let $s_2^{\pm}$ denote the zeros of the denominator $V^-$.

$$s_2^{\pm} = \frac{\beta}{k} \cos\vartheta \pm \frac{\alpha}{k} \sin\vartheta$$

The curves $\{ s_2^{\pm}(\vartheta), \ 0 \leqslant \vartheta \leqslant \Pi \}$ are represented in figures 5a and 5b, depending on the sign of the real part of $\alpha$.

The curve $\{ s_2^{+}(\vartheta), \ 0 \leqslant \vartheta \leqslant \Pi \}$ crosses the contour L when $\vartheta$ is in the interval $]0, \Pi/2[$.

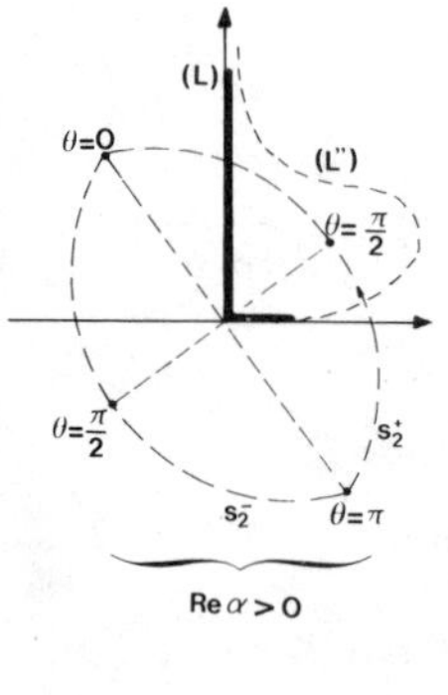


- Figure 5a -

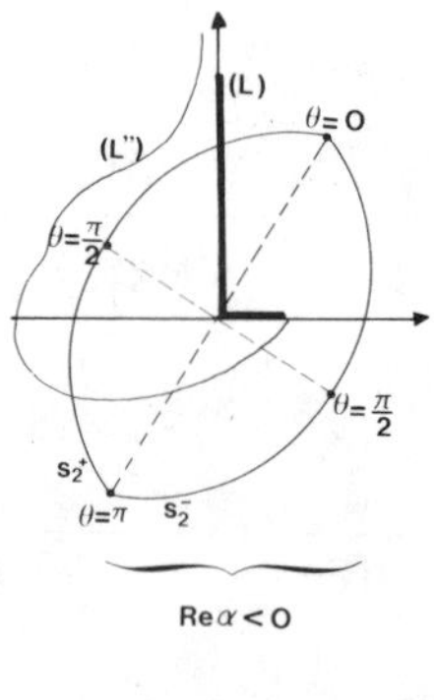


- Figure 5b -

Let (L") be a contour which does not cross the curve $s_2^{+}(\vartheta)$ for $0 \leqslant \vartheta \leqslant \Pi/2$.

Then, using a theorem of analytical continuation,

$$R_2 = \frac{1}{8i\Pi\beta} \int_{L''} \frac{e^{ik\gamma_1 z_1}}{\varepsilon_2 \sqrt{V^-(\gamma_1)}}\, d\gamma_1 + I_2^B \quad \text{if} \quad 0 \leqslant \vartheta \leqslant \Pi/2 \,,$$

if the root $s_2^+$ is located between the two contours L and L".

$I_2^B$ is the branch integral :

$$I_2^B = \frac{1}{8i\Pi\beta} \int_B \frac{e^{ik(s_2^+ + it)z_1}}{\varepsilon_2 \sqrt{it(it + s_2^+ - s_2^-)}}\, d(s_2^+ + it)$$

where B is represented in figure 6

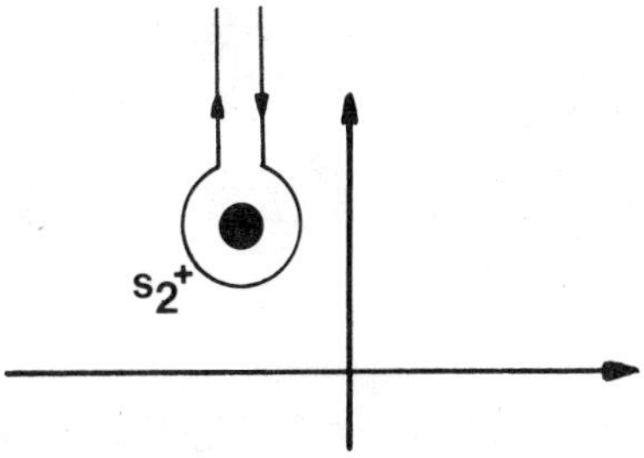


- Figure 6 -

From /26 /, it must be noticed that :

$$I_2^B = \frac{1}{8\beta} e^{i\beta z} H_0^{(1)}(\alpha\rho)$$

Finally, coming back to the contour L,

$$(4\text{-}14) \qquad R_2 = \frac{1}{8i\Pi\beta} \int_L \frac{e^{ik\gamma_1 z_1}}{\varepsilon_2 \sqrt{V^-(\gamma_1)}}\, d\gamma_1$$

$$+ \frac{e^{i\beta z}}{8\beta} H_0^{(1)}(\alpha\rho) \{ Y(\text{Re } s_2^+) + Y(\text{Re } \alpha)\left[1 - 2\, Y(\text{Re } s_2^+)\right] .$$

e/ - Steepest descent contour

The final step consists of replacing the contour L by the steepest descent contour. The variable change is :

$$\gamma_1 = 1 + it \,, \quad 0 \leqslant t < +\infty$$

Let $s_1^\pm$ denote the zeros of the denominator $V^+$ in the integral (4-12) $s_1^\pm = - s_2^\mp$ .

If $s_1^{\pm}$ are located between the two contours, a branch integral appears. Again, from /26/ , it can be shown that it is equal to :

$$-\frac{e^{-i\beta z}}{8\beta} H_0^{(1)}(\alpha\rho)$$

Then, the final expression of $\varphi$ is given by :

$$\varphi(M) = \frac{-1}{8\beta} H_0^{(1)}(\alpha\rho)\ \{\ Y(\mathrm{Re}\,\alpha)\,.\,\mathrm{sgn}\,(\mathrm{Re}\ s_2^+ - 1) + Y(1 - \mathrm{Re}\ s_2^+)\,\}\ e^{i\beta z} \tag{4-15}$$

$$+ \frac{1}{8\beta} H_0^{(1)}(\alpha\rho)\,.\,Y(\mathrm{Re}\,\alpha)\,.\,Y(1 - \mathrm{Re}\ s_1^+)\ e^{-i\beta z}$$

$$+ \frac{e^{ikz_1}}{8\Pi\beta}\int_0^{+\infty}\left\{\frac{-\varepsilon_1\ e^{-ktz_1}}{\sqrt{(\beta/k + (1+it)\cos\vartheta)^2 - t(t-2i)\sin^2\vartheta}} + \frac{\varepsilon_2\ e^{-ktz_1}}{\sqrt{(\beta/k - (1+it)\cos\vartheta)^2 - t(t-2i)\sin^2\vartheta}}\right\} dt$$

This formula shows how a simple layer potential of density $H_0^{(1)}$ can be expressed as a sum of surface wave terms and a Laplace integral.

The same kind of formula is obtained for the z-derivative of this layer potential. Using

$$\frac{\partial}{\partial z}\varphi(M) = i\beta\,.\,OS + i\beta\,(R_2 - R_1) \qquad \text{for} \quad z > 0$$

leads to :

$$\frac{\partial}{\partial z}\varphi(M) = -\frac{i}{8} H_0^{(1)}(\alpha\rho)\ e^{i\beta z}\ \{\ Y(\mathrm{Re}\,\alpha)\,.\,\mathrm{sgn}\,(\mathrm{Re}\ s_2^+ - 1) + Y(1 - \mathrm{Re}\ s_2^+)\ \} \tag{4-16}$$

$$-\frac{i}{8} H_0^{(1)}(\alpha\rho)\ e^{-i\beta z}\ Y(\mathrm{Re}\,\alpha)\ Y(1 - \mathrm{Re}\ s_1^+)$$

$$+ \frac{i e^{ikz_1}}{8\Pi}\int_0^{+\infty}\left\{\frac{\varepsilon_2\ e^{-ktz_1}}{\sqrt{(\beta/k - (1+it)\cos\vartheta)^2 - t(t-2i)\sin^2\vartheta}} + \frac{\varepsilon_1\ e^{-ktz_1}}{\sqrt{(\beta/k + (1+it)\cos\vartheta)^2 - t(t-2i)\sin^2\vartheta}}\right\} dt$$

4.3. - "Local reaction " case

For the "local reaction" model, the pressure is given by :

(4-17)
$$p(M) = -\frac{e^{ikR(S,M)}}{4\Pi R(S,M)} + \frac{e^{ikR(S',M)}}{4\Pi R(S',M)}$$

$$+ \frac{k}{4\zeta}(1 + \mathrm{sgn}\,\hat{\zeta})\ Y(\vartheta - \vartheta_0)\ H_0^{(1)}(\alpha\rho)\ e^{-\frac{ik}{\zeta}(z+s)}$$

$$+ \frac{k}{2\zeta}\ \frac{e^{ikR(S',M)}}{\Pi} \int_0^{+\infty} \frac{e^{-kR(S',M)t}}{\varepsilon\sqrt{W(t)}}\, dt$$

if the impedance $\zeta = \dot{\zeta} + i\hat{\zeta}$ ; $\alpha^2 = k^2(1 - 1/\zeta^2)$, $\mathrm{Im}\ \alpha > 0$

$\vartheta_0$ is defined by : $-\dfrac{\zeta}{|\zeta|^2}\cos\vartheta_0 + \dfrac{\mathrm{Re}\ \alpha}{k}\sin\vartheta_0 = 1$

$$W(t) = \left(\frac{1}{\zeta} + \cos\vartheta\right)^2 + 2it\left(1 + \frac{\cos\vartheta}{\zeta}\right) - t^2$$

$$t_0 = \frac{\hat{\zeta}}{|\zeta|^2}\left\{\frac{\dot{\zeta}}{|\zeta|^2} + \cos\vartheta\right\}\left\{1 + \frac{\dot{\zeta}}{|\zeta|^2}\cos\vartheta\right\}^{-1}$$

and

$$\varepsilon \left|\begin{array}{l} = -1 \text{ if } \mathrm{Re}\ W(t_0) < 0\ ,\quad t > t_0\ ,\ \hat{\zeta} > 0 \\ = +1 \text{ if not.}\end{array}\right.$$

4.4. - Conclusion

In this chapter, we have shown that the pressure can be expressed with a Laplace type integral and surface wave terms. The method can be applied to layer potentials of density $H_0^{(1)}$ , at least. Various examples can be already found in /25/ (diffraction by a finite layer of porous medium, by a plate...).

The surface wave terms can be computed very easily, using Padé techniques (see, for example /27/). The Laplace integral can be computed, using Laguerre polynomials. Both computation techniques are rapid and efficient.

## 5. - CONCLUSION

The three methods proposed here provide three ways to express the acoustic field emitted in the presence of an absorbing plane. The first one seems to be the most general. The other two are interesting for a more restricted number of problems: diffraction by plates, membranes, some ground models. As far as ground models are concerned, these last two methods can be applied for the "locally reaction" case and for the finite thickness layer of porous medium. Fortunately, it seems that these two models are the most convenient for the description of the acoustical diffraction by a ground. Also, it seems that for the model of porous medium with a decreasing porosity, the pressure can be written exactly using a Laplace type integral.

Among these three representations, the last one, proposed in chapter 4, allows a very rapid and efficient computation of the pressure. So, it does not seem necessary to look for analytical approximations anymore, except perhaps for the very far field where the first two terms of the asymptotic series in ( $R^{-n}$) are convenient.

To achieve this study of the diffraction by a flat, homogeneous ground, the next problem is the identification of the parameters characterizing a ground.

The aim is to compute the parameters corresponding to each ground model studied, from sound levels measurements over a given ground. The technique is to minimize a function of the difference between measured and calculated levels, using an iterative method. Then, the parameters are evaluated for each model and the best model is choosen,i.e. the model which provides the best minimization. Using this model and the corresponding parameters, it is then possible to predict the acoustic field everywhere over the ground.

BIBLIOGRAPHIE

1. B. VAN DER POL. 1935. Physica, n° 2, pp 843-853, Theory of the reflection of the light from a point source by a finitely, conducting flat mirror, with an application to radiotelegraphy.

2. I. RUDNICK. 1947. Journal of the Acoustical Society of America, vol 19, n° 2, pp 348-356. The propagation of an acoustic wave along a boundary.

3. U. INGARD. 1951. Journal of the Acoustical Society of America, vol 23, n° 3, pp 329-335. On the reflection of a spherical sound wave from an infinite plane.

4. R.B. LAWHEAD and I. RUDNICK . 1951. Journal of the Acoustical Society of America, vol 23, n° 5, pp 546-549. Acoustic wave propagation along a constant normal impedance boundary.

5. D.I. PAUL. 1959. Journal of Mathematics and Physics , pp 1-15. Wave propagation in acoustics using the saddle point method.

6. S.P. PAO and L.B. EVANS. 1971. Journal of the Acoustical Society of America, vol 49, n° 4, pp 1069-1075. Sound attenuation over simulated ground cover.

7. A.R. WENZEL. 1974. Journal of the Acoustical Society of America, vol 55, n° 5, pp 956-963. Propagation of waves along an impedance boundary.

8. W.K. VAN MOORHEM. 1975. Journal of Sound and Vibration, vol 45, n° 2, pp 201-208. Reflection of a spherical wave from a plane surface.

9. C.F. CHIEN and W.W. SOROKA. 1975. Journal of Sound and Vibration, vol 43, n°1, pp 9-20. Sound propagation along an impedance plane.

10. T.F.W. EMBLETON, J.E. PIERCY, N. OLSON. 1976. Journal of the Acoustical Society of America, vol 59, n° 2, pp 267-277. Outdoor sound propagation over ground of finite impedance.

11. S.I. THOMASSON. 1976. Journal of the Acoustical Society of America, vol 59, n° 4, pp 780-785. Reflection of waves from a point source by an impedance boundary.

12. R.J. DONATO. 1976. Journal of The Acoustical Society of America, vol 60, n°1, pp 34-39. Propagation of a spherical wave near a plane boundary with a complex impedance.

13. S.I. THOMASSON. 1977. Journal of the Acoustical Society of America, vol 61, n° 3, pp 659-674. Sound propagation above a layer with a large refraction index.

14. C.I. CHESSELL. 1977. Journal of the Acoustical Society of America, vol 62, n° 4, pp 825-834. Propagation of noise along a finite impedance boundary.

15. K. ATTENBOROUGH, S.I. HAYEK, J.M. LAWTHER. 1980. Journal of the Acoustical Society of America, vol 68, n° 5, pp 1493-1501. Propagation of source above a porous half-space.

16. L.M. BREKHOVSKIKH. 1960. Waves in layered media, New-York, Academic Press Inc.

17. H. SAADAT and P. FILIPPI. 1981. Journal of the Acoustical Society of America, vol 69, n° 2, pp 397-403. Diffraction of a spherical wave by a thin infinite plate.

18. D. HABAULT. 1981. Note L.M.A. n° 1848, Marseille. Propagation acoustique en présence d'une membrane semi-infinie et d'un demi-plan réfléchissant : cas bidimensionnel.

19. P. FILIPPI. 1982. See lecture : Integral equations in Acoustics.

20. M. BRIQUET and P. FILIPPI. 1977. Journal of the Acoustical Society of America, vol 61, pp 640-646. Diffraction of a spherical wave by an absorbing plane.

21. D. HABAULT, and P.J.T. FILIPPI. 1978. Journal of Sound and Vibration, vol 56, pp 87-95. On the resolvant of the Pekeris operator with a Neumann condition.

22. C. H. WILCOX. 1975. Lecture Notes in Mathematics, Springer Verlag, Berlin-Heidelberg - New York. Scattering theory for the d'Alembert equation.

23. D. HABAULT and P.J.T. FILIPPI. 1981. Journal of Sound and Vibration vol 79, n° 4, pp 529-550. Ground effect analysis : surface wave and layer potential representations.

24. H. WEYL. 1919. Annales der Physik. Leipzig, vol 60, pp 481-500. Ausbreitung elektromagnetisher Weller über einen ebenen Leiter.

25. P.J.T. FILIPPI. 1981. Note L.M.A. n° 1847, Marseille. Rayonnement de sources réparties et intégrales de Laplace : applications aux milieux stratifiés.

26. G.N. WATSON. 1962. A treatise on the theory of Bessel functions, 2nd edition. Cambridge at the University Press.

27. Y.L. LUKE. 1975. Mathematical functions and their applications, New-York : Academic Press Inc.

# BOUNDARY ELEMENT METHODS AND THEIR ASYMPTOTIC CONVERGENCE

W.L. WENDLAND
Fachbereich Mathematik
Technische Hochschule Darmstadt
D-6100 Darmstadt, Fed. Rep. Germany

## INTRODUCTION

Nowadays the most popular numerical methods for solving elliptic boundary value problems are finite differences, finite elements and, more recently, boundary element methods. The latter are numerical methods for solving integral equations (or their generalizations) on the boundary $\Gamma$ of the given domain. The reduction of interior or exterior stationary boundary value problems as well as transmission problems to equivalent boundary integral equations is by no means unique, the two most popular reductions are the "direct method" and the "method of potentials". In all these cases one needs a fundamental solution of the differential equations explicitly since it will be used in numerical computations. This restricts the boundary integral methods to cases of simple computability of a fundamental solution, i.e. essentially to differential equations with constant coefficients. The formulation on the boundary surface $\Gamma$ reduces the dimensions of the original problem by one. For the computational treatment the boundary surface is decomposed into a finite number of segments and the boundary functions are approximated by corresponding finite elements, the boundary elements. The appropriately discretized version of the boundary integral equation then provides a finite system of linear approximate equations whose coefficient matrix, the influence matrix is fully distributed. Therefore the latter are usually resolved by direct methods. Consequently, a careful comparison of computational expenses of the finite element domain method for interior problems with the corresponding boundary element method shows that they are both of the same magnitude, i.e. proportional to $N^3$ where $N$ denotes the number of boundary elements [113]. (This comparison does not take into account fast solvers which are yet to be developed for the boundary element methods). One should also notice that the computing time for the influence matrix is usually rather high, namely up to 80% of the total time for solving one boundary value problem. Here a careful design of the numerical implementation could save much time. On the other hand there are several advantages of boundary integral methods: (i) Experiments show reasonable numerical results already for small $N$. (ii) The method is applicable to interior as well as <u>exterior</u> boundary value problems without any modifications. (iii) The desired fields away from the boundary $\Gamma$ are given by boundary potentials and, hence, also the computed approximations can be differentiated analytically. (iv) Often the interesting physical quantities are to be computed on the boundary

$\Gamma$ and this is just the solution of an appropriate boundary integral equation.

All these arguments show that an error analysis for an improvement of the boundary element methods is indispensible. Here an asymptotic error analysis in terms of orders of diminishing grid size can be developed similarly to that for finite element domain methods. One finds the least difficulties in this analysis for Galerkin procedures although computations mostly use collocation methods. For two-dimensional problems, however, the collocation method can be treated as a modified Galerkin method which also here provides sharp asymptotic convergence results in the sense that the order of convergence is optimal. For all Galerkin methods including the above mentioned cases of collocations our error analysis rests on <u>strong ellipticity</u> of the <u>boundary integral operators</u> which is formulated in the frame of pseudodifferential operators yielding coerciveness in the form of Gårding inequalities. Since Michlin's fundamental work [ 73 ] and the constructive proof of the Lax-Milgram theorem by Hildebrandt and Wienholtz [ 49 ] it is well known that Gårding's inequality implies asymptotic convergence of the Galerkin method in an appropriate energy norm. This approach can be extended to projection methods in Banach spaces yielding e.g. some convergence results for collocation methods in $\mathbb{R}^2$ and $\mathbb{R}^3$ for Fredholm integral equations of the second kind. But for the more general strongly elliptic problems such results can only be obtained by applying the Aubin-Nitsche duality arguments to Galerkin methods in dual pairs of Hilbert spaces, i.e. here in boundary Sobolev spaces.

The numerical implementation of Galerkin's method requires for the influence matrix two numerical quadratures on $\Gamma$ and for ordinary collocation one. If these are done accurately enough then the first Strang lemma provides the same optimal asymptotic orders of convergence as for the theoretical Galerkin or collocation method, respectively. If these numerical quadratures are done by Gaussian formulas then the collocation method is superior to Galerkin's procedure but both methods are rather time consuming and rather inefficient concerning the computing time. In two dimensions this can be overcome by treating the principal part and smooth remainders differently and the latter by using integration formulas with regular grid points only. Here the function values can be stored and partly be used again for mesh refinements. Moreover the computing time for the influence matrix is essentially the least possible. Even the transition to higher degree splines then can be done extremely fast improving the accuracy of the numerical results tremendously (see the table [113 p. 442]).

For three dimensional problems the situation is different and up to now standard programs use an additional approximation of $\Gamma$ . This approximation can also be treated with the first Strang lemma.

For me it was a great surprise to see how well numerical experiments revealed the orders of asymptotic convergence in [ 51 ], [ 53 ], [ 67 ]. This shows that our asymptotic error analysis provides useful informations for the numerical work, in particular how to design and to improve sensibly the boundary element programs to be used.

The four lectures are organized as follows. Section 1 is devoted to strong ellipticity, Gårding's inequality and the proof of Céa's lemma along the lines of Michlin's treatment of variational problems. As examples we present four different types of boundary integral equations. Three of them model the two and three-dimensional classical scattering problems in acoustics. These formulations are done in such a way that the boundary integral equations are perturbations of the case of wave number $k = 0$ , i.e. the low frequency expansions are immediately available. In Section 2 we present the asymptotic error analysis for boundary element approximations in the framework of variational formulations in Sobolev spaces on $\Gamma$ . In particular we prove the super approximation results by [ 55 ] for Galerkin's method applied to strongly elliptic systems. Then we present the recent results by [ 7 ] for ordinary collocation with odd degree splines for two-dimensional problems. We also review a comparison of the Galerkin method with collocation on curves and recent (unpublished) results on the collocation using even degree splines. Section 3 deals with Fredholm integral equations of the second kind. First we apply our general approach obtaining superconvergence for smoothed approximations and for Galerkin's solution at nodal points. Then we apply convergence results in Banach spaces to the special equations of classical scattering if the scatterer has corners and edges. In Section 4 we apply numerical integration to the foregoing procedures. For two-dimensional problems with the boundary equations on a curve $\Gamma$ we present the Galerkin collocation for equations with convolutional kernels in the principal part which combines the high accuracy of Galerkin's method with the efficiency of ordinary collocation. The latter is also implemented in the same framework. For three-dimensional problems we present the isoparametric boundary element approximation by Nedelec [ 78 ]. The error estimates for the numerical quadratures are incorporated via the first Strang lemma leading to asymptotic errors for the fully discretized schemes.

## § 1 STRONGLY ELLIPTIC INTEGRAL EQUATIONS AND PROJECTION METHODS

In the following we shall consider boundary integral equations of the form

$$Au + B\omega = f \ , \quad \Lambda u = b \tag{1.1}$$

where $A$ is a $p\times p$ matrix of linear operators mapping the p-vector valued functions $u = (u_1,..,u_p)$ on $\Gamma$ into p-vector valued functions on $\Gamma$ . $B$ is a given $p\times q$ matrix of suitably smooth functions on $\Gamma$ and $\Lambda$ is a given $q\times p$ matrix of linear functionals.
$f = (f_1,...,f_p)$ a p-vector valued function and $b = (b_1,...,b_q)\in \mathbb{R}^q$ are given, $u$ and $\omega\in \mathbb{R}^q$ are the unknowns. $\Gamma$ is a given compact $n-1$ dimensional manifold in $\mathbb{R}^n$ $n=2$ or $3$ and at least piecewise Ljapounov. We shall specify the mapping properties of $A$, $B$, $\Lambda$ later on.

As simple examples let us consider some classical scattering problems with an obstacle $\Omega$ whose boundary is $\Gamma$ . Then, if we denote the velocity potential of the total field by $U$ and if an incident plane wave is moving in the $x_1$-direction, we have from Green's formula the representation $U$ for two-dimensional problems

$$U(x) = e^{ikx_1} - \frac{i}{4}\int_\Gamma \{H_o^{(1)}(kr)\,\frac{\partial U}{\partial \nu}(y) - U(y)(\frac{\partial}{\partial \nu}\,H_o^{(1)}(kr))\}\,ds_y \tag{1.2}$$

for $x \in \mathbb{R}^2\setminus(\Omega\cup\Gamma)$ . Here $H_o^{(1)}$ denotes the Hankel function of the first kind [84], $r = |x-y|$ , $\nu$ the exterior normal vector to $\Gamma$ and $ds_y$ the arc length element at the integration point $y$ . Following [70] a simple computation yields on $\Gamma$ the relation between the Cauchy data $U|_\Gamma$ and $\frac{\partial U}{\partial \nu}\Big|_\Gamma$ ,

$$\begin{aligned} & U(x) + \frac{1}{2\pi}\int_\Gamma (U(y) - U(x)) \frac{\partial}{\partial \nu_y}(\log r)\, ds_y \\ & - \int_\Gamma U(y) \frac{\partial}{\partial \nu_y}\{\frac{i}{4} H_o^{(1)}(kr) + \frac{1}{2\pi}\log r\}\, ds_y \\ & = \frac{1}{2\pi}\int_\Gamma \log r \frac{\partial U}{\partial \nu}\, ds - \int_\Gamma \{\frac{i}{4} H_o^{(1)}(kr) + \frac{1}{2\pi}\log r\}\frac{\partial U}{\partial \nu}\, ds + e^{ikx_1} . \end{aligned} \tag{1.3}$$

In case of a <u>hard obstacle</u> we have $\frac{\partial U}{\partial \nu}\Big|_\Gamma = 0$ and (1.3) gives the <u>Fredholm integral equation of the second kind</u> for $u := U|_\Gamma$ ,

$$\begin{aligned} Au(x) := u(x) &+ \frac{1}{2\pi}\int_\Gamma (u(y) - u(x)) \frac{\partial}{\partial \nu_y}(\log r)\, ds_y \\ &- \int_\Gamma u(y) \frac{\partial}{\partial \nu_y}\{\frac{i}{4} H_o^{(1)}(kr) + \frac{1}{2\pi}\log kr\} ds_y = e^{ikx_1} \end{aligned} \tag{1.4}$$

for $x\in\Gamma$ . Provided $k$ is not a critical value, the equation (1.4) is uniquely solvable in suitable function spaces. In $\mathbb{R}^3$ , the corresponding integral equation reads as

$$\begin{aligned} Au(x) := u(x) &- \frac{1}{4\pi}\int_\Gamma (u(y) - u(x)) \frac{\partial}{\partial \nu_y}\frac{1}{r}\, ds_y \\ &+ \frac{1}{4\pi}\int_\Gamma u(y) \frac{\partial}{\partial \nu_y}\{\frac{1}{r}(e^{ikr} - 1)\}\, ds_y = e^{ikx_1} \end{aligned} \tag{1.5}$$

(see [60]). Note that these formulations with more general boundary data can be used to compute the low frequency asymptotics for $k \to 0$ by following MacCamy [70] since the integral operators in (1.4), (1.5) define regular perturbations to the harmonic case $k = 0$.

Also note that these equations remain valid if the normal derivatives are defined in the distributional sense as boundary flows and if the boundary is not necessarily smooth but provides the condition

$$\sup_{x \in \mathbb{R}^2 \setminus \Gamma} \int_\Gamma \left|\frac{\partial}{\partial \nu_y} \log r\right| ds_y < \infty \quad \text{respectively} \quad \sup_{x \in \mathbb{R}^3 \setminus \Gamma} \int_\Gamma \left|\frac{\partial}{\partial \nu_y} \frac{1}{r}\right| ds_y < \infty \tag{1.6}$$

(see [60]).

For a soft scatterer we have $U|_\Gamma = 0$ and (1.3) becomes a Fredholm integral equation of the <u>first kind</u> for $u = \frac{\partial U}{\partial \nu}\Big|_\Gamma$. In view of low frequency asymptotics we now modify the equation by setting

$$u = \alpha u_o + u_1 \tag{1.7}$$

where $\alpha$, $u_o$ and $u_1$ to be determined, and solve the two systems

$$\begin{aligned} Au_j + \omega_j &:= -\frac{1}{2\pi}\int_\Gamma \log r \; u_j ds + \int_\Gamma \{\frac{i}{4} H_o^{(1)}(kr) + \frac{1}{2\pi} \log kr\} u_j ds + \omega_j \\ &= \delta_{1j}\, e^{ikx_1} \end{aligned} \tag{1.8}$$

$$\Lambda u_j := \int_\Gamma u_j ds = \delta_{oj} \quad , \quad j = 0,1 .$$

Here $\delta_{kj}$ denotes the Kronecker and

$$\alpha = -\omega_1 / (\frac{1}{2\pi} \log k + \omega_o) .$$

Since the system (1.8) is a regular perturbation of the case $k = 0$ (see [113 (1.5), (1.6)]) it can be used to compute again the low frequency expansion of the Dirichlet problem [70].

The equations in three dimensions corresponding to (1.8) read as

$$\begin{aligned} Au_j + \omega_j &:= \frac{1}{4\pi}\int_\Gamma \frac{1}{r} u_j \, ds + \frac{1}{4\pi}\int_\Gamma \frac{1}{r}\{e^{ikr}-1-ikr\}u_j \, ds + \omega_j = -\delta_{1j} e^{ikx_1}, \\ \Lambda u_j &= \int_\Gamma u_j \, ds = \delta_{oj} \quad , \quad j = 0,1 \end{aligned} \tag{1.9}$$

and

$$\alpha = -\omega_1 \; / \; (\frac{ik}{4\pi} + \omega_o) \; .$$

A further type of boundary integral equations arises from the double layer approach for a hard scatterer [33 p. 163],[37 p. 25] (see also [11]) and the screen problem [29],[116] (for $k = 0$ see also [38]). Here we use the idea of Brakhage and Werner [16] to avoid the exceptional frequencies by the representation

$$U(x) = e^{ikx_1} + \frac{i}{4}\int_\Gamma u(y)\{i\eta \frac{\partial}{\partial \nu_y} + 1\} H_o^{(1)}(kr) \, ds_y \quad \text{for} \quad x \in \mathbb{R}^2 \setminus \bar{\Omega} \tag{1.10}$$

where $\eta > 0$ is a sufficiently large real constant and $u$ a complex density to be determined from the equation

$$Au(x) := -\frac{1}{2\pi}\int_\Gamma (u(y)-u(x))\frac{\partial}{\partial\nu_x}\frac{\partial}{\partial\nu_y}(\log r)ds_y + \frac{i}{2\eta}u(x)$$

$$+\int_\Gamma u(y)\frac{\partial}{\partial\nu_x}[\{\frac{\partial}{\partial\nu_y}-\frac{i}{\eta}\}\{\frac{1}{4i}H_o^{(1)}(kr)+\frac{1}{2\pi}\log(kr)\}+\frac{i}{2\pi\eta}\log r]ds_y$$

$$= -\frac{k}{\eta}e^{ikx_1}\frac{\partial x_1}{\partial\nu_x}\Big|_\Gamma \quad \text{for } x\in\Gamma . \tag{1.11}$$

The corresponding equations in three dimensions read as

$$U(x) = e^{ikx_1} + \frac{1}{4\pi}\int_\Gamma u(y)\{i\eta\frac{\partial}{\partial\nu_x}+1\}\frac{1}{r}e^{ikr}ds_y \quad \text{for } x\in\mathbb{R}^3\setminus\overline{\Omega} \tag{1.12}$$

and

$$Au(x) := \frac{1}{4\pi}\int_\Gamma (u(y)-u(x))\frac{\partial}{\partial\nu_x}\frac{\partial}{\partial\nu_y}(\frac{1}{r})ds_y$$

$$+\frac{i}{2\eta}u(x)+\frac{1}{4\pi}\int_\Gamma u(y)\frac{\partial}{\partial\nu_x}[\{\frac{\partial}{\partial\nu_y}-\frac{i}{\eta}\}\{e^{ikr}-1-ikr\}\frac{1}{r}-\frac{i}{\eta r}]ds_y$$

$$= -\frac{k}{\eta}e^{ikx_1}\frac{\partial x_1}{\partial\nu}\Big|_\Gamma . \tag{1.13}$$

It can easily be shown that the respective solution of these equations is unique including the case $k = 0$ and that the above operators are regular pertrubations to the case $k = 0$ .

Besides the above three types of boundary integral equations there also arise singular integral equations with Cauchy kernels in two dimensions, i.e. operators of the form

$$Au(x) = a(x)u(x) + \frac{1}{\pi i}\int_{y\in\Gamma} \frac{b(x,y)u(y)}{(y_1-x_1)+i(y_2-x_2)}\, d(y_1+iy_2) \qquad (1.14)$$

and in three dimensions singular integral equations with Giraud kernels

$$Au(x) = a(x)u(x) + \frac{1}{2\pi}\int_{y\in\Gamma} \frac{b(x,y)}{r^2}\, u(y)\, ds_y \qquad (1.15)$$

where $b = b_o(x, \frac{y-x}{r}) + rb_1(x,y)$ satisfying $\int_{|y-x|=r} b_o(x, \frac{y-x}{r}) d(\frac{y-x}{r}) = 0$ for all $r > 0$ and with $b_1$ smooth for $x \neq y$ and bounded. These equations occur for vibrations of elastic and thermoelastic materials [56],[57],[63],[64],[65].

For further integral equations in acoustics we refer to [22],[34],[39], [40], [58], [59], [74], [75], [83], [86], [97],[106] and references given there.

Although these four types of equations have different properties in the classical theory of integral equations it turns out that if they are considered as pseudodifferential operators they have a very strong common property. Namely the equations of practical interest are strongly elliptic. In order to formulate this property we need the Sobolev spaces $H^s(\Gamma)$ of generalized functions on $\Gamma$ , their interpolation and their dual spaces. For the Sobolev spaces we shall assume that the boundary manifold $\Gamma$ is sufficiently smooth. For the definition of the Sobolev spaces see [ 1 ]. The pairs $H^{j+s}(\Gamma)$ and $H^{j-s}(\Gamma)$ are dual spaces with respect to the j-scalar product $(f,g)_j$ on $\Gamma$ , for $j = 0$ we have

$$(f,g)_o = \int_\Gamma f\bar{g}\, ds \ .$$

Then each of the above operators $A$ defines a continuous linear mapping $A : H^s \to H^{s-2\alpha}$ for a whole scale of real $s$ (depending on the smoothness of $\Gamma$ ). $A$ is in each case a pseudodifferential operator of order $2\alpha$ [19 ], [105]. For our examples we have

$2\alpha = 0$ for (1.4), (1.5), (1.14) and (1.15),

$2\alpha = -1$ for (1.8), (1.9) and

$2\alpha = 1$ for (1.11) and (1.13) .

The announced common property is the <u>Gårding inequality</u>

$$\text{Re } (Av,v)_j \geq \gamma \ \|v\|^2_{j+\alpha} - \text{Re } C[v,v] \text{ for all } v \in H^{j+\alpha}(\Gamma) \qquad (1.16)$$

where $\gamma > 0$ is a constant independent of $v$ and where $C[u,v]$ denotes a compact bilinear form on $H^{j+\alpha} \times H^{j+\alpha}$ .

In some cases (1.16) can be proved directly as in [93],[81],[54] or it is equivalent to the energy estimates for the corresponding solution of the original boundary value problem and the Gårding inequality holds even with $C = 0$ as in [31],[38],[80],[81] .

In general (1.16) is a consequence of a simple property of the principal part of $A$ which has a principal symbol $a_o(x,\xi)$ for $x \in \Gamma$ and $\xi \in R^{n-1}$. $a_o$ is not uniquely determined, it depends on the local representation of $\Gamma$ [99]. If the principal part of $A$ is given by a convolution as in (1.8), (1.9), (1.11), (1.13) one may replace $\Gamma$ by its tangential plane in $x$ and find $a_o$ by the 1 - respectively 2 - dimensional Fourier transformed kernel function. For our example the principal symbols are given by

$$a_o = 1 \quad \text{for (1.4), (1,5),}$$

$$a_o = \frac{1}{2|\xi|} \quad \text{for (1.8), (1.9),}$$

$$a_o = \frac{1}{2}|\xi| \quad \text{for (1.11), (1.13) and} \tag{1.17}$$

$$a_o = a(x) + b_o(x,x)\frac{\xi}{|\xi|} \quad \text{for (1.14)},$$

$$a_o = a(x) + \sum_{\ell=1}^{\infty}\sum_{k=1}^{2\ell(\ell+1)} \frac{\pi\, i^{\ell}\,\Gamma(\frac{\ell}{2})}{\Gamma(1+\ell/2)}\, a_\ell^{(k)}(x)\, Y_{\ell,2}^{(k)}\left(\frac{\xi}{|\xi|}\right) \quad \text{for (1.15)}$$

where $b_o(x,\frac{y-x}{r}) = \sum\limits_{\ell=1}^{\infty}\sum\limits_{k=1}^{2\ell(\ell+1)} a_\ell^{(k)}(x)\, Y_{\ell,2}^{(k)}(\frac{y-x}{r})$

is the expansion with respect to the two-dimensional spherical harmonics of order $\ell$, see [74].

Note that the principal symbol is a positive-homogeneous function of $\xi$ for $|\xi| \geq 1$ [105 p. 32] whereas for $|\xi| \leq 1$ it is arbitrarily extended to a $C^\infty$ function of $\xi$ [19 ,Chap. IV].

Here the principal symbol in (1.17) for (1.8) follows directly from the convolution in

$$-\frac{1}{2\pi}\int_\Gamma u\log r\, ds = -\frac{1}{2\pi}\int_{\mathbb{R}} \log|t-s|\,(\chi(s)u(s))\,ds\ +$$

$$+\int_{\mathbb{R}} \log\left|\frac{x(t)-x(s)}{t-s}\right| \chi u ds \quad \text{for} \quad t\in[o\,,\int_\Gamma ds]$$

by Fourier transformation[3]. The principal symbol of (1.11) then is obtained from the identity

$$\frac{\partial}{\partial\nu_x}\frac{1}{2\pi}\int_\Gamma u(y)\frac{\partial}{\partial\nu_y}(\log r)\,ds_y = \frac{1}{2\pi}\int_\Gamma \frac{du}{ds_y}\,(\frac{d}{ds_x}\,(\log|x-y|)\,ds_y\,.$$

The principal symbols for (1.9) and (1.13) can be obtained by the use of surface polar coordinates $r$ and $\varphi$ introduced in [72 p. 76 ff.] and local coordinates $u_1 = r\cos\varphi$, $u_2 = r\sin\varphi$ in the neighbourhood of $x$. Inserting [72 (2.63), (2.68)] in (1.9) and [72 (2.63)-(2.68)] into the kernel of (1.13) one finds a representation with pseudo-homogeneous kernels [99 , Chap. III] whose principal symbols (1.17) follow by Fourier transformation (see also[3],[35],[71]and [101] for symbols corresponding to fundamental boundary integral operators.)

A pseudodifferential operator of order $2\alpha$ is called <u>strongly elliptic</u>, if there exists a positive constant $\gamma^1$ and a complex function $\Theta(x) \neq 0$ such that

$$\mathrm{Re}\ \Theta(x)\quad a_o(x,\xi) \geq \gamma' \quad \text{for all} \quad |\xi| = 1 \ . \tag{1.18}$$

Obviously, the operators in (1.4), (1.5), (1.8), (1.9), (1.11), (1.13) are all strongly elliptic with $\Theta \equiv 1$ whereas for (1.14) and (1,15) strong ellipticity (1.18) is an additional requirement. For (1.14) one can show [ 87 ] that strong ellipticity holds if and only if

$$a(x) + \lambda\, b(x,x) \neq 0 \quad \text{for all} \quad \lambda \in [-1,1] \ .$$

Due to Kohn and Nirenberg [61] and Hörmander [50] strong ellipticity (1.18) implies Gårding's inequality (1.16) for the operator $\Theta A$ .

For most boundary integral equations for stationary problems in applications it turns out that they are strongly elliptic (see e.g. [112]). However, most problems of practical applications yield more than just one boundary integral equation but systems, i.e. $u$ has to be replaced by a vector of unknowns $u(x) = \{u_1(x),\ldots,u_p(x)\}$ . Then the boundary integral equations form a system with

$$Au = \sum_{k=1}^{P} A_{\ell k}\, u_k$$

and our concept of strong ellipticity must be modified. To this end we assume that to $A$ there exist two index vectors $\alpha = (\alpha_1,\ldots,\alpha_p) \in \mathbb{R}^P$ and $\beta = (\beta_1,\ldots,\beta_p) \in \mathbb{R}^P$ such that the numbers

$$\alpha_{\ell k} = \alpha_\ell + \alpha_k - 2\beta_\ell \tag{1.19}$$

are the orders of $A_{\ell k}$ with corresponding principal symbols $a_{o\ell k}(x,\xi)$ , (see also [ 2 ]). Then the system (1.1) is called strongly elliptic if there exists a regular complex matrix function $\Theta(x)$ such that the following quadratic form is positive definite:

$$\operatorname{Re} \sum_{r,\ell,k=1}^{p} \zeta_r \, \Theta_{r\ell}(x) \, a_{o\ell k}(x,\xi) \, \bar{\zeta}_k \geq \gamma' \, |\zeta|^2 \tag{1.20}$$

for all $\zeta \in \mathbb{C}^p$ , all $x \in \Gamma$ and all $\xi \in \mathbb{R}^{n-1}$ where $\gamma' > 0$ is independent of $\zeta$, $x$ and $\xi$ . Now set for fixed $\gamma \in \mathbb{R}^p$ and $j \in \mathbb{R}$

$$\mathcal{H}^{\gamma+j} := \prod_{\ell=1}^{p} H^{\gamma_\ell + j}(\Gamma) \times \mathbb{R}^q \tag{1.21}$$

and define the operator associated to (1.1) ,

$$_\Theta\mathcal{A} := \begin{pmatrix} \Theta A \, , & B \\ \Lambda \, , & 0 \end{pmatrix} \, . \tag{1.22}$$

For the numerical boundary element approximation we shall restrict us to two proper subclasses of the strongly elliptic systems providing a single form of Gårding's inequality for systems [102].

1. The case of equal orders

Here we require

$$\alpha_\ell = \alpha \quad \text{for} \quad \ell = 1,..,p \, . \tag{1.23}$$

In this case strong ellipticity provides a Gårding inequality of the form

$$\mathrm{Re}\,({}_{\Theta}A(v,\omega),\ (v,\omega))_{\beta+j} := \mathrm{Re}\ \{\sum_{r,\ell=1}^{p} (\Theta_{r\ell}(\sum_{k=1}^{p} A_{\ell k}v_k + B_\ell\omega), v_r)_{j+\beta_\ell} + \Lambda v\cdot\omega\}$$

$$\geq \gamma\ \{\|v\|^2_{\alpha+j} + |\omega|^2\} - \mathrm{Re}\ C[(v,\omega),\ (v,\omega)] \tag{1.24}$$

for any $(v,\omega)\in H^{\alpha+j}$ with $\gamma > 0$ and $C$ a compact bilinear form. The proof follows from the Gårding inequality shown in [32 ], [36 , Sec. 12], [ 68 ], [107]. (For special boundary integral equations see [35 ] and for $\beta_\ell = 0$ , $C = 0$, $\Theta_{r\ell} = \delta_{r\ell}$ see [31 , p. 207 and Chap. VII ).

To the above case belong the systems of singular integral equations (1.14) and (1.15) where $\alpha = 0 = \beta_\ell$ . For (1.14), i.e. $n = 2$ the strong ellipticity (1.20) is equivalent to

$$\det\ (a(x) + \lambda\ b(x,x)) \neq 0 \quad \text{for all}\ \ \lambda\in[-1,1] \tag{1.25}$$

see [ 88 ], there one also finds the construction of $\Theta(x)$ .)

2. The case of positive definite principal symbol

In this case let in (1.20) be

$$\Theta_{r\ell}(x) = \delta_{r\ell}\ , \quad \text{the Kronecker symbol.} \tag{1.26}$$

Then strong ellipticity, i.e. in this case positive definiteness of the principal symbol again provides a Gårding inequality [36],[68],[107] in the form

$$\mathrm{Re}\ (A\ (v,\omega),\ (v,\omega))_{j+\beta} :=$$

$$\mathrm{Re}\ \{\sum_{\ell=1}^{p} (\sum_{k=1}^{p} A_{\ell k}\ v_k + B_\ell\omega,\ v_\ell)_{j+\beta_\ell} + \Lambda v\cdot\omega\} \tag{1.27}$$

$$\geq \gamma\{\|v\|^2_{\alpha+j} + |\omega|^2\} - \mathrm{Re}\ C\ [(v,\omega),\ (v,\omega)]\ .$$

Strongly elliptic systems providing a Gårding inequality of this type with various orders, e.g. $\alpha_1 = 0$, $\alpha_2 = -\frac{1}{2}$, $j = \beta_\ell = 0$ arise on closed subspaces of $H^\alpha$ e.g. for mixed boundary value problems in [23],[24],[25],[66],[67],[115] or in [29],[116].

For the approximation of equations (1.1) let us introduce a family of finite dimensional nested subspaces for each component

$$\tilde{H}_{h_k} \subset \tilde{H}_{\frac{h}{2^k}} \subset \ldots \subset H^{\alpha_k + j} \cap H^{\beta_k + j}$$

where $h$ denotes a parameter associated to the dimension $N$ of $\tilde{H}_{hk}$ by $h = c\,N^{-(n-1)}$. Here and the sequel $c$ will denote a generic constant independent of $h$. By $\tilde{H}$ let us denote the approximating product space $\tilde{H} = \tilde{H}_h = \prod_{k=1}^{p} \tilde{H}_{hk}$.

Since $\tilde{H} \times \mathbb{R}^q$ ought to approximate $H^{\alpha+j}$ we require that $\bigcup_{o<h} \tilde{H}_h \times \mathbb{R}^q$ is dense in $H^{\alpha+j}$. Let $\mu_\ell$, $\ell = 1,\ldots, N$ denote a basis of $\tilde{H}$. Then the well known Galerkin procedure for (1.1) is to find the coefficients $\gamma_\ell$ of the approximate solution

$$v_G(x) = \sum_{\ell=1}^{N} \gamma_\ell \, \mu_\ell(x) \quad , \quad x \in \Gamma \, , \quad \omega_G \in \mathbb{R}^q \tag{1.28}$$

by solving the finite system of linear equations for $\gamma_\ell$, $\omega_G$,

$$\begin{aligned} &\sum_{\ell=1}^{N} (\Theta A\mu_\ell, \mu_k)_{\beta+j}\, \gamma_\ell + (\Theta B, \mu_k)_{\beta+j} \cdot \omega_G = (\Theta f, \mu_k)_{\beta+j} \\ &\sum_{\ell=1}^{N} (\Lambda \mu_\ell)\gamma_\ell = b \, , \quad k = 1,\ldots,N \, . \end{aligned} \tag{1.29}$$

For the convergence of this procedure we have well known results going back to Michlin [73] and Hildebrandt Wienholtz [49]. Here we use the version known as Cea's lemma [19]. For its application to (1.29) we need an additional property of our approximation. Let $P_h$ denote the $H^{\beta+j}$ orthogonal projection onto $\tilde{H} \times \mathbb{R}^q$. Then we require for the representation of $P_h$ in $H^{\alpha+j}$,

$$\| P_h g - g \|_{H^{\alpha+j}} \to 0 \quad \text{for} \quad h \to 0 \text{ and any } g \in H^{\alpha+j} . \tag{1.30}$$

This assumption implies with the Banach-Steinhaus theorem the stability

$$\| P_h \|_{H^{\alpha+j}, H^{\alpha+j}} \le c \tag{1.31}$$

and by duality

$$\| P_h \|_{H^{j+2\beta-\alpha}, H^{j+2\beta-\alpha}} \le c . \tag{1.32}$$

This requirement is satisfied for the Fourier expansion with trigonometric polynomials, spherical harmonics and also with regular finite elements. Now we are in the position to state the convergence result for equations satisfying the general assumption

(A) *Let* (1.1) *with* A *strongly elliptic of order* $2\alpha$ *providing* (1.24) *or* (1.27) *be uniquely solvable for any* $(f,b) \in H^{2\beta-\alpha+j}$ .

THEOREM 1.1: *Suppose* (A).
*Then there exists* $h_o > 0$ *such that* (1.29) *is uniquely solvable for every* $0 < h \le h_o$ . *Moreover there exists a constant* c *independent of*

h *and* (f,b) *such that*

$$|\omega-\omega_G| + \|u - v_G\|_{\alpha+j} \le c \inf_{\chi \in \tilde{H}} \|u - \chi\|_{\alpha+j} \,. \tag{1.33}$$

*Proof:* Since A is strongly elliptic we have Gårding's inequality. That means that ${}_{\Theta}A$ can be decomposed such that

$$\begin{aligned}({}_{\Theta}A(v,\omega),\ (g,\eta))_{\beta+j} &= (\mathcal{A}(v,\omega),\ (g,\eta))_{\alpha+j} \\ &= (\mathcal{D}(v,\omega),\ (g,\eta))_{\alpha+j} + (C(v,\omega),\ (g,\eta))_{\alpha+j}\end{aligned} \tag{1.34}$$

for all $(v,\omega),\ (g,\eta) \in H^{\alpha+j}$ since the bilinear form on the left of (1.34) is continuous on $H^{\alpha+j} \times H^{\alpha+j}$ and $H^{\alpha+j}$ is a Hilbert space. There $\mathcal{D}$ is positive definite,

$$\mathrm{Re}\ (\mathcal{D}(v,\omega),\ (v,\omega))_{\alpha+j} \ge \gamma\ \{\|v\|^2_{\alpha+j} + |\omega|\ \} \tag{1.35}$$

and $C : H^{\alpha+j} \to H^{\alpha+j}$ is compact. Equations (1.29) are equivalent to finding $\tilde{v} := (v_G,\ \omega_G) \in \tilde{H} \times \mathbb{R}^q$ from the projection equations

$$P_h \mathcal{A} P_h \tilde{v} = P_h \mathcal{D} P_h \tilde{v} + P_h C P_h \tilde{v} = P_h \mathcal{A}(u,\omega)\ . \tag{1.36}$$

From the definiteness of $\mathcal{D}$ we have

$$\mathrm{Re}\ (\mathcal{D}\tilde{v},\ \tilde{v})_{\alpha+j} \ge \gamma\ \|\tilde{v}\|^2_{\alpha+j}$$

and also with continuity and $\tilde{v} = P_h \tilde{v}$

$$\gamma \|\tilde{v}\|^2_{\alpha+j} \le \mathrm{Re}\,(P_h \mathcal{D} P_h \tilde{v}, \tilde{v})_{\alpha+j} \le c\, \|(P_h \mathcal{D} P_h)\tilde{v}\|_{\alpha+j}\, \|\tilde{v}\|_{\alpha+j}$$

which yields the stability estimate on $\tilde{H} \times \mathbb{R}^q$ ,

$$\| (P_h \mathcal{D} P_h)^{-1} \|_{H^{\alpha+j},\, H^{\alpha+j}} \le c / \gamma \tag{1.37}$$

where $c/\gamma$ is independent of $h$ . Now we have

$$P_h \mathcal{A} P_h = P_h \mathcal{D} P_h (I + (P_h \mathcal{D} P_h)^{-1} P_h C P_h) . \tag{1.38}$$

The sequence of operators

$$(P_h \mathcal{D} P_h)^{-1} P_h C P_h$$

is the composition of a stable sequence (1.37) with $P_h$ and the compact operator $C$ . Because of (1.37) and $P_h g \to g$ for any $g \in H^{\alpha+j}$ with $h \to 0$ we also have

$$(P_h \mathcal{D} P_h)^{-1} g \to \mathcal{D}^{-1} g$$

for any $g \in H^{\alpha+j}$ . Hence, by the compactness of $C$ we find the convergence

$$\| (P_h \mathcal{D} P_h)^{-1} P_h C P_h - \mathcal{D}^{-1} C \|_{H^{\alpha+j},\, H^{\alpha+j}} \to 0$$

with respect to the operator norm [17 , Hilfssatz 3] [ 6 , Chap. 1].

Hence

$$(I + (P_h \mathcal{D} P_j)^{-1} P_h C P_h)^{-1} g \to (I + \mathcal{D}^{-1} C)^{-1} g = \mathcal{A}^{-1} \mathcal{D} g \quad \text{as} \quad h \to 0$$

for any $g \in H^{\alpha+j}$, and the approximating inverses exist for $h \le h_o$ with an appropriate $h_o > 0$ [ 6 , p. 10]. This implies the uniform boundedness

$$\| (I + (P_h \mathcal{D} P_h)^{-1} P_h C P_h)^{-1} \|_{H^{\alpha+j},\, H^{\alpha+j}} \le c$$

for all $h \le h_o$ . With (1.37), (1.38) we find the uniform boundedness of the Galerkin projection,

$$G_h := (P_h \mathcal{A} P_h)^{-1} P_h \mathcal{A} , \qquad \| G_h \|_{H^{\alpha+j},\, H^{\alpha+j}} \le c \tag{1.39}$$

for all $0 < h \le h_o$ and corresponding solvability of (1.36). Since

$$G_h(X,\eta) = (P_h \mathcal{A} P_h)^{-1} P_h \mathcal{A} P_h (X,\eta) = (X,\eta) \quad \text{for all} \quad X \in \tilde{H}_h, \eta \in \mathbb{R}^q ,$$

the stability (1.39) implies by

$$\| \tilde{v} - (u,\omega) \|_{\alpha+j} = \| G_h(u,\omega) - (u,\omega) + (X,\omega) - G_h(X,\omega) \|_{\alpha+j}$$

$$\le (c+1) \, \| u - X \|_{\alpha+j}$$

the proposed estimate (1.33). □

## § 2 CONVERGENCE RESULTS FOR BOUNDARY ELEMENT METHODS

In the following we shall approximate functions on the boundary $\Gamma$ by finite elements called boundary elements. To this end we assume that $\Gamma$ is given by local representations such that regular partitions in the parameter domains are mapped onto a corresponding partition of $\Gamma$. On the partitions in the parameter domains we use a regular $(m+1,m)$ system of finite elements (e.g. piecewise polynomials) [ 12 ]. (In general one uses $(\ell,m)$ systems rather than specifying $\ell = m+1$ . We avoid these details here.) Then the local representation of $\Gamma$ transplants these finite element functions on to $\Gamma$ . For calculations,the integrals can be evaluated by using the local coordinates. In these the finite elements appear as simple functions over the parameter domains. This construction of finite elements on $\Gamma$ requires that the parameter representations are fully available. For two dimensions, i.e. if $\Gamma$ is a curve, this is a sensible requirement whereas in higher dimensions, as $n=3$ , for the boundary surface $\Gamma$ the parameter representations can become rather involved. Hence in this case an additional approximation of the surface representation becomes necessary leading to isoparametric elements as well as to a boundary approximation. These approximations have been investigated by Nedelec [ 78 ], [ 80 ] for special cases of equations (1.9) and (1.13) and by Giroire [ 37 ] for (1.13).

But this approach is yet to be generalized to our more general class of problems. If we use the transplanted finite elements subject to regular $(m+1,m)$ systems in the parameter domains in connection with regular representations of the surface then they provide the approximation property.

*Let* $-\infty < t \leq s \leq m+1$, $t \leq m$, *then there exists a constant* $c$ *depending only on* $t$, $s$ *and* $m$ *and to any given* $u \in H^s(\Gamma)$ *and any* $\tilde{H}_h$ *of our family there exists a finite element function* $u_h \in \tilde{H}_h$ *such that*

$$\| u-u_h \|_t \leq c\, h^{s-t} \, \| u \|_s \tag{2.1}$$

[12, Theorem 4.1.2], [46].

*Remark:* For $n=2$ and one-dimensional piecewise polynomials of degree $m$, the approximation property (2.1) also holds for $m < t < m + \frac{1}{2}$. This has been obtained independently by D. Arnold and G. Schmidt and to be published soon.

Sometimes we shall additionally use the inverse property which holds for regular families $\tilde{H}_h$ subject to quasiuniform families of meshes:

*For* $t \leq s \leq m$ *there holds an estimate*

$$\| \mu \|_s \leq c\, h^{t-s} \, \| \mu \|_t \quad \text{for} \quad \mu \in \tilde{H}_h \tag{2.2}$$

*where the constant* $c$ *is independent of* $\mu$ *and* $h$.

For $s = t + 1 = m$ see [82] and for the general case see [7]. Again (2.2) also holds up to $s < m + \frac{1}{2}$ for one-dimensional splines on quasi-uniform families of meshes.

Since we also consider systems with $p$ unknown functions to be approximated we shall consider $p$ individual finite element approximations of $(m_k + 1, m_k)$ systems to $u_k$, $k = 1,\dots,p$, providing (2.1), respectively (2.2), if necessary, for each component. For the respective approximating spaces we write $\tilde{H}_{hk}$, $k = 1,\dots,p$ and again denote by $\tilde{\mathcal{H}}$ the approximating product space

$$\tilde{\mathcal{H}} := \prod_{k=1}^{p} \tilde{H}_{hk} \quad .$$

## 2.1. Galerkin's procedure

Let us first consider Galerkin's procedure (2.29) for (1.1). In order to be able to implement the Galerkin equations computationally we further restrict the class of boundary equations by the requirements

$$\beta_k + j \in \mathbb{N}_0 \quad \text{for every} \quad k = 1,\dots,p \, . \tag{2.3}$$

Note that the simplest case is $\beta_k + j = 0$, then (1.29) becomes the usual Galerkin procedure that has been considered in [37],[38],[54], [55],[79],[80],[93],[102],[113],[114].

For the application of the Aubin-Nitsche lemma we shall need the $\beta+j$-adjoint ${}_\Theta A^*$ to ${}_\Theta A$ which is defined via duality and the Riesz representation theorem by

$$(g, {}_\Theta A^* \phi)_{\mathcal{H}^{\beta+j}} = ({}_\Theta A g, \phi)_{\mathcal{H}^{\beta+j}} \quad \text{for all} \quad g, \phi \in \mathcal{H}^{\alpha+j} \quad . \tag{2.4}$$

Recalling the result of [ 55 ] we find for (1.29) the following asymptotic error estimate.

THEOREM 2.1: [102], [ 55 ], [ 7 ]: *Let* (1.1) *satisfy* (A) . *Then, in addition to Theorem 1.1 we have for* $0 \le \tau \le \min_{k=1,..p} \{m_k+1-\alpha_k-j\}$ *and*

$\alpha_k + j \le s_k \le m_k + 1$ , $\alpha_k + j \le m_k$ , $k = 1,..p$ *the asymptotic error estimate*

$$|\omega-\omega_G| + \|u-v_G\|_{\alpha+j-\tau} \le c \sum_{k=1}^{p} h^{\tau+s_k-j-\alpha_k} \|u_k\|_{s_k} \tag{2.5}$$

*provided* ${}_\Theta A^* : H^{\alpha+j+\tau} \to H^{2\beta-\alpha+j+\tau}$ *is an isomorphism.*

*Proof:* For $\tau = 0$ , (2.5) follows from (1.33) with (2.1),

$$|\omega-\omega_G| + \|u-v_G\|_{\alpha+j} \le c \sum_{k=1}^{p} \| u_k-u_{kh}\|_{\alpha_k+j} \le c \sum_{k=1}^{p} h^{s_k-\alpha_k-j} \|u_k\|_{s_k} \tag{2.6}$$

For $0 < \tau$ we use that $H^{\alpha+j-\tau}$ and $H^{2\beta-\alpha+j+\tau}$ are dual spaces with respect to the $H^{\beta+j}$- scalar product,

$$\|e\|_{\alpha+j-\tau} \le c \sup_{\|\phi\|_{2\beta-\alpha+j+\tau} \le 1} |(e, \phi)_{H^{\beta+j}}|$$

where $e = (u-v_G, \omega-\omega_G)$ . Set $w \in H^{\alpha+j+\tau}$ with

$${}_\Theta A^* w = \phi$$

and use (1.29) in the form $({}_\Theta Ae, X)_{\beta+j} = 0$ for all $X \in \tilde{H}$ obtaining

$$\| e \|_{\alpha+j-\tau} \le c \sup_{\|\phi\|_{2\beta-\alpha+j+\tau} \le 1} |(e, {}_{\Theta}A^* w)_{H^{\beta+j}}| = c \sup |({}_{\Theta}Ae, w)_{H^{\beta+j}}|$$

$$= c \sup |({}_{\Theta}Ae, w-X)_{H^{\beta+j}}|$$

$$\le c \| e \|_{\alpha+j} \sup \| {}_{\Theta}A^* (w-X) \|_{2\beta-\alpha+j}$$

$$\le c \| e \|_{\alpha+j} \sup \| w-X \|_{\alpha+j}$$

for all $X \in \widetilde{H}$ . The choice $X = w_h$ in (2.1) yields

$$\| e \|_{\alpha+j-\tau} \le c \| e \|_{\alpha+j} \sup \{c\, h^{\tau} \| w \|_{\alpha+j+\tau}\}$$

$$\le c \| e \|_{\alpha+j} h^{\tau} \sup \| \phi \|_{2\beta-\alpha+j+\tau} = c\, h^{\tau} \| e \|_{\alpha+j} \ .$$

Inserting (2.6) gives (2.5). □

By using the inverse property (2.2) we are also able to establish

THEOREM 2.2: [ 55 ] *Let* (1.1) *satisfy* (A) *and let* $\widetilde{H}$ *provide the inverse property* (2.2). *Then for*

$0 \le \tau \le m_k - \alpha_k - j$ , $\alpha_k + j \le s_k \le m_k + 1$ , $\alpha_k + j \le m_k$ , $k = 1,..p$

*we find*

$$\| u - v_G \|_{\alpha+\tau+j} \le c \sum_{k=1}^{p} h^{s_k - \tau - j - \alpha_k} \| u_k \|_{s_k} \ . \tag{2.7}$$

*For one-dimensional splines,* i.e. $n = 2$ , (2.7) *also holds if*

$0 \le \tau < m_k - \alpha_k - j + \frac{1}{2}$ *and* $\alpha_k + j < m_k + \frac{1}{2}$ [7] .

Since (2.7) follows immediately from (2.2) with (2.5) $(\tau=0)$ we omit the elementary proof.

For estimating the errors due to numerical integration we also need the following

COROLLARY 2.3: [ 7 ], [ 55], [112] *Suppose that all assumptions of Theorems* 2.1 *and* 2.2 *are fulfilled. Then we have the following conditioning estimates:*

$$|\omega_G| + \|v_{Gk}\|_o \le c\,\{\sum_{\ell=1}^{p} h^{(\alpha_\ell+\alpha_k+2j)'} \|w_\ell\|_{2j+2\beta_\ell} + |b|\}\ ,\ k = 1,..,p,$$

$$|b| + \|w_\ell\|_{2j+2\beta_\ell} \le c\,\{\sum_{k=1}^{p} h^{(-2j-\alpha_\ell-\alpha_k)'} \|v_{Gk}\|_o + |\omega_G|\}\quad \ell = 1,..,p \tag{2.8}$$

*where* $w_\ell = P_h f_\ell \in \tilde{H}_{h\ell}$ *and* $t' = \min\{0,t\}$ .

*Proof:* For the proof let us indicate just one case, the others follow by suitable modifications.

First note that the Galerkin solution $(v_G, \omega_G)$ to $(u,\omega)$ respectively $(f,b)$ is the same as to the solution $(v,\omega')$ belonging to $(P_h f,b)$ due to (1.29). Then (2.5) or (2.7) implies with

$$\tau = \begin{cases} j + \alpha_k & \text{for } j + \alpha_k \ge 0 \text{ and (2.5)},\\ -j - \alpha_k & \text{for } j + \alpha_k \le 0 \text{ and (2.7)} \end{cases}$$

the estimate

$$\|v_{Gk}\|_o + |\omega_G| \le c\,\{\sum_{\ell=1}^{p} \|v_\ell\|_{\alpha_\ell-\alpha_k} + |\omega'|\}$$

$$\le c\,\{\sum_{\ell=1}^{p} \|P_h f_\ell\|_{2\beta_\ell+2j-(\alpha_k+\alpha_\ell+2j)} + |b|\}\ .$$

For the terms on the right hand side we use continuous imbedding for $2j + \alpha_\ell + \alpha_k \ge 0$ , and the inverse property (2.2) for $\alpha_k + \alpha_\ell + 2j < 0$ to find the first inequality in (2.8).

For the second inequality in (2.8) we use continuity of $A$ and again continuous imbedding, respectively (2.2) to obtain

$$|b| + \|w_\ell\|_{2j+\beta_\ell} \le \|w\|_{2\beta+2j-\alpha+\alpha_\ell} + |b|$$

$$\le c\,\{\sum_{k=1}^{p} \|v_{Gk}\|_{\alpha_k+(\alpha_\ell+2j)^+} + |\omega_G|\}$$

$$\le c\,\{\sum_{k=1}^{p} h^{(-2j-\alpha_\ell-\alpha_k)'} \|v_{Gk}\|_o + |\omega_G|\}\ .$$

All the above estimates require that the corresponding Sobolev indices satisfy appropriate inequalities subject to $m_k$ . In case that these are violated one needs modifications as in [52 , p. 109] or [ 7 ,(2.1.37)].□

Besides the error estimates in Sobolev spaces we propose also almost quasioptimal pointwise estimates.

PROPOSITION 2.4: *For* $u \in \prod_{k=1}^{p} W^{\alpha_k+\tau,\infty} =: W^{\alpha+\tau,\infty}$ *and*

$\tau \geq -\alpha_\ell$ , $\ell = 1,..,p$ , $\alpha_k - m_k \leq \alpha_\ell \leq m_k + 1 - \alpha_k - 2j$ *for all* $\ell, k = 1,..p$ *we have the pointwise error estimates*

$$|\omega-\omega_G| + \|u_\ell - v_{G\ell}\|_{L^\infty} \leq c\, h^{\alpha_\ell+\tau} |\log h|^t \|u\|_{C^{\alpha+\tau},\infty} \tag{2.9}$$

*with* $t \geq 0$ *an appropriate integer,* $\ell = 1,..,p$.

This proposition has yet been proved only for equations of the type (1.8) respectively (1.9) in [91].

## 2.2 Ordinary collocation

The ordinary collocation method is that one mostly used in engineering computations. Here let $x_\ell \in \Gamma$ be $N$ appropriately given collocation points such that $\{x_\ell\}$ form a unisolvent set with respect to every $\tilde{H}_{hk}$, $k = 1,...,p$ [21]. Then the collocation procedure for (1.1) is to find the vector coefficients $\gamma_j$ of the approximate solution

$$v_c(x) = \sum_{m=1}^{N} \gamma_m \mu_m(x) \ , \ x \in \Gamma \ , \quad \omega_c \in \mathbb{R}^q \tag{2.10}$$

by solving the collocation equations

$$\sum_{m=1}^{N} (A\mu_m)(x_\ell)\gamma_m + B(x_\ell)\omega_c = f(x_\ell) \ , \quad \ell = 1,..,N \ , \ \sum_{m=1}^{N} \Lambda\mu_m \gamma_m = b \ . \tag{2.11}$$

But in contrary to Galerkin's procedure, for the collocation (2.11) convergence results are known yet only in the special case of Fredholm integral equations of the second kind as for (1.4), (1.5) which will be treated in Chapter 3 and for one-dimensional boundary integral equations, i.e. $n = 2$ . In the latter case we shall consider here only polynomial splines and distinguish the cases of odd and even degree splines. For these two-dimensional cases we are able to prove optimal order convergence by converting the collocation equations into modified Galerkin equations. For them we can use our already developed analysis including the Aubin-Nitsche duality arguments. It should be pointed out that in contrary to the finite element methods these results cannot be obtained from treating collocation as a one point Gaussian quadrature version of the usual Galerkin method for which one never gains enough powers of $h$ (except for Fredholm equations of the second kind). Our approach is however restricted to the use of spline functions. For other trial functions as for the collocation of singular integral equations as (1.14) using trigonometric polynomials and Fourier series we refer to [89] .

$n = 2$ and odd degree splines

In the following we essentially resemble the presentation in [7] .

To the increasing sequence of collocation points $\Delta = \{x_\ell\}$ let us choose the space of all $m_k - 1$ times continuously differentiable splines of degree $m_k$ subordinate to the partition $\Delta$ . Note that $\varphi^{(m_k)}$, the $m_k$-th derivative with respect to the arc length $s$ on $\Gamma$ (or an equivalent regular parameter) is a step function for $\varphi \in \tilde{H}_{hk}$ . Moreover we require that $m_k$ be an odd integer and be related to the indices $\beta_k$ in (1.19) by the equation

$$m_k = 2(j+\beta_k) - 1 \quad \text{for} \quad k = 1,..,p \tag{2.12}$$

where $j \in \mathbb{R}$ is a <u>fixed number</u>. Since $Au_c$ must be continuous at the nodes for (2.11) we further assume that

$$2j > 1 + (\alpha_k + \alpha_\ell) - 2(\beta_\ell + \beta_k) \quad \text{for} \quad \ell, k = 1,..,p . \tag{2.13}$$

If the equations (2.11) are considered to be Galerkin-Bubnov equations with Dirac functions as test functions, integration by parts yields modified Galerkin equations. To this end we define mappings $J$ and $J_\Delta$ by

$$Ju := \int_\Gamma u\,ds \quad \text{and} \quad J_\Delta u = \sum_{\ell=1}^{N} \delta_\ell\, u(x_\ell) \tag{2.14}$$

where the $\delta_\ell$ are the weights of the trapezoidal rule subject to the nodal points $x_\ell \in \Gamma$. Then we have from [7 , Theorem 2.1.1]:

THEOREM 2.5: *For smooth* $w$ *the equations*

$$w(x_i) = 0 , \quad i = 1,..,N$$

*hold if and only if*

$$(w - Jw + J_\Delta w, \varphi)_{j+\beta_k} = 0 \quad \text{for all} \quad \varphi \in \tilde{H}_{hk} .$$

Therefore the collocation equations (2.11) are equivalent to the modified Galerkin equations

$$((I - J + J_\Delta)\Theta\{A(v_c - u) + B(\omega_c - \omega)\}, \chi)_{\beta+j} = 0 \quad \text{for all} \quad \chi \in \tilde{H}_h ,$$

$$\Lambda\,(v_c - u) = 0 . \tag{2.15}$$

This formulation now allows to apply our stability result (1.39) obtained in Theorem 1.1 yielding the following theorem.

THEOREM 2.6 [ 7 , Theorem 2.1.4]: *Let* (1.1) *satisfy* (A). *Then under the foregoing assumptions there exist positive constants* $\gamma$ *and* $h_o$ *such that if* $\Delta$ *is a partition for which* $h = \max_{i=1,..,N} |x_i - x_{i-1}| \le h_o$, *then there holds the Babuška - Brezzi stability condition*

$$\inf_{\substack{v\in\tilde{H}_h\,,\ \omega\in\mathbb{R}^q \\ \|(v,\omega)\|_{H^{\alpha+j}}=1}} \ \sup_{\substack{z\in\tilde{H}_h\,,\ \eta\in\mathbb{R}^q \\ \|(z,\eta)\|_{H^{\alpha+j}}=1}} ((I-J+J_\Delta)\,{}_\Theta A(v,\omega),(z,\eta))_{\tilde{H}^{\beta+j}} \ge \gamma \tag{2.16}$$

*Proof:* For showing (2.16) we reformulate (2.15) for $v_c$ in the form of the equations (1.29),

$$\begin{aligned}&({}_\Theta A(v_c,\omega_c),(z,\eta))_{H^{\beta+j}}\\ &= ({}_\Theta A[(u,\omega)-{}_\Theta A^{-1}\{(J_\Delta-J)\,{}_\Theta A(v_c-u,\omega_c-\omega)\}],\ (z,\eta))_{H^{\beta+j}}\end{aligned} \tag{2.17}$$

for all $(z,\eta)\in\tilde{H}\times\mathbb{R}^q$ . This yields via (1.33), respectively (1.39) the estimate

$$\|v_c\|_{\alpha+j} + |\omega_c| \le c\ \|[(u,\omega)-{}_\Theta A^{-1}\{(J_\Delta-J)\,{}_\Theta A(v_c-u,\omega_c-\omega)]\|_{\alpha+j}$$

where the constant $c$ is independent of $h$ for $0 < h \le h_1$ with a suitable $h_1 > 0$ . Since the images of $J_\Delta$ and $J$ are constants we further have

$$\|v_c\|_{\alpha+j} + |\omega_c| \le c\,\{\|u\|_{\alpha+j} + |\omega| + |(J_\Delta - J)\,{}_\Theta A(v_c-u,\omega_c-\omega)|\}$$

and from the integration error of the trapezoidal rule

$$\|v_c\|_{\alpha+j} + |\omega_c| \le c\,\{\|u\|_{\alpha+j} + |\omega|\} + c'h^\mu \|{}_\Theta A(v_c-u,\omega_c-\omega)\|_\mu$$

where $\mu := \min\,\{2,\ j-\alpha_k+2\beta_k\} > \frac{1}{2}$ due to (2.13). Hence we find with continuity of ${}_\Theta A$

$$\|v_c\|_{\alpha+j} + |\omega_c| \le c\,\{\|u\|_{\alpha+j} + |\omega|\} + c_1 h^\mu\{\|v_c\|_{\alpha+j} + |\omega_c| +$$

$$+ \|u\|_{\alpha+j} + |\omega|\}$$

which yields stability

$$\|v_c\|_{\alpha+j} + |\omega_c| \le c\,\{\|u\|_{\alpha+j} + |\omega|\} \le c'\,\{\|f\|_{2\beta-\alpha+j} + |b|\} \quad (2.18)$$

for all $0 < h \le h_o := \min\,\{h_1, (2c_1)^{-1/\mu}\}$. This stability estimate implies uniqueness and, hence, solvability of (2.11). Moreover, it can easily be shown that (2.18) implies (2.16). □

Now quasioptimal convergence in the $H^{\alpha+j}$ norm of the collocation method is a simple consequence of (2.16). Moreover, we can apply the Aubin-Nitsche trick to obtain

THEOREM 2.7 [ 7 , Theorem 2.1.6]: *Let* (1.1) *satisfy* (A) *and let*

$0 \le \tau \le \min\limits_{k=1,..,p} \{m_k+1-\alpha_k-j\}$ *and* $\alpha_k + j \le s_k \le m_k + 1$ , $k = 1,..,p$

*and suppose that* ${}_\Theta A^* : H^{\alpha+j+\tau} \to H^{2\beta-\alpha+j+\tau}$ *is an isomorphism. Then the collocation method converges asymptotically as*

$$|\omega-\omega_c| + \|u-v_c\|_{\alpha+j-\tau} \le c \sum_{k=1}^{p} h^{\tau+s_k-j-\alpha_k} \|u_k\|_{s_k} \qquad (2.19)$$

*for* $0 < h \le h_o$ *with some appropriate* $h_o$ .

*Proof:* In the case $\tau = 0$ , (2.16) is equivalent to (2.18) implying the existence and stability of the collocation projection $C_h$ which is defined by solving the collocation equations (2.11). $C_h$ satisfies

$$(v_c,\omega_c) := C_h(u,\omega) \ , \quad C_h : H^{\alpha+j} \to \tilde{H} \times \mathbb{R}^q \subset H^{\alpha+j}$$

with

$$C_h\Big|_{\tilde{H}\times\mathbb{R}^q} = \text{identity and } \|C_h\|_{\alpha+j,\ \alpha+j} \le c \ .$$

Hence, as in the proof of Theorem 1.1 we find with (2.1)

$$|\omega-\omega_c| + \|u-v_c\|_{\alpha+j} \le c \inf_{X\in\tilde{H}} \|u-X\|_{\alpha+j} \le c \sum_{k=1}^{p} h^{s_k-j-\alpha_k} \|u_k\|_{s_k} \ ,$$

i.e. (2.19) with $\tau = 0$ .

For $\tau > 0$ define $\phi \in H^{2\beta-\alpha+j+\tau}$ by duality and the Riesz representation theorem via

$$\|u-v_c\|^2_{\alpha+j-\tau} + |\omega-\omega_c|^2 = ((u-v_c,\ \omega-\omega_c),\phi)_{H^{\beta+j}} = \|\phi\|^2_{H^{2\beta-\alpha+j+\tau}}. \tag{2.20}$$

To $\phi$ define

$(w,\eta) := {}_{\Theta}A^{*-1}\ \phi \in H^{\alpha+j+\tau}$ and

$(y,\eta)$ by

$y_k = w_k$ for all $k \in K_1 := \{k$ with $2\beta_k + j - \alpha_k \le \tau + \frac{1}{2}\}$

$y_k = (I+J-J_\Delta^*)w_k$ for all $k \notin K_1$, $k = 1,..,p$ .

Note that $J^2 = J_\Delta\ J = J$ , $J_\Delta^2 = JJ_\Delta = J_\Delta$ and therefore

$w_k = (I-J+J_\Delta^*)y_k$ .

Moreover note that $J_\Delta : H^s \to H^t$ is uniformly bounded for $s > \frac{1}{2}$ only and, hence, $J_\Delta^* : H^{2\beta_k+2j-t} \to H^{-s+2\beta_k+2j}$ . Thus, $J_\Delta^*:H^{\alpha_k+j+\tau} \to H^{\alpha_k+j+\tau}$ only in the case $k \notin K_1$ .

Now let us proceed with the proof first in case $0 < \tau \le 2$ . Then

$$\|u-v_c\|^2_{\alpha+j-\tau} + |\omega-\omega_c|^2 = ((u-v_c,\ \omega-\omega_c),\ {}_{\Theta}A^*(w,\eta))_{\beta+j}$$

$$= ({}_{\Theta}A(u-v_c,\ \omega-\omega_c),\ (\omega,\eta))_{\beta+j}$$

$$= ((I-J+J_\Delta)\,{}_{\Theta}A(u-v_c,\ \omega-\omega_c),\ (y,\eta))_{\beta+j} + \sum_{k\in K_1} ((J_\Delta-J)({}_{\Theta}A(u-v_c))_k, y_k)_{\beta_k+j}$$

and by equations (2.15)

$$\|u-v_c\|^2_{\alpha+j-\tau} + |\omega-\omega_c|^2 = ((I-J+J_\Delta)\Theta\{A(u-v_c)+B(\omega-\omega_c)\},\ y-\chi)_{\beta+j}$$

$$+ \sum_{k\in K_1} ((J_\Delta-J)\Theta_{k\cdot}A_\cdot(u-v_c),\ y_k)_{\beta_k+j}$$

with any $\chi \in \widetilde{H}$. That implies with (2.1) and the error of the trapezoidal rule

$$\|u-v_c\|^2_{\alpha+j-\tau} + |\omega-\omega_c|^2 \le c\ (\|u-v_c\|_{\alpha+j} + |\omega-\omega_c|)\cdot \inf_{\chi\in\widetilde{H}} \|y-\chi\|_{\alpha+j}$$
$$+ c\,h^\tau \|A(u-v_c)\|_\tau \sum_{k\in K_1} |Jy_k|$$

$$\le c_1 h^\tau (\|u-v_c\|_{\alpha+j} + |\omega-\omega_c|)\ \|y\|_{\alpha+j+\tau}$$

since $\tau \le 2\beta_k - \alpha_k + j = m_k + 1 - \alpha_k - j$. Our definition of $y$ implies

$$\|y\|_{\alpha+j+\tau} \le c\ \|(w,\eta)\|_{\alpha+j+\tau} \le c\ \|\phi\|_{2\beta-\alpha+j+\tau} \le c(\|u-v_c\|_{\alpha+j-\tau} + |\omega-\omega_c|)$$

which assures (2.19) with the above estimate. In the remaining case $\tau > 2$ we proceed in the same manner to find

$$\|u-v_c\|^2_{\alpha+j-\tau} + |\omega-\omega_c|^2 \le c\,h^\tau (\|u-v_c\|_{\alpha+j} + |\omega-\omega_c|)\ \|y\|_{\alpha+j+\tau}$$

$$+ c\,h^2 \sum_{k\in K_1} \|u-v_c\|_{\alpha+j-\tau_k} \|y\|_{\alpha+j+\tau}$$

where $\tau_k = j - \alpha_k + 2\beta_k - 2$ and $k \in K_1$. If we restrict $\tau$ to $2 < \tau \leq \frac{7}{2}$ then $\tau_k \leq 2$ and we can use (2.19) with $\tau = \tau_k$ yielding

$$\|u-v_c\|_{\alpha+j-\tau} \leq c \sum_{\ell=1}^{p} h^{\tau+s_\ell-j-\alpha_\ell} \|u_\ell\|_{s_\ell}$$

$$+ c \sum_{k\in K_1}^{p} \sum_{\ell=1}^{p} h^{\tau_k+2-s_\ell-j-\alpha_\ell} \|u_\ell\|_{s_\ell}$$

$$\leq c \sum_{\ell=1}^{p} h^{\tau+s_\ell-j-\alpha_\ell} \|u_\ell\|_{s_\ell}$$

since $\tau_k + 2 = m_k + 1 - \alpha_k - j \geq \tau$.

If $\tau > \frac{7}{2}$ use the same approach to show (2.19) for $\frac{7}{2} < \tau \leq 5$. Repeating the above arguments, (2.19) is proved after finitely many steps. □

Since Theorem 2.2 depended only on the additional inverse property we deduce from Theorem 2.7 the

CONCLUSION 2.8: [ 7 ] *If the inverse property* (2.2) *holds in addition to the assumptions of Theorem* 2.7 *then Theorem* 2.2 *remains valid also for the collocation procedure.*

In the same manner as for Corollary 2.3 one shows here

COROLLARY 2.9: [ 7 ] *If the inverse property* (2.2) *holds in addition to the assumptions of Theorem 2.7 then we have*

$$|\omega_c| + \|v_{ck}\|_o \le c \sum_{\ell=1}^{p} h^{(\alpha_\ell+\alpha_k-2\beta_\ell)'} \{\|w_\ell\|_o + |b|\} , \; k = 1,\ldots,p ,$$

$$|b| + \|w_\ell\|_o \le c \{\sum_{k=1}^{p} h^{(2\beta_\ell-\alpha_\ell-\alpha_k)'} \|v_{ck}\|_o + |\omega_c|\} , \; \ell=1,\ldots,p \tag{2.21}$$

*where* $w_\ell = P_h f_\ell \in \tilde{H}_{h,\ell}$ .

For the special case of singular integral equations with $\alpha_\ell = \beta_\ell = 0$ and $j = m_\ell = 1$ , i.e. piecewise linear trial functions the convergence result follows from [88],[89]. Here the strong ellipticity condition is even nécessary for the convergence. For all the other cases these results have been obtained for the first time in [ 7 ]. There one can also find several references to numerical treatments of boundary integral methods to which our results apply.

Note that our error estimates can be used to compare the Galerkin method with the collocation method [ 8 ]. For simplicity let us consider only the case $\beta_\ell = 0$ and $\alpha_\ell = \alpha$ . Let us first consider the case of using the <u>same degree splines</u> $S_m(\Delta)$ for <u>both</u> methods. Further let us restrict to <u>quasiuniform families of meshes</u> and smooth solutions. Then the highest rate of convergence achieved by the collocation method is $O(h^{m+1-2\alpha})$ in $H^{2\alpha}$ , whilst the Galerkin method converges with rate $O(h^{2m+2-2\alpha})$ in $H^{2\alpha-m-1}$ . This situation is summarized in Figure 1. Hence, theoretically the Galerkin method seems to be superior to the collocation as long as the same kind of trial functions is used. To obtain the <u>same order</u> of superapproximation for <u>both</u> methods, one must use splines of <u>different</u> orders $m_G$ for the Galerkin method and $m_c$ for the collocation method which are related by $m_c = 2m_G + 1$ . Then for both methods we have the <u>same</u> rate $O(h^{2m_G+2-2\alpha}) = O(h^{m_c+1-2\alpha})$ of convergence in $H^{-m_G-1+2\alpha}$ and $H^{2\alpha}$ , respectively. On the other hand, the construction of the stiffness matrix for the Galerkin method requires the evaluation of double integrals whilst the collocation method only requires single integrals. A comparison of these computational expenses has been done in [8] .

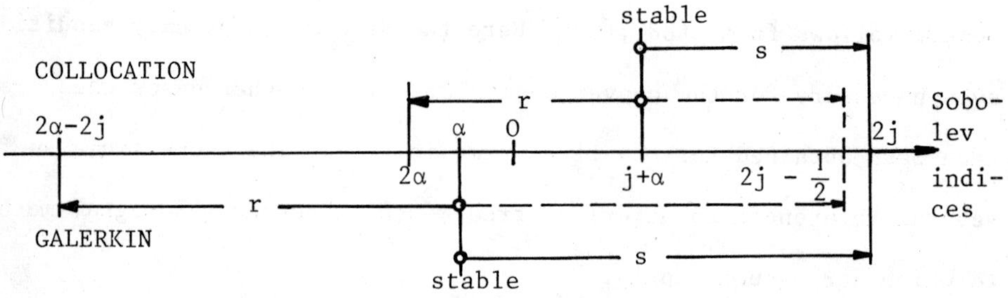


The case $\alpha \le 0$ ($j+\alpha > 0$ pictured, $j+\alpha \le 0$ also possible)

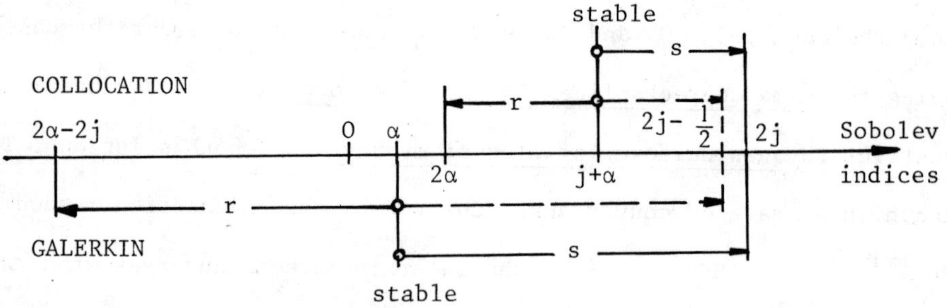


The case $\alpha \ge 0$

<u>Figure 1.</u> The indices $r \le s$ for which $\| u-v \|_r \le ch^{s-r} \|u\|_s$ .
Dashed lines indicate estimates requiring a quasiuniform mesh family.

<u>$n = 2$ and even degree splines</u>

In addition to the collocation points $\Delta$ let us choose $N$ meshpoints $\{\tilde{x}_\ell\} \subset \Gamma$ such that for corresponding arc length (or parameter values of a regular parameter representation) we have $s(x_\ell) = s(\tilde{x}_{\ell-1}) + \frac{1}{2}(s(\tilde{x}_\ell) - s(\tilde{x}_{\ell-1}))$ i.e. $x_\ell$ lies in the "middle" between $\tilde{x}_{\ell-1}$ and $\tilde{x}_\ell$. Now choose $\tilde{H}_{hk}$ as the space of $m_k - 1$ times continuously differentiable splines of degree $m_k$ subordinate to the meshpoints $\{\tilde{x}_\ell\}$. In particular the choice $m_k = 0$, i.e. piecewise constant trial functions is very often used in practical computations [13],[14],[45],[76],[104].

From results for more specific equations (1.1) [95] it can be conjectured that Theorem 2.7 remains valid also in this case provided

$$m_k = 2(j+\beta_k)-1 > \alpha_k + \alpha_\ell - 2\beta_\ell \quad \text{for} \quad k,\ell = 1,..,p \ .$$

But the proof for this case is yet to be done.

It should also be mentioned that the collocation with even order splines subordinate to the nodal points (i.e. nodal points and brake points coincide) in general will <u>not</u> converge for the strongly elliptic equations anymore but for a completely different class as was shown recently by G. Schmidt [98].

For $n = 3$, i.e. integral equations on surfaces, convergence results for the collocation method are yet available only for the Fredholm integral equations of the second kind. For our fairly more general class of strongly elliptic equations the mathematical analysis is yet to be done.

Note that all the foregoing error estimates hinge crucially on the smoothness of $u$ on one hand and the Gårding inequality on the other hand.

If $\Gamma$ is not smooth any more, e.g. $\Gamma$ contains corners or the boundary conditions change then the desired layers $u$ will contain singular functions. These have been augmented to the trial (and test) functions in [23], [24], [25], [28], [66], [67], [115] where the analysis again rests on Gårding's inequality for the remaining smooth parts of $u$. The latter actually also is the basic property in [29].

## § 3 FREDHOLM INTEGRAL EQUATIONS OF THE SECOND KIND

In this section we consider the special case of linear equations of the second kind, i.e.

$$u + Qu + Cu + B\omega = g\ , \quad \Lambda u = b \tag{3.1}$$

in a Banach space $X \times \mathbb{R}^q = X^p \times \mathbb{R}^q$ where $Q$ is a contraction mapping, i.e.

$$\| Qv \|_X \leq q_o \| v \|_X \quad \text{with} \quad q_o < 1 \tag{3.2}$$

and where $C : X \to X$ is compact. Although (3.1) seems to be rather specific, this case corresponds to strong ellipticity.

LEMMA 3.1 [111], [21, p. 9] : *Let* $X = H^j$ , $\beta = \alpha = 0$ . *Then Gårding's inequality* (1.24) *implies that* (1.1) *is equivalent to* (3.1) *with*

$$Qu + Cu = \frac{\gamma}{\|\Theta A\|} \Theta A u - u \ , \quad g = \frac{\gamma}{\|\Theta A\|} f \ ,$$

*and* $0<q_o = 1 - \frac{\gamma^2}{\|\Theta A\|^2} < 1$ . *Conversely,* $A = I + Q + C$ *provides the Gårding inequality* (1.24) *with* $\Theta = I$ .

Note that the quations (1.4) and (1.5) (after multiplication by 2) are of the form (3.1). Here let us consider the case of smooth $\Gamma$ first, i.e. $\Gamma \in C^{r+2}$ , $r \in \mathbb{N}_o$ (or $\Gamma \in C^{1+\alpha}$ a Ljapounov curve, respectively surface). Then for $n = 2$ , i.e. $\Gamma$ a curve, in (2.14) set $Q = 0$ , and there

$$Cu := -\frac{1}{2i} \int_\Gamma u(y) \left(\frac{\partial}{\partial \nu_y} H_o^{(1)}(kr)\right) ds_y \tag{3.3}$$

defines compact mappings

$$H^t \to H^{t+r} \ , \quad C^t \to C^{t+r} \ , \quad H^t \to C^{t+r} \quad \text{for} \quad |t| \le r + 2$$

where $C^t$ denotes for integer $t$ the space of t-times continuously differentiable functions with corresponding maximum norms, or for non integer $t$ the usual Hölder space. For $\Gamma \in C^\infty$ , $C$ is a pseudodifferential operator of order $-\infty$ since it has a $C^\infty$-kernel [99, Chap. III]. For a Ljapounov curve $\Gamma \in C^{1+\alpha}$ , $C : C^o \to C^\varepsilon$ is compact for any $0 \le \varepsilon \le \alpha < 1$ .

In the three-dimensional case $n = 3$ , i.e. $\Gamma$ a smooth closed surface, again $Q = 0$ and

$$Cu := \frac{1}{2\pi} \int_\Gamma u(y) \frac{\partial}{\partial \nu_y} \left(\frac{e^{ikr}}{r}\right) ds_y \tag{3.4}$$

is a weakly singular integral operator. For $\Gamma \in C^\infty$ the kernel of $C$ is pseudohomogeneous [99, Chap. III] whose expansion with respect to kr can be found in [60, (1.14)]. Choosing surface polar coordinates r and $\varphi$, the expansions given by Martensen [72,(2.63), (2.65)] yield the principal symbol

$$a_o = -\frac{1}{2} \sum_{i,j=1}^{2} b_{ij}(x)\xi_i\xi_j \,/\, |\xi|^3$$

where $b_{ij}(x)$ denote the coefficients of the second fundamental tensor associated to $\Gamma$ and the local coordinates $u_1 = r\cos\varphi$, $u_2 = r\sin\varphi$ on $\Gamma$.

Hence C defines a pseudodifferential operator of order $-1$ [99, Chap. III]. For $r \geq 1$ we find from differentiating (3.4) (and integrating by parts) [74, p. 249 ff.] and the Giraud theorem [74, p. 239] that $C : C^{\varepsilon} \to C^{1+\varepsilon}$ is continuous for any $0 < \varepsilon < 1$ and with [74, Theorem 5.1, p. 266] that $C : H^t \to H^{t+1}$ is continuous if $|t| \leq r-1$. For $r \geq 0$ we find in [74, p. 213] Sobolev's theorem,

$$C : L_p \to L_q \text{ is compact (continuous) for } \begin{cases} 2 \leq p \underset{(-)}{<} q \leq \infty & \text{or} \\ 1 \leq p \leq q \underset{(-)}{<} 2p/(2-p) & \text{if } p<2 . \end{cases}$$

For a Ljapounov surface $\Gamma \in C^{1+\alpha}$ the estimates [41, p. 54] yield that $C : C^o \to C^{\varepsilon}$ is compact for any $0 \leq \varepsilon < 1$ and $\varepsilon \leq \alpha$.

## 3.1 Smoothing and superconvergence

If $Q = 0$ in (3.1) then the compactness of C can be used to improve the quality of an approximation [15] including the order of convergence [100]. To this end let us define by

$$v_{(\ell+1)} := -Cv_{(\ell)} + f - B\omega_G \quad \text{with} \quad v_{(o)} := \begin{cases} v_G & \text{for Galerkin's procedure and} \\ v_c & \text{for collocation} \end{cases} \tag{3.5}$$

$\ell = 0,\ldots,\rho-1$ , the $\rho$-times smoothed approximation.

THEOREM 3.2: *Let* $C : H^t \to H^{t+\kappa}$ *with* $\kappa>0$ *be continuous where*
$-m - 1 \le t \le \frac{n}{2} + \varepsilon - \kappa - \frac{1}{2}$ *for Galerkin's procedure and*
$0 \le t \le \frac{n}{2} + \varepsilon - \kappa - \frac{1}{2}$ *for collocation with any* $\varepsilon > 0$ . *Let* $v_{(\rho)}$ *be the* $\rho$*-times smoothed approximation with* $\rho > (\frac{n}{2} + m + \frac{1}{2})\kappa^{-1}$ *for Galerkin's procedure and* $\rho > (\frac{n}{2} - \frac{1}{2})\kappa^{-1}$ *for collocation. Then we have superconvergence*

$$|\omega - \omega_G| + \| u - v_{(\rho)} \|_{L_\infty} \le ch^\tau \|u\|_{m+1} \tag{3.6}$$

*with* $\tau = 2m+2$ *for the Galerkin procedure and* $\tau = m+1$ *for collocation.*

*Proof.* With the same idea as in [55] we have via the Sobolev imbedding theorem

$$|\omega-\omega_G| + \| u-v_{(\rho)} \|_{L_\infty} \le c\, \{|\omega-\omega_G| + \| u-v_{(\rho)} \|_{\frac{n}{2} + \varepsilon - \frac{1}{2}}\}$$

and with

$$v_{(\rho)} - u = (-C)^\rho \, (u-v_G) + \sum_{\ell=0}^{\rho-1} (-C)^\ell \, B(\omega_G-\omega) ,$$

the estimate

$$|\omega-\omega_G| + \| u-v_{(\rho)} \|_{L_\infty} \le c\, \{|\omega-\omega_G| + \| u-v_G \|_{-m-1}\}$$

since $\frac{n}{2} + \varepsilon - \frac{1}{2} - \kappa\rho \le -m - 1$ for $\varepsilon > 0$ small enough. Now Theorem 2.1

yields (3.6) in case of Galerkin's procedure.

For the remaining case of collocation proceed in the corresponding manner. □

For a smooth kernel, say $\kappa \geq m+1$ one has superconvergence even <u>without</u> smoothing.

THEOREM 3.3: *Let* $C : H^{-m-1} \to L_2$ *be continuous. Then for Galerkin's procedure we have superconvergence at the nodal points* $x_i = z_i$ *for* $m$ *odd or at the mesh points* $\tilde{x}_i = z_i$ *for* $m$ *even* ,

$$|u(z_i) - v_G(z_i)| \leq c h^{m+2} \| u \|_{2m+3} \tag{3.7}$$

For the proof and for further results on superconvergence see [20], [44], [55], [92], [96].

## 3.2 The case $Q \neq 0$ and results in Banach spaces

If $\Gamma$ is not smooth any more then our model equations (1.4) and (1.5) are still of the form (3.1), now containing a contraction $Q \neq 0$ .

i) For $n = 2$ let us consider $\Gamma$ being piecewise $C^2$ and let $\Omega(x) = \int_{|y-x|>0} \frac{\partial}{\partial \nu_y} (\log r) ds_y$ denote the inner corner angle at $x \in \Gamma$ . (For $\Gamma$ smooth at $x$ note $\Omega(x) = \pi$ .) Now suppose

$$\sup_{x\in\Gamma} |\Omega(x) - \pi| = \tilde{q}\pi \quad \text{with} \quad \tilde{q} < 1 , \tag{3.8}$$

i.e. <u>exclude</u> <u>spines</u>. Then there exist $q_o < 1$, $q_o > \tilde{q}$ and $\delta > 0$ such that the operator

$$Qu(x) := \frac{1}{\pi}\{\int_{0<|y-x|\le\delta} u(y)(\frac{\partial}{\partial\nu_y}\log r)ds_y + (\pi-\Omega(x))u(x)\} \tag{3.9}$$

defines a contraction in $C^o \to C^o$ [90] and also in $L_2 \to L_2$ [23],[24]. For the compact term set

$$Cu(x) := \frac{1}{\pi}\int_{\delta\le|x-y|} u(y)(\frac{\partial}{\partial\nu_y}\log r)ds_y \tag{3.10}$$

$$- \int_\Gamma u(y)\frac{\partial}{\partial\nu_y}\{\frac{1}{2i}H_o^{(1)}(kr) + \frac{1}{\pi}\log r\}ds_y .$$

Then $C$ maps $L_2$ and $C^o$ compactly into $C^{2-\varepsilon}$ with any $\varepsilon > 0$ .

ii) Let $\Gamma$ be a rectifyable continuous curve satisfying (1.6) and suppose

$$\lim_{\delta\to 0}\frac{1}{\pi}\sup_{x\in\Gamma}\{\int_{0<|x-y|\le\delta}|\frac{\partial}{\partial\nu_y}\log r|\,ds_y + |\pi-\Omega(x)|\} = \tilde{q} < 1 . \tag{3.11}$$

Then $Q$ in (3.9) defines a contraction $C^o \to C^o$ and $C$ in (3.10) is compact in $C^o$ [18] , [60] , [62] .

iii) For $n = 3$ let $\Gamma$ either be a closed piecewise Ljapounov or a rectifyable surface satisfying (1.6). Let

$$\Omega(x) = -\int_{|x-y|>0}\frac{\partial}{\partial\nu_y}(\frac{1}{r})\,ds_y .$$

Suppose

$$\lim_{\delta\to 0}\frac{1}{2\pi}\sup_{x\in\Gamma}\{\int_{0<|x-y|<\delta}|\frac{\partial}{\partial\nu_y}\frac{1}{r}|\,ds_y + |2\pi-\Omega(x)|\} = \tilde{q} < 1 \tag{3.12}$$

and set

$$Qu(x) := \frac{1}{2\pi}\{-\int_{0<|x-y|\le\delta} u(y)(\frac{\partial}{\partial\nu_y}\frac{1}{r})\,ds_y + (2\pi-\Omega(x))u(x)\} . \tag{3.13}$$

Then there exist $q_o$ and $\delta>0$, $\tilde{q} < q_o < 1$ such that $Q : C^o \to C^o$ is a contraction [18], [60], [110]. The remaining operator

$$Cu(x) := -\frac{1}{2\pi} \int_{\delta \le |x-y|} u(y) \frac{\partial}{\partial \nu_y} \left(\frac{1}{r}\right) ds_y \tag{3.14}$$

$$- \frac{1}{2\pi} \int_\Gamma u(y) \frac{\partial}{\partial \nu_y} \left\{ \frac{1}{r} (e^{ikr} - 1)\right\} ds_y$$

is compact in $C^o$ [60].

*Remark 3.4:* Note that if $Q : L_2 \to L_2$ is a contraction as is (3.9) for a piecewise $C^2$-curve $\Gamma$ with (3.8) then our asymptotic estimates in Theorems 2.1 and 2.2 remain valid for Galerkin's procedure (with $\alpha=\beta=j=0$). However, $u$ will in general not be smooth at the corner points any more. Therefore then appropriate singular functions must be introduced which can be treated as additional trial and test functions (see [23], [24]).

Now let us consider the approximation of (3.1) in a Banach space $X \times \mathbb{R}^q$ following closely [17] (see also [6] for $Q = 0$ and [108]).

Suppose that there are given a closed subspace $X_o = \overline{X_o} \subset X$,

$$Q : X_o \to X_o \quad \text{and} \quad C : X \to X_o \quad \text{compact}, \tag{3.15}$$

a family of projections $\pi_h : X \to \tilde{H}_h$ with the properties

$$\| \pi_h f - f \|_X \to 0 \quad \text{for} \quad h \to 0 \quad \text{and for any} \quad f \in X_o, \tag{3.16}$$

$$\| Q\pi_h \|_{X,X} \le q_o < 1 \quad \text{uniformly for all} \quad 0 < h, \tag{3.17}$$

and a family of compact operators $C_h : X \to X_o$ satisfying

$$\| C_h \|_{X,X} \le c\,, \quad \| C_h v - Cv \|_X \to 0 \quad \text{for} \quad h \to 0 \quad \text{and any} \quad v \in X_o\,,$$

in addition (3.18)

$$\bigcup_{0<h} \{ w = (C_h - C)\varphi \mid \varphi \in X \text{ and } \| \varphi \| \le 1 \}$$

is relatively compact in $X_o$, i.e. $(C_h - C)$ is a collectively compact operator family.

Consider the approximating equations

$$v_h + (Q\pi_h + C_h)v_h + B\omega_h = f\,, \quad \Lambda v_h = b\,. \tag{3.19}$$

<u>THEOREM 3.5:</u> *Let* (3.1) *be uniquely solvable and let* $f \in X_o$. *Suppose* (3.15) - (3.18). *Then there exist positive constants* $h_o$ *and* $c_1, c_2$ *such that* (3.19) *has a unique solution* $(v_h, \omega_h)$ *for any* $0 < h \le h_o$. *We also have the asymptotic estimates*

$$\| v_h \| + |\omega_h| \le c_1 \{ \| u \| + |\omega| \} \le c_2 \{ \| f \| + |b| \} \tag{3.20}$$

*and*

$$\| u - v_h \| + |\omega - \omega_h| \le c_2 \{ \| Q\pi_h u - Qu \| + \| (C_h - C)u \| \}\,. \tag{3.21}$$

*Proof*. [103] The equations (3.19) and (3.1) yield

$$(I + Q\pi_h)v_h + Cv_h + B\omega_h = (I + Q)u + Cu + B\omega + (C - C_h)v_h\,,$$
$$\Lambda v_h = \Lambda u\,. \tag{3.22}$$

If (3.20) were not true then there would exist a sequence $f_h, b_h$ with corresponding $u_h, \eta_h \in X_o \times \mathbb{R}^q$ with $\| u_h \| + |\eta_h| = 1$ and corresponding $\| v_h \| + |\omega_h| \to \infty$ . Deviding (3.22) by $\| v_h \| + |\omega_h|$ and introducing

$$\psi_h = v_h \,/\, (\| v_h \| + |\omega_h|) \ , \quad \alpha_h = \omega_h \,/\, (\| v_h \| + |\omega_h|)$$

would yield

$$(I + Q\pi_h)\psi_h + C\psi_h + B\alpha_h \ = \ \chi_h + (C - C_h)\psi_h \ ,$$

$$\Lambda_h \psi_h = \Lambda u \ / \ (\| v_h \| + |\omega_h|)$$

where $\chi_h \to 0$ . Now we could choose a subsequence with converging $\alpha_h \to \alpha$ and $(C - C_h)\psi_h \to \lambda \in X_o$ . But this would imply the convergence of a subsequence of

$$\psi_h \ = \ (I + Q\pi_h)^{-1} \{\chi_h + (C - C_h)\psi_h - C\psi_h - B\alpha_h\} \to \psi \in X_o$$

with $\| \psi \| + |\alpha| = 1$ and satisfying

$$(I + Q)\psi + C\psi + B\alpha = 0 \ , \quad \Lambda\psi = 0$$

implying $\psi = 0$ , $\alpha = 0$ , a contradiction.

Inequality (3.21) now follows from

$$(u-v_h) + Q\pi_h(u-v_h) + C_h(u-v_h) + B(\omega-\omega_h) \ = \ Q\pi_h u - Qu + C_h u - Cu \ ,$$

$$\Lambda(u-v_h) \ = \ 0$$

with (3.20) since for $u \in X_o$ the right hand side also belongs to $X_o$ .

□

Note that one of the most important applications of Theorem 3.5 is that to the Nyström method. There $Q = 0$ and

$$Cu(x) = \int k(x,y)u(y)ds_y ,$$

$$C_h u(x) = \sum_{\ell} b_{\ell}\, k(x,y_{\ell})u(y_{\ell}) ,$$

with a smooth kernel function $k(x,y)$ (see e.g. [ 9 ] ) . But for non-smooth $k(x,y)$ the method must be modified as for discretizations of (1.4) and (1.5) involving corners and edges.

For two-dimensional problems with $\Gamma$ piecewise $C^{1+\alpha}$ and corners we choose a family of boundary segments $F_j \subset \Gamma$ with $\Gamma = \bigcup_{j=1}^{N} F_j$ , $F_j \cap F_{\ell} = \emptyset$ for $j \neq \ell$ and the associated family $p_j$ of respective midpoints of $F_j$ with $h = |p_j - p_{j-1}| \to 0$ .

First we consider an approximation related to piecewise constant trial functions by defining

$$\pi_h g(x) := g(p_j) \quad \text{for all} \quad x \in F_j \ , \quad j = 1,\dots,N \tag{3.23}$$

for $g$ continuous on $F_j$ and replace (1.4) by the approximation

$$\begin{aligned} v_h(x) &+ \frac{1}{2\pi}\int_{\Gamma} (\pi_h v_h(y) - v_h(x))\Big(\frac{\partial}{\partial \nu_y}\log r\Big)ds_y \\ &- \int_{\Gamma} (\pi_h v_h(y)) \frac{\partial}{\partial \nu_y}\Big\{ \frac{1}{4i} H_o^{(1)}(kr) + \frac{1}{2\pi}\log kr\Big\}\, ds_y = f(x) \end{aligned} \tag{3.24}$$

which is equivalent to the system of linear equations for $v_h(p_j)$ ,

$$\begin{aligned} v_h(p_j) &+ \frac{1}{2\pi}\sum_{\ell\neq j} (v_h(p_{\ell}) - v_h(p_j)) \int_{F_{\ell}} \Big(\frac{\partial}{\partial \nu_y}\log r\Big)(p_j)ds_y \\ &- \sum_{\ell} v_h(p_{\ell}) \int_{F_{\ell}} \frac{\partial}{\partial \nu_y}\Big\{\frac{1}{4i} H_o^{(1)}(kr) + \frac{1}{2\pi}\log kr\Big\}(p_j)ds_y = f(p_j) \ , j=1,\dots,N. \end{aligned} \tag{3.25}$$

Here we set:

$X = B^1(\Gamma)$ the functions of the first Baire class on $\Gamma$ equipped with the norm $\|f\|_X := \sup_{x\in\Gamma} |f(x)|$,

$X_o$ = space of continuous functions on $\Gamma$.

Then $X$ is a Banach space and $X_o$ a closed subspace (see [77]). An extension of $\pi_h$ (3.23) from $X_o$ to $X$ is given by

$$\pi_h f(x) := f(p_j) \quad \text{for} \quad x \in \bar{F}_j, \quad j = 1, \dots, N . \tag{3.26}$$

Now $\|\pi_h\|_{X,X} = 1$ and (3.17) follows from (3.11) where $Q$ is defined by (3.9). $C_h$ is defined by (3.10) and $C_h = C\pi_h$. Since $C$ has a continuous kernel we have $C, C_h : X \to X_o$ and (3.18) follows immediately. Hence, the collocation method (3.24), (3.25) converges uniformly for $h \to 0$ due to (3.21).

Moreover, since for (1.4) the boundary values of $u$ belong to the space $C^\varepsilon$ with $\varepsilon = \min\{\pi/(2\pi - \Omega(x)), 1\} > 0$ (this can be obtained from [69]) the estimate (3.21) implies in this case the asymptotic convergence estimate

$$\|u - v_h\|_{C^o} \le c \|\pi_h u - u\|_{C^o} \le ch^\varepsilon \|u\|_{C^\varepsilon} . \tag{3.27}$$

*Remark:* $C_h$ could also have been defined by numerical quadrature as in Nyström's method.

Now let us consider the corresponding three-dimensional case following [60] with $\Gamma = \bigcup_{\ell=1}^{L} S_\ell$ where every $S_\ell$ is a segment of a Ljapounov surface. Each $S_\ell$ can be considered to be the image of a

$C^{1+\alpha}$-application $x(t)$ defined on a polygonal parameter domain $U_\ell \subset \mathbb{R}^2$, $S_\ell = x(U_\ell)$. Then a regular family of triangular partitions of $U_\ell$ with triangles $T_j$ defines a corresponding family of partitions of each $S_\ell$ and, eventually, on $\Gamma$, $\Gamma = \bigcup_{j=1}^{N} F_j$, $F_j = x(T_j)$. By $p_j \in F_j$ let us denote the images of the centers of gravity in the corresponding triangle $T_j$ of the parameter domain. To $\{p_j\}$ we associate the one point Gaussian formula

$$\int_\Gamma f\, ds \simeq \sum_{j=1}^{N} f(p_j) g_j$$

for numerical quadrature. Since $\Gamma$ may have corners and edges we define to the set $\gamma \subset \Gamma$ of all corner and edge points

$$\Sigma := \{x \in \Gamma \mid \text{distance } (x-\gamma) < d\} \quad \text{and}$$

$$\Sigma_\ell := \Sigma \cap \{x \mid |x-p_\ell| < d\}$$

where $d > 0$ is a quanitity to be chosen fixed and beforehand.

Then the integral equation (1.5) is replaced by the following system of linear equations.

$$v_h(p_\ell) = \frac{1}{2} v_h(p_\ell) - \frac{1}{2}\Big\{ v_h(p_\ell) \frac{1}{2\pi} (2\pi - \Omega(p_\ell)) \qquad (3.28)$$

$$- \sum_{\substack{0<|p_j-p_\ell|\le\delta \\ p_j \in \Sigma_\ell}} v_h(p_j) \frac{1}{2\pi} \int_{F_j} \frac{\partial}{\partial \nu_y}\left(\frac{1}{r}\right)(p_\ell)\, ds_y$$

$$+ \sum_{\substack{0<|p_j-p_\ell|\le\delta \\ p_j \notin \Sigma_\ell}} v_h(p_j) \frac{1}{2\pi} \frac{n(p_j)\cdot(p_j-p_\ell)}{|p_j - p_\ell|^3} g_j$$

$$- v_h(p_\ell) \frac{1}{2\pi} [-\Omega(p_\ell) - \sum_{\substack{p_j \in \Sigma_\ell \\ j \neq \ell}} \int_{F_j} \frac{\partial}{\partial \nu_y} (\frac{1}{r})(p_\ell) ds_y$$

$$+ \sum_{\substack{p_j \notin \Sigma_\ell \\ j \neq \ell}} \frac{n(p_j) \cdot (p_j - p_\ell)}{|p_j - p_\ell|^3} g_j ]\Big\}$$

$$- \frac{1}{2}\Big\{\frac{1}{2\pi} \sum_{j \neq \ell} v_h(p_j) [e^{ik|p_\ell - p_j|} - 1 - ik|p_j - p_\ell| e^{ik|p_j - p_\ell|}]$$

$$\times \frac{n(p_j)\cdot(p_j - p_\ell)}{|p_j - p_\ell|^3} g_j - \sum_{\substack{|p_j - p_\ell| > \delta \\ p_j \in \Sigma_\ell}} v_h(p_j) \frac{1}{2\pi} \int_{F_j} \frac{\partial}{\partial \nu_y} (\frac{1}{r})(p_\ell) ds_y$$

$$+ \sum_{\substack{|p_j - p_\ell| > \delta \\ p_j \notin \Sigma_\ell}} v_h(p_j) \frac{1}{2\pi} \frac{n(p_j)\cdot(p_j - p_\ell)}{|p_j - p_\ell|^3} g_j \Big\} + f(p_\ell) . \qquad (3.28)$$

The first curly bracketed term on the right-hand side of (3.28) represents a discretized form $Q_h u$ of $Qu$ and the second term in curly brackets is the discretization $C_h u$ of $Cu$. The last, apparently additional, term in the discretized $Qu$ is included (even though it vanishes as $h \to 0$) in order to obtain the discretized version of (1.5) namely

$$v_h(p_\ell) = - \frac{1}{4\pi} \sum_{j \neq \ell} \alpha_{j\ell} [v_h(p_j) - v_h(p_\ell)] + f(p_\ell)$$

$$- \frac{1}{4\pi} \sum_{j \neq \ell} v_h(p_j) \{e^{ik|p_j - p_\ell|} - 1 - ik|p_j - p_\ell| e^{ik|p_j - p_\ell|}\} g_j \qquad (3.29)$$

$$\times \frac{n(p_j)\cdot(p_j - p_\ell)}{|p_j - p_\ell|^3} .$$

where the coefficients $\alpha_{j\ell}$ are given by

$$\begin{aligned}\alpha_{j\ell} &= -\int_{F_j} \frac{\partial}{\partial \nu_y}\left(\frac{1}{r}\right)(p_\ell)ds_y \qquad \text{for } p_j \in \Sigma_\ell\,, \\ &= \frac{n(p_j)\cdot(p_j-p_\ell)}{|p_j-p_\ell|^3}\, g_j \qquad \text{for } p_j \notin \Sigma_\ell\,.\end{aligned} \tag{3.30}$$

Again we choose $X$ to be the first Baire class of functions on $\Gamma$, $X_o = C^o(\Gamma)$ and $\pi_h$ as in (3.23), (3.26). Here $C_h$ is defined differently to $C\pi_h$ and therefore the justification of (3.18) is more tedious. We refer for the proof to [60], [110]. There we also present a procedure to compute the weights $\alpha_{j\ell}$ explicitly via its geometric meaning.

Again, (3.21) implies the asymptotic convergence of this method. Note that the methods used in [42], [47], [48], [86] correspond to the choice $C_h = C\pi_h$ (they are similar to (3.28), (3.29)) and that (3.21) (together with the estimates in Chapter 4) assures their convergence.

For $\Gamma \subset C^2$ <u>without</u> corners and edges, i.e. $\Sigma = \Sigma_\ell = \emptyset$, $Q = 0$ and the choice $C_h = P_h\, C\, P_h$ with $P_h$ the $L_2$-projection onto the piecewise constant functions on $\{F_j\}$ one finds equations similar to (3.28). Then (3.29) can be understood as a corresponding numerically integrated version (see Chapter 4). In this case we have the following estimate due to V. Thomée,

$$\| Cu - CP_h u \|_{C^o} + \| Cu - P_h Cu \|_{C^o} \le ch \log \frac{1}{h} \| u \|_{C^o} \tag{3.31}$$

(oral communication,will be published soon) and (3.21) yields the asymptotic estimate

$$\| v_h - u \|_{C^o} \leq ch \log \frac{1}{h} \| u \|_{C^o} . \tag{3.32}$$

The convergence of the above methods can again be improved by the use of higher order splines. In the case $n = 2$ the definition of $\pi_h$ by

$$\pi_h u(x) := u(x_j) \frac{s_{j+1} - s}{s_{j+1} - s_j} + u(x_{j+1}) \frac{s - s_j}{s_{j+1} - s_j} \qquad \text{for} \quad s_j \leq s \leq s_{j+1}$$

where $x = x(s)$ , $x_j = x(s_j)$ yields a modification of (3.24), (3.25) with piecewise linear functions [17] , [94] . Here choose $X = X_o = C^o(\Gamma)$ and find $\| \pi_h \|_{C^o,C^o} = 1$ . Again we have the asymptotic convergence

$$\| u - v_h \|_{C^o} \leq c \| \pi_h u - u \|_{C^o} \leq ch^{\varepsilon} \|u\|_{C^{\varepsilon}} , \tag{3.33}$$

now with $\varepsilon = \min \{\pi/(2\pi-\Omega(x)),2\} > 0$ . But this estimate improves (3.27) only for corners with $2\pi-\Omega(x) < \pi$ , otherwise one again needs to incorporate the singular functions.

For smooth $\Gamma$ we have further improvements by using higher order splines and $\pi_h$ the interpolation operator. For orders exceeding one, however, the $C^o$-norms of $\pi_h$ exceed 1 [85] and (3.17) becomes more restrictive. This is in contrast to our approach in Section 2.2 where we could interpret the collocation as a modified Galerkin method in appropriate Sobolev spaces.

For the spacial problems, $n = 3$ , the use of higher order isoparametric elements would also improve the rate of convergence. Here we introduce the piecewise linear Courant elements on the partitions $\{T_j\}$ of $U_\ell$ and lift these functions by $x(t) : U_\ell \to S_\ell$ on $\Gamma$ . On joint boundary curves $S_\ell \cap S_m \neq \emptyset$ we require the traces to coincide. This defines a boundary element space $\tilde{H}_h$ . Now $\pi_h$ again can be chosen to be the collocation operator with the corner points of $F_j$ as collocation points $p_\ell$ . Since then we have $\| \pi_h \|_{C^o,C^o} = 1$ and the Taylor formula provides the estimate

$$\| \pi_h u - u \|_{C^o} \leq ch^2 \|u\|_{C^2} ,$$

the approximation of (1.5) with $Q = 0$ and $C_h = C\pi_h$ via (3.21) yields the asymptotic convergence

$$\| u - v_h \|_{C^o} \leq c \, \| C(\pi_h u - u) \|_{C^o} \leq ch^2 \|u\|_{C^2} \tag{3.34}$$

which is of optimal order.

*Remark 3.6:* As was shown in [ 4] , [ 5 ] , [60] , the equations (1.4), (1.5) as well as (3.25), (3.29) can be solved by successive approximation via C. Neumann's method provided the wave number k is small enough. The rate of convergence of that iteration is independent of h . For large systems of discretized equations as (3.29) this is of great advantage compared with direct methods as have been used in [47] , [48] , [86] .

## § 4 FULL DISCRETIZATION AND NUMERICAL INTEGRATION

For the numerical implementation of the Galerkin equations (1.29) or the collocation equations (2.11) or (3.25), respectively, the entries of the influence matrix and right hand sides have to be evaluated numerically. For the Galerkin procedure (1.29) these are

$$a_{mk} := (\Theta A\mu_m,\mu_k)_{\beta+j}\ ,\ b_k := (\Theta B,\mu_k)_{\beta+j}\ ,\ \lambda_m := \Lambda\mu_m \quad \text{and}$$
$$f_k := (\Theta f,\mu_k)_{\beta+j}\ ,\qquad m,k = 1,\dots,N \tag{4.1}$$

and for the collocation,

$$a_{m\ell} := A\mu_m(x_\ell)\ ,\ \ b_\ell := B(x_\ell)\ ,\ \ \lambda_m := \Lambda\mu_m\ ,\ \ m,\ell = 1,\dots,N, \tag{4.2}$$

respectively. By $\tilde{a}_{mk}$, $\tilde{b}_k$, $\tilde{\lambda}_m$ and $\tilde{f}_k$ let us denote the corresponding erroneous numerical values. For related error estimates we introduce for Galerkin's method by the equations

$$(\widetilde{\Theta A}_h\mu_m,\mu_k)_{\beta+j} = \tilde{a}_{mk}\ ,\ (\widetilde{\Theta B}_h,\mu_k)_{\beta+j} = \tilde{b}_k\ ,\ \tilde{\Lambda}_h\mu_m = \tilde{\lambda}_m\ ,\quad m,k = 1,\dots,N \tag{4.3}$$

new operators $\widetilde{\Theta A}_h$ , $\widetilde{\Theta B}_h$ , $\tilde{\Lambda}_h$ on the finite dimensional space $\tilde{H} \to \tilde{H}$ and a corresponding operator ${}_\Theta\tilde{A}_h$ on $\tilde{H} \times \mathbb{R}^q \to \tilde{H} \times \mathbb{R}^q$ and also by

$$({}_\Theta\tilde{f}_h,\mu_k)_{\beta+j} = \tilde{f}_k\ ,\quad k = 1,\dots,N \tag{4.4}$$

a function ${}_\Theta\tilde{f}_h \in \tilde{H}$ .

Correspondingly we define $\tilde{A}_h$, $\tilde{B}_h$, $\tilde{\Lambda}_h$ for the collocation method by

$$\tilde{A}_h\mu_m(x_\ell) = \tilde{a}_{m\ell}\ , \ \tilde{B}_h(x_\ell) = \tilde{b}_\ell\ , \ \tilde{\Lambda}_h\mu_m = \tilde{\lambda}_m \quad \text{and}$$
$$\tilde{\mathcal{A}}_h : \tilde{H} \times \mathbb{R}^q \to \tilde{H} \times \mathbb{R}^q\ . \tag{4.5}$$

The coefficient errors result in corresponding errors for $P_{h\Theta}\mathcal{A} - {}_\Theta\tilde{\mathcal{A}}_h$ or $I_h\mathcal{A} - \tilde{\mathcal{A}}_h$, respectively, on $\tilde{H} \times \mathbb{R}^q$. Here $P_h$ denotes the orthogonal projection $P_h : H^{2\beta+2j} \to \tilde{H} \times \mathbb{R}^q$ and $I_h$ the interpolation operator with $\tilde{H} \times \mathbb{R}^q$-functions at the collocation points. For these error estimates we restrict the finite elements to regular families providing the inverse property (2.2). Moreover one needs the equivalence between the discrete and continuous norms:

$$\|\sum_{j=1}^{N}\gamma_j\mu_j\|_{L_2} \le c_1\{h^{n-1}\sum_{j=1}^{N}\gamma_j^2\}^{\frac{1}{2}} \le c_2\|\sum_{j=1}^{N}\gamma_j\mu_j\|_{L_2} \tag{4.6}$$

and

$$\|\sum_{j=1}^{N}\gamma_j\mu_j\|_{C^o} \le c_1\{\max_{j=1,\dots,N}|\gamma_j|\} \le c_2\|\sum_{j=1}^{N}\gamma_j\mu_j\|_{C^o}\ . \tag{4.7}$$

LEMMA 4.1: *Let the entries of the influence matrix and the right hand sides, respectively, satisfy*

$$|a_{mk} - \tilde{a}_{mk}| + h^{n-1}|b_k - \tilde{b}_k| + |\lambda_m - \tilde{\lambda}_m| \le ch^{L+1+\varepsilon(n-1)}\ ,$$
$$|f_k - \tilde{f}_k| \le c_f h^{L+n}\ , \qquad k,m = 1,\dots,N \tag{4.8}$$

*where* $\varepsilon = 2$ *for the Galerkin method and* $\varepsilon = 1$ *for collocation. Then we have the following consistencies on* $\tilde{H} \times \mathbb{R}^q$ :

$$\|(P_{h\,\Theta}\mathcal{A} - {}_\Theta\tilde{\mathcal{A}}_h)(w,\eta)\|_{H^{2\beta+2j}} \le ch^{L+1}\{\|w\|_{L_2} + |\eta|\}\ , \tag{4.9}$$

$$\|{}_\Theta\tilde{f}_h - P_h\Theta f\|_{2\beta+2j} \le c\, c_f\, h^{L+1} \tag{4.10}$$

*for the Galerkin procedure and*

$$\|(I_h A - \mathcal{A}_h)(w,\eta)\|_{L_2} \leq ch^{L+1}\{\|w\|_{L_2} + |\eta|\} \tag{4.11}$$

*respectively,*

$$\|(I_h A - \mathcal{A}_h)(w,\eta)\|_{C^o} \leq ch^{L+1}\{\|w\|_{C^o} + |\eta|\}$$

*for collocation.*

The proof of (4.9), (4.10) follows from the duality between $H^o$ and $H^{2\beta+2j}$ with respect to the $H^{\beta+j}$-scalar product by the use of (4.6) and the Schwarz inequality. (4.11) is correspondingly obtained and (4.12) follows directly with (4.7). We omit the details [114] . □

Clearly, the fully discretized numerical equations are defined by $\tilde{a}_{m,k}$ , $\tilde{b}_k$ , $\tilde{\lambda}_m$ and $\tilde{f}_k$ and read with our notations as to compute $(\tilde{v},\tilde{\omega}) \in \tilde{H} \times \mathbb{R}^q$ by solving

$${}_\Theta\mathcal{A}_h(\tilde{v},\tilde{\omega}) = ({}_\Theta\tilde{f},b) \tag{4.13}$$

(with $\Theta = I$ for collocation.) With the consistency estimates of Collary 2.3 we now have by the first Strang lemma the following error estimates [114].

THEOREM 4.2: *Let* $L > -(2j + \alpha_\ell + \alpha_k)' - 1$ , $\ell,k = 1,\dots,p$ . *Then there is* $h_o > 0$ *such that the fully discretized Galerkin equations* (4.13) *are uniquely solvable for any* $0 < h \leq h_o$ . *There we have the asymptotic error bounds*

$$|\omega_G - \tilde{\omega}_G| + \| v_{Gk} - \tilde{v}_{Gk} \|_{L_2} \leq c h^{\{\min_{\ell=1,\dots,p} \{(\alpha_\ell + \alpha_k + 2j)'\} + L + 1\}} \cdot \{ \| v_G \|_{L_2} + |\omega_G| + c_f \} . \tag{4.14}$$

For the collocation method let $L > -(\alpha_\ell + \alpha_k - 2\beta_\ell)' - 1$, $\ell, k = 1, \dots, p$. Then we have correspondingly

$$|\omega_c - \tilde{\omega}_c| + \| v_{ck} - \tilde{v}_{ck} \|_{L_2} < c h^{\min_{\ell=1,\dots,p} \{(\alpha_\ell + \alpha_k - 2\beta_\ell)'\} + L + 1} \cdot \{ \| v_c \|_{L_2} + |\omega_c| \}$$

*Proof.* Since

$$_\Theta\tilde{A}_h = P_h \, {}_\Theta A P_h \{ I + (P_h \, {}_\Theta A P_h)^{-1} ({}_\Theta\tilde{A}_h - P_h \, {}_\Theta A P_h) \} ,$$

with Corollary 2.3 and estimate (4.9) we find

$$\| (P_h \, {}_\Theta A P_h)^{-1} ({}_\Theta\tilde{A}_h - P_h \, {}_\Theta A)(v, \eta) \|_o \leq c h^{\min\{L+1+(\alpha_\ell+\alpha_k+2j)'\}} \cdot \{ \| v \|_o + |\eta| \}$$

for all $v \in \tilde{H}$. Hence there is an $h_o > 0$ such that ${}_\Theta\tilde{A}_h^{-1}$ exists for every $0 < h \leq h_o$ and the estimates (2.8) hold also for all solutions of (4.13).

For the estimate (4.14) subtract (4.13) from the Galerkin equations to find

$$_\Theta\tilde{A}_h (v_G - \tilde{v}_G, \omega_G - \tilde{\omega}_G) = ({}_\Theta\tilde{A}_h - P_h \, {}_\Theta A)(v_G, \omega_G) + (P_h f - \tilde{f}_h, 0) .$$

The above secured estimate in the form (2.8) yields

$$|\omega_G - \tilde{\omega}_G| + \| v_{Gk} - \tilde{v}_{Gk} \|_o \leq c h^{\min_{\ell=1,\dots,p} \{(\alpha_\ell + \alpha_k + 2j)'\}} \cdot \{ \| ({}_\Theta\tilde{A}_h - P_h \, {}_\Theta A)(v_G, \omega_G) \|_{2\beta+2j} + \| P_h f - \tilde{f}_h \|_{2\beta+2j} \} .$$

Inserting (4.9) and (4.10) gives (4.14). For collocation proceed in the same manner. □

## 4.2 Two-dimensional problems

Here $\Gamma$ is a curve which is given by a regular parameter representation

$$\Gamma : x = x(t) \;, \quad 0 \le t \le 1 \tag{4.15}$$

with $x(t)$ a 1-periodic sufficiently smooth vector valued function satisfying $|dx/dt| = \rho(t) \ge \rho_o > 0$ for all $t$. Then the principal part of $A$ takes the form

$$A_o v(\tau) = A_o * v(\tau) = \text{p.v.} \int \{p_1(\tau,t-\tau)+\log|t-\tau|\, p_2(\tau,t-\tau)\}\rho(t)v(t)dt \tag{4.16}$$

where the convolution must be understood in the distributional sense, i.e. it may also contain Dirac functions and their derivatives. For fixed $\tau$ the kernels $p_1(\tau,\zeta)$, $p_2(\tau,\zeta)$ in (4.16) are matrices of positive homogeneous functions of $\zeta \neq 0$ of degrees $-1-\alpha_{\ell k}$. Since it is evident from Lemma 4.1 and Theorem 4.2 that the accuracy of the final numerical results depends significantly on how to compute the weights of the principal part, we shall treat $A_o$ and the remaining smoother terms of $A$ differently. For $A_o$ let us assume essentially that the corresponding weights

$$a_{opq} = (\Theta A_o \mu_p, \mu_q)_{\beta+j} \quad \text{or} \quad a_{opq} = (A_o \mu_p)(x_q) \;, \tag{4.17}$$

respectively can be computed <u>exactly</u> up to the available number of digits. In the following two cases we are able to propose a scheme how to perform these corresponding computations.

Galerkin weights for convolutions and regular meshes

Here let us assume that the functions $\Theta p_1 = \pi_1(\zeta)$ and $\Theta p_2 = \pi_2(\zeta)$ become independent of $\tau$, i.e. the principal part of $\Theta A$ is a pure convolution. Therefore we shall introduce $(\rho u)$ to be the new unknowns, and (1.1) may be considered on the circle corresponding to $x = \frac{1}{2\pi} e^{2\pi i t}$. Note that the principal part $\Theta A_o$ then becomes independent of the special choice of the curve $\Gamma$.

In order to utilize the convolution in the principal part we use regular finite elements on a uniform grid of [0.1] defined with shifts and stretched variables from one shape function $\mu(\eta)$. The latter we define as in [10, Chap. 4] by suitable piecewise polynomials of order $m$ with $\mu \in C^{m-1}$. For $m = 0,1,2$ e.g. we have

| | $m = 0$ | $m = 1$ | $m = 2$ | for |
|---|---|---|---|---|
| | $1$ | $\eta$ | $\frac{1}{2}\eta^2$ | $0 \le \eta < 1$ |
| $\mu(\eta) =$ | $0$ | $2-\eta$ | $-\eta^2 + 3\eta - 3/2$ | $1 \le \eta < 2$ |
| | $0$ | $0$ | $\frac{1}{2}\eta^2 - 3\eta + 9/2$ | $2 \le \eta < 3$ |
| | $0$ | $0$ | $0$ | elsewhere . |

With $\mu$ we define a basis of $\tilde{H}_h$ by

$$\mu_q(t) := \frac{1}{\rho}\, \mu\!\left(\frac{t}{h} - q + 1\right) \quad \text{for} \quad h(q-1) \le t \le 1 + h\cdot q \,,$$
$$q = 1,\ldots,N \,, \quad h = 1/N \tag{4.18}$$

and their 1-periodic extensions. Now insert (4.18) into (4.17) with (4.16) and introduce coordinates $t' = \frac{t}{h} - q + 1$ to find the weights having the form

$$a_{opq} = h^{1-\alpha_{\ell k}-2(\beta_k+j)} \{ \int_{\tau'=0}^{m+1} \text{p.v.} \int_{t'=0}^{m+1} [\pi_1(t'-\tau'+(p-q)) + \pi_2 \log|t'-\tau'+(p-q)|] \cdot \mu(t')\mu(\tau')dt'd\tau'$$

$$+ \log h \int_{t=0}^{m+1} \text{p.v.} \int_{t'=0}^{m+1} \pi_2(t'-\tau'+(p-q))\mu(t')\mu(\tau')dt'd\tau' \tag{4.19}$$

$$= h^{1-\alpha_{\ell k}-2(\beta_k+j)} \{W_{1\lambda} + W_{2\lambda} \log h\} \quad , \quad \lambda = p-q \in \mathbb{Z} \, .$$

for every single operator in the matrix of operators forming the principal part $\Theta A_o$. Note that the principal part entries (4.19) form a Toeplitz matrix. Moreover, the two vectors of weights $W_{1\lambda}$ and $W_{2\lambda}$ can be computed once for all independent of $\Gamma$ and $h$ for each type of shape functions $\mu$ and then be used for further computations. For equations with $\beta = j = 0$, $\alpha_\ell = -\frac{1}{2}$, $\pi_1 = 0$ and $\pi_2 = 1$ this has been done in [52],[53] and with $\alpha_1 = -\frac{1}{2}$ and $\alpha_2 = 0$ in [66], [67].

Smooth remainders

Let us assume that the remaining weights

$$a_{1pq} = (\Theta(A-A_o)\mu_p, \mu_q)_{\beta+j}$$

can be written (after integration by parts, if necessary) in the form

$$a_{1pq} = \iint L(\tau,t)\mu_p(t)\mu_q(\tau)dtd\tau \tag{4.20}$$

with a smooth kernel $L$. Otherwise we again split according to the asymptotic expansion of the complete symbol of $\Theta A$ into two terms where the first contains the singularities and has to be treated similarly to the principal part. For the weights $f_k$, $b_k$, $a_{lpk}$ we now use a numerical quadrature formula with degree of precision $L$ [30] on every patch being an interval of length $h$ or a product of such intervals, respectively.

Most numerical implementations in engineering are done with Gaussian quadrature on the patches [109]. But this requires the evaluation of the integrand at $([\frac{L}{2}] + 1)^2$ Gaussian nodal points for product integration or a little less for two-dimensional formulas [30, Chap. 9.4.2] and, hence, excessive computing time for the influence matrix. Moreover for a mesh refinement the function values on the coarser grid cannot be used on the refined grid again since all Gaussian nodal points are placed differently. Therefore we try to use a more economical numerical integration.

## Galerkin collocation

To reduce the computing time significantly, we have developed numerical integration formulas with the shape functions as weights using regular grid points only. As in [52] , [112], [114] this approach yields formulas

$$\tilde{f}_k = h \sum_{\ell=-M}^{M} b_\ell \, f(h(k + \ell\cdot\gamma + \frac{m+1}{2})) \tag{4.21}$$

where $\gamma = 1$ or $\frac{1}{2}$ or $\frac{1}{4}$ ,..., depending on the degree $m$ of splines and where the degree of precision is $L = 2M + 1$ . The corresponding formulas for the numerical weights are given by

$$\tilde{a}_{lpq} = h^2 \sum_{\ell,i=-M}^{M} b_\ell b_i L(h(q+\ell\gamma+ \frac{m+1}{2}) , h(p+i\gamma+ \frac{m+1}{2})) . \tag{4.22}$$

This method we have called <u>Galerkin collocation</u> [52], [112] , where also the derivation of weights $b_\ell = b_{-\ell}$ is performed for various cases. Some of the weights [8] we present here.

| $\gamma$ | M | m=0 | | m=1 | | | m=2 | | | |
|---|---|---|---|---|---|---|---|---|---|---|
| | | $b_o$ | $b_1$ | $b_o$ | $b_1$ | $b_2$ | $b_o$ | $b_1$ | $b_2$ | $b_3$ |
| 1 | 0 | 1 | – | 1 | – | – | 1 | – | – | – |
| 1 | 1 | $\frac{11}{12}$ | $\frac{1}{24}$ | $\frac{5}{6}$ | $\frac{1}{12}$ | – | $\frac{3}{4}$ | $\frac{1}{8}$ | – | – |
| $\frac{1}{2}$ | 2 | | | $\frac{13}{30}$ | $\frac{4}{15}$ | $\frac{1}{60}$ | $\frac{2}{5}$ | $\frac{7}{30}$ | $\frac{1}{15}$ | – |
| $\frac{1}{2}$ | 3 | | | | | | $\frac{358}{945}$ | $\frac{157}{630}$ | $\frac{19}{315}$ | $\frac{1}{945}$ |

<u>Table</u> Weights of numerical integrations against splines.

If the computation of the values of the kernel functions at the grid points is executed in advance and these values then are stored for further use then the transition of the method to different shape functions can be done in a most efficient way (see [52, Table 4]) .

Here the computational expense per element of the influence matrix is only proportional to $\gamma^{-1}$ or just 1 and the method becomes optimally efficient.

The method also provides optimal order convergence (4.14) if $L = 2M + 1$ satisfies the assumptions $2M + 2 > -(2j + \alpha_\ell + \alpha_k)'$ of Theorem 4.2. This is in amazingly good agreement with some numerical experiments in [53].

Collocation weights for regular meshes

For the collocation method we need not to require the principal part to be a convolution but use only the general form (4.16). On the regular mesh we again use the basis (4.18) and find

$$a_{opq} = h^{-\alpha_{\ell k}}\{\text{p.v.} \int_{t'=0}^{m+1} p(\tau_q, t' - \frac{m+1}{2} + (q-p))\mu(t')dt'$$

$$+ \log h \ \text{p.v.} \int_{t'=0}^{m+1} p_2(\tau_q, t' - \frac{m+1}{2} + (q-p))\mu(t')dt'\} , \tag{4.22}$$

where $\tau_q = h(q + \frac{m+1}{2})$ and $p = p_1 + \log|\cdot|p_2$. If $p_1, p_2$ are independent of $\tau_p$, then (4.22) again is a Toeplitz matrix whose entries can be evaluated in terms of two weight vectors which are independent of the special choice of $\Gamma$ and of $h$.

Smooth remainders

Let us again assume that the remaining weights have the form

$$a_{1pq} = (A-A_o)\mu_p(\tau_q) = \int L_c(\tau_q, t)\mu_p(t)dt \tag{4.23}$$

with a smooth kernel $L_c$. For the numerical integration with (4.8) we again need formulas with degree of precision $L$ on every patch.

With Gaussian quadrature we need the evaluation of $L_c$ at $[\frac{L}{2}] + 1$ Gaussian nodal points. For $L$ satisfying the assumptions of Theorem 4.2 and Symm's integral equation our estimate (4.14) in connection with the estimates (2.19) have been confirmed with high accuracy by numerical computations in [51]. Although here the computing time is less than for Galerkin's method with Gaussian quadrature, we still have excessive computing times combined with the disadvantages in connection with mesh refinements. A more detailed comparison of the corresponding computing expenses can be found in [8].

Collocation with grid point quadrature

Again we get rather efficient methods if we use the quadrature formulas against splines given by (4.21) computing the weights by

$$\tilde{a}_{lpq} = h \sum_{i=-M}^{M} b_i L_c(\tau_q, h(p+i\gamma+\frac{m+1}{2})) , \tag{4.24}$$

$L = 2M + 1$ . The computational expense per element of the influence matrix is now proportional to $\gamma^{-1}$ or just $1$ and, hence, the least possible.

## 4.3 Three-dimensional problems

As we already have indicated in Chapter 3 , representations $x = x(t_1,t_2)$ of the surface $\Gamma$ over polygonal parameter domains which are partitioned by a regular family of triangulations can be used to define a corresponding family of patches $\{F_\ell\}$ on the surface $\Gamma$ . The approximating finite element spaces can be obtained by lifting appropriate finite elements $\varphi_h(t)$ from the parameter domain onto $\Gamma$ by $w_h(p) := \varphi_h(x^{-1}(p))$. Then the operations $\Theta A w_h$ as well as the scalar products in Galerkin's method (1.29) can all be expressed in terms of integrals having integration domains in the parameter domains. Now Gaussian quadrature on the four-dimensional product domains of the triangles for Galerkin's method or on the two-dimensional triangles for collocation with degree of precision $L$

leads to methods governed by Theorem 4.2 provided the principal parts are computed accurately and the remainders have smooth kernels. The boundary integral methods on two-dimensional surfaces in $\mathbb{R}^3$, however, mostly contain <u>singular</u> integrals as in (1.5), (1.9), (1.13) or (1.15). Here even an approximation with piecewise constant trial functions confronts us with a large amount of computational difficulties, even in the case of Fredholm integral equations of the second kind as in (1.5) (see [110]) in connection with exact integration of the principal part and one point Gaussian quadrature. For higher order methods the computational expense will grow tremendeously due to higher degree Gaussian quadratures. Here formulas using the grid points of the triangulations only, i.e. a surface Galerkin collocation method should also be developed.

Therefore most numerical computations [26],[37],[38],[42],[43],[47],[48],[78],[79],[80],[81],[109] are based on simultaneous approximations of the geometry as well as of the desired densities similarly to the finite element treatment of shell problems [21, Chap. 8] . The corresponding error analysis for this approximation to (1.9) can be found in [37] and is based on the fundamental paper [78], the approximation to (1.13) is treated in [38],[43] and [79].

Let us explain these methods and briefly review the corresponding error estimates. To the triangles $T_\ell$ of the parameter triangulations we associate a $C^o$ finite element $S_h^{\kappa+1,1}$ system of Lagrange type, $\kappa \geq 1$, containing piecewise polynomials of degree $\kappa$ and also associated with a unisolvent set of grid points $\{p_{\ell i}\}$ such that the interpolation problem

$\Phi_h(p_{\ell i}) = \Phi(p_{\ell i})$ for $\Phi_h \in S_h$ is uniquely solvable. The corresponding interpolation operator denote by $I_h$. Let $X_j(t)$ be the parameter representation of $\Gamma$ over the parameter domain $U_j$. Then the approximate surface is defined by $\Gamma_h$ : $X_{hj} = I_h X_j(t)$, $t \in U_j$. It is further assumed that along the curves $\overline{F}_j \cap \overline{F}_i \subset \Gamma$ adjacent to $F_j$ and $F_i$ we have the coincidence $X_{hj} = X_{hi}$. For the construction of $H_h$ we choose a regular $S^{m+1,1}$ system, $m \leq \kappa$ of finite elements $\varphi_h$ associated with the regular triangulations in the parameter polygonal domains and a corresponding interpolation operator $\pi_h$. On the boundary of $U_j$ we require via $\overline{F}_j \cap \overline{F}_i$ identical interpolations. The finite elements $\varphi_h$ are lifted with $X_h$ onto $\Gamma_h$ by

$$\chi_h(x) = \varphi_h(X_h^{-1}(x)) \quad \text{for } x \in \Gamma_h$$

defining $H_h$ on $\Gamma_h$. Now let $\Gamma \in C^{\kappa+2}$.

For $x \in \Gamma$ let the straight line through $x$ in the $\pm$ directions of the normal vector $\nu_x$ hit $\Gamma_h$ at $\psi^{-1}(x)$ being the nearest hit. For $h$ small enough this mapping and its inverse $\psi$ exist. Then $\tilde{H}_h$ on $\Gamma$ is defined by the functions

$$w_h(x) := \varphi_h(X_h^{-1}(\psi^{-1}(x))) \quad \text{where } \varphi_h \in S^{m+1,1}.$$

For any of the operators in (1.5), (1.9), (1.13) or (1.15) replace $\Gamma$ by $\Gamma_h$ and $x,y$ by corresponding points on $\Gamma_h$ as well as $r = |x-y|$, $\nu_x, \nu_y$ and the surface elements. With the basis of $\tilde{H}_h$, respectively $H_h$ on $\Gamma_h$ this gives rise to coefficients

$$\tilde{a}_{pq} = (A_h \mu_p, \mu_q)_{L_2(\Gamma_h)} \tag{4.25}$$

which are defined by integrals on the parameter domains. A particularly simple choice of approximation is $\kappa = 1$ and $m = 0$ , i.e. $\Gamma_h$ is piecewise constant. For this approximation one finds in [27] a method to compute (4.25) for the first kind equation (1.9) numerically. For (1.13) and $m = 1$ one finds explicit numerical treatments in [43] and in [29].

The case $\kappa = 1$ and $m = 0$ for the collocation method has been worked out in [48] and [86] for the second kind equations (1.5). The explicit solid angle formula in [110] can be also be used to compute the weights

$$\frac{1}{4\pi} \int_{F_p \subset \Gamma_h} \frac{\partial}{\partial \nu_{yh}} \frac{1}{r} \, ds_{yh}$$

in (1.5) exactly up to round off errors. (A further explicit method to compute $\tilde{a}_{pq}$ for collocation by using Gaussian quadrature can be found in [109].)

In order to prove estimates (4.9), (4.10), in [78] and [38] estimates are given for the differences between $\Gamma$ and $\Gamma_h$ and the derived quantities.

<u>LEMMA 4.3:</u> *Let* $\Gamma \in C^{\kappa+2}$ . *Then one has the estimates*

$$\sup |\psi \circ X_h(t) - X_h(t)| + \sup_{x \in \Gamma_h} |J_h(x) - J \circ \psi(x)| + \sup_{x \in \Gamma_h} |\nu_h(x) - \nu \circ \psi(x)| \leq c h^{\kappa+1} \tag{4.26}$$

*and*

$$0 < c_1 \leq |\psi(x) - \psi(y)| \,/\, |x-y| \leq c_1^{-1} \tag{4.27}$$

*for all* $0 \leq h \leq h_o$ *and* $x,y \in \Gamma_h$ *where the constants* $c, c_1$ *and* $h_o$

*are inedependent of* $h,x,y$ .

For the boundary element spaces one obtains from [78, Lemma 4] that the approximation property (2.1) holds for $-m-1 \leq t \leq s \leq m+1$ and $t \leq 1$ with $u_h = P_o u$ where $P_o$ is the $L_2$ orthogonal projection onto $\tilde{H}_h$ . One also has (2.2) for $-1 \leq t \leq s \leq 1$ . Collecting the results, one has

LEMMA 4.4: *Let* $\Gamma \in C^{\kappa+2}$ . *Then for the approximation of*

i) (1.5) *hold* (4.9), (4.11) *and* (4.12) *with* $\kappa = L$ [37],

ii) (1.9) *holds* (4.9) *with* $\kappa = L$ [78] ,

iii) (1.13) *holds* (4.9) *with* $\kappa = L + 1$ [38].

CONCLUSION 4.5: *For the above approximations we find*

i) *for* (1.5) *an order* $h^{m+1}$ *of* $L_2$*-convergence from* (2.5) *and* (4.14) *if* $\kappa = m$ . *The same order in* $C^o(\Gamma)$ *is obtained for* $\kappa = m$ *from* (2.5) *and* (3.21);

ii) *for* (1.9) *an order* $h^{m+1}$ *of* $L_2$*-convergence from* (2.7) *(with* $\alpha = -\frac{1}{2} = -\tau$ , $j = \beta = 0$*) and* (4.14) *and the optimal choice is* $\kappa = m+1$;

iii) *for* (1.13) *an order* $h^{m+1}$ *of* $L_2$*-convergence from* (2.5) *(with* $\alpha = \frac{1}{2} = \tau$ , $j = \beta = 0$) *and* (4.14) *and the optimal choice is* $\kappa = m$ .

Further estimates (in particular for problems involving irregularities) need still to be analized in view of the numerical integrations.

*References:*


[1] Adams, R.A., Sobolev Spaces, Academic Press, New York, 1975.

[2] Agmon, A., Douglis, A. and Nirenberg, L., Estimates near the boundary for solutions of elliptic partial differential equations satisfying general boundary conditions II, Comm. Pure Appl. Math. 17 (1964) 35-92.

[3] Agranovich, M.S., Spectral properties of diffraction problems, in The General Method of Natural Vibrations in Diffraction Theory by N.N. Voitovic, B.Z. Katzenellenbaum and A.N. Sivov (Russian) Izdat. Nauka, Moscow 1977.

[4] Ahner, J.F. and Hsiao, G.C., On the two-dimensional exterior boundary-value problems of elasticity, SIAM J. Appl. Math. 31 (1976) 677-685.

[5] Ahner, J.F. and Kleinman, R.E., The exterior Neumann problem for the Helmholtz equation, Arch. Rat. Mech. Anal. 52 (1973) 26-43.

[6] Anselone, P.M., Collectively Compact Operator Approximation Theory, Prentice Hall, London, 1971.

[7] Arnold, D.N. and Wendland, W.L., On the asymptotic convergence of collocation methods, Math. Comp., to appear. (Preprint Nr. 665, Technical Univ. Darmstadt, Dept. Mathematics, D-61 Darmstadt, Fed. Rep. Germany 1982).

[8] Arnold, D. and Wendland, W.L., Collocation versus Galerkin procedures for boundary integral methods, in Fourth Int. Sem. Boundary Element Methods (ed. C.A. Brebbia), to appear, (Preprint Nr. 671, Technical Univ. Darmstadt, Dept. Mathematics, D-61 Darmstadt 1982)

[9] Atkinson, K.E., A Survey of Numerical Methods for the Solution of Fredholm Integral Equations of the Second Kind, Soc. Ind. Appl. Math. Philadelphia, 1976.

[10] Aubin, J.P,. Approximation of Elliptic Boundary-Value Problems, Wiley-Interscience, New York, 1972.

[11] Aziz, A.K., Dorr, M.R. and Kellogg, R.B., Calculation of electromagnetic scattering by a perfect conductor, Naval Surface Weapons Center Report TR80-245, 1980, Silver Spring, Maryland 20910.

[12] Babuška, I. and Aziz, A.K., Survey lectures on the mathematical foundations of the finite element method, in the Mathematical Foundation of the Finite Element Method with Applications to Partial Differential Equations (ed. A.K. Aziz), Academic Press, New York (1972) 3-359.

[13] Blue, J., Boundary integral solutions of Laplace's equation, The Bell System Tech. J. 57 (1978) 2797-2822.

[14] Bolteus, L, and Tullberg, O., BEMSTAT-A new type of boundary element program for two-dimensional elasticity problems, in Boundary Element Methods (ed. C.A. Brebbia), Springer, Berlin, Heidelberg, New York (1981) 518-537.

[15] Brakhage, H., Über die numerische Behandlung von Integralgleichungen nach der Quadraturformelmethode, Num. Math. 2 (1960) 183-196.

[16] Brakhage, H. and Werner, P., Über das Dirichletsche Außenraumproblem für die Helmholtzsche Schwingungsgleichung, Arch. Math. 16 (1965) 325-329.

[17] Bruhn, G. and Wendland, W.L., Über die näherungsweise Lösung von linearen Funktionalgleichungen, in Funktionalanalysis, Approximationstheorie, Numerische Mathematik (ed. L. Collatz and H. Ehrmann) Intern. Ser. Numer. Math. 7 Birkhäuser Basel (1967) 136-164.

[18] Burago, Y.D., Maz'ja, V.G. and Sapozhnikova, V.D., On the theory of simple and double-layer potentials for domains with irregular boundaries, in Problems in Math. Analysis Vol. 1, Boundary Value Problems and Integral Equations (ed. V.I. Smirnov), Consultats Bureau, New York (1968) 1-30.

[19] Cea, J., Approximation variationelle des problèmes aux limites, Ann. Inst. Fourier, Grenoble, 14 (1964) 345-444.

[20] Chandler, G.A., Superconvergence of numerical solutions to second kind integral equations, Ph. D. Thesis, Australian National University, 1979.

[21] Ciarlet, P.G., The Finite Element Method for Elliptic Problems, North Holland, Amsterdam, New York, Oxford, 1978.

[22] Colton, D.L., Analytic Theory of Partial Differential Equations, Pitman Publ. London, 1980.

[23] Costabel, M. and Stephan, E., Boundary integral equations for mixed boundary value problems in polygonal domains and Galerkin approximation, Banach Center Publications, Warsaw, to appear (Preprint Nr. 593, Technical Univ. Darmstadt, Dept. Mathematics, D-61 Darmstadt, Fed. Rep. Germany 1981).

[24] Costabel, M. and Stephan, E., Curvature terms in the asymptotic expansions for solutions of boundary integral equations on curved polygons, to appear (Preprint Nr. 673, Technical Univ. Darmstadt, Dept. Mathematics, D-61 Darmstadt, Fed. Rep. Germany 1982).

[25] Costabel, M., Stephan, E. and Wendland, W.L., On boundary integral equations of the first kind for the bi-Laplacian in a polygonal plane domain, to appear (Preprint Nr. 670, Technical Univ. Darmstadt, Dept. Mathematics, D-61 Darmstadt, Fed. Rep. Germany 1982).

[26] Cruse, T.A., Application of the boundary integral equation method to three-dimensional stress analysis, Comp. Struct. 3 (1973) 309-369.

[27] Djaoua, M., Méthode d'éléments finis pour la résolution d'un problème extérieur dans $\mathbb{R}^3$, Centre de Math. Appl., École Polytechniques, Rapport Int. 3, Palaiseau, France, 1975.

[28] Djaoua, M., A method of calculation of lifting flows around 2-dimensional corner shaped bodies, Centre de Math. Appl, École Polytechniques, Rapport Int. 34, Palaiseau, France, 1978.

[29] Durand, M., Diffraction d'ondes acoustiques par un écran mince, Publ. de Math. Appl. Marseille - Toulon 80/3 Université de Provence, France, 1980.

[30] Engels, H., Numerical Quadrature and Cubature, Academic Press, London, New York, Toronto, Sydney, San Francisco, 1980.

[31] Eskin, G.I., Boundary Value Problems for Elliptic Pseudodifferential Equations, AMS, Transl. Math. Mon. 52, Providence, Rhode Island, 1980.

[32] de Figueiredo, D.G., The coerciveness problem for forms over vector valued functions, Comm. Pure Appl. Math. 16 (1963) 63-94.

[33] Filippi, P., Potentiels de couche pour les ondes mécaniques scalaires, Révue du Cethedec 51 (1977) 121-175.

[34] Filippi, P., Layer potentials and acoustic diffraction, J. Sound and Vibration 54 (1977) 473-500.

[35] Fischer, T., An integral equation procedure for the exterior three-dimensional viscous flow, Integral Equations and Operator Theory 5 (1982) 490-505.

[36] Friedrichs, K.O., Pseudo-Differential Operators, Lecture Notes, Courant Inst. New York University 1970.

[37] Giroire, J., Integral equation methods for exterior problems for the Helmholtz equation, Centre de mathématiques appliquées, École Polytechnique, Palaiseau, France, Rapport interne N$^{\underline{o}}$ 40 (1978).

[38] Giroire, J. and Nedelec, J.C., Numerical solution of an exterior Neumann problem using a double layer potential, Math. Comp. 32 (1978) 973-990.

[39] Goldstein, C., Numerical Methods for Helmholtz Type Equations in Unbounded Domains, BNL-26543, Brookhaven Lab., Brookhaven, N.Y. 1979.

[40] Gregoire, J.P., Nedelec, J.C. and Planchard, J., Problèmes relatifs à l'équation d'Helmholtz, Serv. Inf. et Math. Appl. Bull. Direction des Études et Recherches, Ser. C, 2 (1974) 15-32.

[41] Günter, N.M., Die Potentialtheorie, Teubner, Leipzig 1957.

[42] Ha Duong, T., La méthode de Schenck pour la résolution numérique du problème de radiation acoustique, E.D.F., Bull. Dir. Études Recherches, Ser. C, Math., Informatique, Service Informatique et Mathématiques Appl. 2 (1979) 15-50.

[43] Ha Duong, T., A finite element method for the double-layer potential solutions of the Neumann exterior problem, Math. Meth. Appl. Sci. 2 (1980) 191-208.

[44] Hämmerlin, G. and Schumaker, L.L., Procedures for kernel approximation and solution of Fredholm integral equations of the second kind, Numer. Math. 34 (1980) 125-141.

[45] Hayes, J.K., Kahaner, D.K. and Kellner, R.G., An improved method for numerical conformal mapping, Math. Comp. 26 (1972) 327-334.

[46] Helfrich, H.P., Simultaneous approximation in negative norms of arbitrary order, R.A.I.R.O. Num. Analysis 15 (1981) 231-235.

[47] Hess, J.L., Calculation of acoustic fields about arbitrary three-dimensional bodies by a method of surface source distributions based on certain wave number expansions, Report DAC 66901, Mc Donnell Douglas, 1968.

[48] Hess, J.L. and Smith, A.M.O., Calculation of potential flow about arbitrary bodies, in Progress in Aeronautical Sciences (ed. D. Kuchemann) Pergamon, Oxford 8 (1967) 1-138.

[49] Hildebrandt, St. and Wienholtz, E., Constructive proofs of representation theorems in separable Hilbert space, Comm.Pure Appl. Math. 17 (1964) 369-373.

[50] Hörmander, L., Pseudo-differential operators and non-elliptic boundary problems, Annals Math. 83 (1966) 129-209.

[51] Hoidn, H.-P., Die Kollokationsmethode angewandt auf die Symmsche Integralgleichung, ETH Zürich, Switzerland, in preparation.

[52] Hsiao, G.C., Kopp, P. and Wendland, W.L., A Galerkin collocation method for some integral equations of the first kind, Computing 25 (1980) 89-130.

[53] Hsiao, G.C., Kopp, P. and Wendland, W.L., Some applications of a Galerkin collocation method for integral equations of the first kind, in preparation.

[54] Hsiao, G.C. and Wendland, W.L., A finite element method for some integral equations of the first kind, J. Math. Anal. Appl. 58 (1977) 449-481.

[55] Hsiao, G.C. and Wendland, W.L., The Aubin-Nitsche lemma for integral equations, J. Integral Equations 3 (1981) 299-315.

[56] Jentsch, L., Über stationäre thermoelastische Schwingungen in inhomogenen Körpern, Math. Nachr. 64 (1974) 171-231.

[57] Jentsch, L., Stationäre thermoelastische Schwingungen in stückweise homogenen Körpern infolge zeitlich periodischer Außentemperatur, Math. Nachr. 69 (1975) 15-37.

[58] Jones, D.S., Integral equations for the exterior acoustic problem, Quart. J. Mech. Appl. Math. 27 (1974) 129-142.

[59] Kleinman, R.E. and Roach, G.F., Boundary integral equations for the three-dimensional Helmholtz equation, SIAM Review 16 (1974) 214-236.

[60] Kleinman, R. and Wendland, W.L., On Neumann's method for the exterior Neumann problem for the Helmholtz equation, J. Math. Anal. Appl. 57 (1977) 170-202.

[61] Kohn, J.J. and Nirenberg, L., On the algebra of pseudo-differential operators, Comm. Pure Appl. Math. 18 (1965) 269-305.

[62] Kral, J., Integral Operators in Potential Theory, Lecture Notes Math. 823, Springer Berlin, Heidelberg, New York 1980.

[63] Kupradze, W.D., Randwertaufgaben der Schwingungstheorie und Integralgleichungen, Dt. Verlag d. Wissenschaften, Berlin, 1956.

[64] Kupradze, V.D., Potential Methods in the Theory of Elasticity, Israel Program Scientific Transl., Jerusalem, 1965.

[65] Kupradze, V.D., Gegelia, T.G., Basheleishvili, M.O., Burchuladze, T.V., Three-Dimensional Problems of the Mathematical Theory of Elasticity and Thermoelasticity, North Holland, Amsterdam, 1979.

[66] Lamp, U.,Schleicher, T., Stephan, E. and Wendland, W.L., The boundary integral method for a plane mixed boundary value problem, in Advances in Computer Methods for Partial Differential Equations - IV (ed. R. Vichnevetsky and R.S. Stepleman) IMACS Dept. Computer Science, Rutgers Univ. New Brunswick, N.Y. 08903 U.S.A. (1981) 223-229.

[67] Lamp, U., Schleicher, T., Stephan, E. and Wendland, W.L., Theoretical and experimental asymptotic convergence of the boundary integral method for a plane mixed boundary value problem, Proc. Fourth Intern. Seminar on Boundary Element Methods (ed. C.A. Brebbia), Springer, Berlin, Heidelberg, New York, to appear 1982 (Preprint Nr. 667, Technical Univ. Darmstadt, Dept. Mathematics, D-61 Darmstadt Fed. Rep. Germany 1982).

[68] Lax, P.D. and Nirenberg, L., On stability for difference schemes; a sharp form of Gårding's inequality, Comm. Pure Appl. Math. 19 (1966) 473-492.

[69] Lehmann, R., Developments at an analytic corner of solutions of elliptic partial differential equations, J. Math. Mech. 8 (1959) 727-760.

[70] MacCamy, R.C., Low frequency acoustic oscillations, Quaterly Appl. Math. 23 (1965) 247-265.

[71] MacCamy, R.C. and Stephan, E., A boundary element method for an exterior problem for three-dimensional Maxwell's equations, to appear (Preprint Nr. 681 , Technical Univ. Darmstadt, Dept. Mathematics, D-61 Darmstadt, Fed. Rep. Germany 1982).

[72] Martensen, E., Potentialtheorie, B.G. Teubner, Stuttgart, 1968.

[73] Michlin, S.G., Variationsmethoden der Mathematischen Physik, Akademie-Verlag, Berlin 1962.

[74] Michlin, S.G. and Prössdorf, S., Singuläre Integraloperatoren, Akademie-Verlag, Berlin, 1980.

[75] Müller, C., Foundations of the Mathematical Theory of Electromagnetic Waves, Springer, Berlin • Heidelberg • New York , 1969.

[76] Mustoe, G.G. and Mathews, I.C., Direct boundary integral methods, point collocation and variational procedures, to appear.

[77] Natanson, E., Theory of Functions of a Real Variable, Ungar Publ., New York, 1955.

[78] Nedelec, J.C., Curved finite element methods for the solution of singular integral equations on surfaces in $\mathbb{R}^3$ , Comp. Math. Appl. Mech. Eng. 8 (1976) 61-80.

[79] Nedelec, J.C., Approximation par potentiel de double cuche du problème de Neumann extérieur, C.R. Acad. Sci. Paris, Sér. A 286 (1977) 616-619.

[80] Nedelec, J.C., Formulations variationelles de quelques équations integrales faisant intervenir des parties finies, in Innovative Numerical Analysis for the Engineering Sciences (R. Shaw ed.), Univ. Press of Virginia, Charlottesville (1980) 517-524.

[81] Nedelec, J.C. and Planchard, J., Une méthode variationnelle d'éléments finis pour la résolution numérique d'un problème extérieur dans $\mathbb{R}^3$ , R.A.I.R.O. 7 (1973) R3, 105-129.

[82] Nitsche, J.A., Zur Konvergenz von Näherungsverfahren bezüglich verschiedener Normen, Num. Math. 15 (1970) 224-228.

[83] Panič, I.J., On the solubility of exterior boundary value problems for the wave equation and for a system of Maxwell's equations (Russian), Uspehi Mat. Nauk 20 (1965) 221-226.

[84] Petiau, G., La théorie des fonctions de Bessel, Centre National de la Rech. Scientifique, Paris 1955.

[85] Phillips, J.L., The use of the collocation as a projection method for solving linear operator equations, SIAM J. Numer. Anal. 9 (1972) 14-28.

[86] Poggio, A.J. and Miller, E.K., Integral equation solutions of three-dimensional scattering problems, in Computer Techniques for Electromagnetics (R. Mittra ed.) Pergamon, Oxford, 1973.

[87] Prössdorf, S. and Schmidt, G., A finite element collocation method for singular integral equations, Math. Nachr. 100 (1981) 33-60.

[88] Prössdorf, S. and Schmidt, G., A finite element collocation method for systems of singular integral equations, Preprint P - MATH - 26/81, Akademie d. Wissenschaften DDR, Inst. Math., DDR-1080 Berlin, Mohrenstr. 39, 1981.

[89] Prössdorf, S. and Silbermann, B., Projektionsverfahren und die näherungsweise Lösung singulärer Gleichungen, Teubner, Leipzig, 1977.

[90] Radon, J., Über die Randwertaufgaben beim logarithmischen Potential, Sitz.ber. Akad. Wiss. Wien, Math.-nat. Kl. IIa 128 (1919) 1123-1167.

[91] Rannacher, R. and Wendland, W.L., On the order of pointwise convergence of some boundary element methods, in preparation.

[92] Richter, G.R., Superconvergence of piecewise polynomial Galerkin approximations for Fredholm integral equations of the second kind, Numer. Math. 31 (1978) 63-70.

[93] Le Roux, M.N., Résolution numérique du problème du potential dans le plan part une méthode variationelle d'éléments finis, These, L'Université de Rennes, Ser. A, No. D'ordre 347, No. Ser. 38, 1974.

[94] Ruland, C., Ein Verfahren zur Lösung von $(\Delta+k^2)u = 0$ in Aussengebieten mit Ecken, Appl. Analysis 7 (1978) 69-79.

[95] Saranen, J. and Wendland, W.L., On the asymptotic convergence of collocation methods with spline functions of even degree, to appear (Preprint Nr. 690, Technical Univ. Darmstadt, Dept. Mathematics, D-61 Darmstadt, Fed. Rep. Germany 1982).

[96] Schäfer, E., Fehlerabschätzungen für Eigenwertnäherungen nach der Ersatzkernmethode bei Integralgleichungen, Numer. Math. 32 (1979) 281-290.

[97] Schenck, H.A., Improved integral formulation for acoustic radiation problems, J. Acoustic Soc. Amer. 44 (1968) 41-58.

[98] Schmidt, G., On spline collocation for singular integral equations, Math. Nachr., to appear.

[99] Seeley, R., Topics in pseudo-differential operators, in Pseudo-Differential Operators (L. Nirenberg ed.) C.I.M.E., Cremonese, Roma, 1969.

[100] Sloan, I., Improvement by iteration for compact operator equations, Math. Comp. 30 (1976) 758-764.

[101] Stephan, E., Solution procedures for interface problems in acoustics and electromagnetics, in these Lecture Notes.

[102] Stephan, E. and Wendland, W.L., Remarks to Galerkin and least squares methods with finite elements for general elliptic problems, Springer Lecture Notes Math. 564 (1976) 461-471 and Manuscripta Geodaetica 1 (1976) 93-123.

[103] Stummel, F., Diskrete Konvergenz linearer Operatoren I and II, Math. Zeitschr. 120 (1971) 231-264.

[104] Symm, G.T., Integral equation methods in potential theory, II, Proc. Royal Soc. London A275 (1963) 33-46.

[105] Treves, F., Introduction to Pseudodifferential and Fourier Integral Operators I. Plenum Press, New York and London, 1980.

[106] Ursell, F., On the exterior problems of acoustics, Proc. Camb. Phil. Soc. 74 (1973) 117-125.

[107] Vaillancourt, R., A simple proof of Lax-Nirenberg theorems, Comm. Pure Appl. Math. 23 (1970) 151-163.

[108] Vainikko, G., Funktionalanalysis der Diskretisierungsmethoden, G.B. Teubner, Leipzig, 1976.

[109] Watson, J.O., Advanced implementation of the boundary element method for two- and three-dimensional elastostatics, in "Developments in Boundary Element Methods - 1" (ed. P.K. Banerjee and R. Butterfield), Appl. Science Publ. LTD, London (1979) 31-63.

[110] Wendland, W.L., Die Behandlung von Randwertaufgaben im $\mathbb{R}_3$ mit Hilfe von Einfach- und Doppelschichtpotentialen, Num. Math. 11 (1968) 380-404.

[111] Wendland, W.L., Über Galerkin-Verfahren zur Lösung des Dirichlet-Problems in $L_p$-Räumen für gleichmäßig stark elliptische Differentialgleichungen, Abh. Math. Seminar Univ. Hamburg, 36 (1971) 185-197.

[112] Wendland, W.L., On Galerkin collocation methods for integral equations of elliptic boundary value problems, in Numerical Treatment of Integral Equations (ed. J. Albrecht and L. Collatz), Intern. Ser. Num. Math. 53, Birkhäuser Basel (1980) 244-275.

[113] Wendland, W.L., Asymptotic convergence of boundary element methods, in Lectures on the Numerical Solution of Partial Differential Equations (ed. I. Babuška, T.-P. Liu and J. Osborn) Lecture Notes # 20, Univ. of Maryland, College Park Md. (1981) 435-528.

[114] Wendland, W.L., Asymptotic accuracy and convergence, in Progress in Boundary Element Methods (ed. C.A. Brebbia), Pentech Press, London, Plymouth, 1 (1981) 289-313.

[115] Wendland, W.L., Stephan, E. and Hsiao, G.C., On the integral equation method for the plane mixed boundary value problem of the Laplacian, Math. Meth. in the Appl. Sci., 1 (1979) 265-321.

[116] Wolfe, P., An integral operator connected with the Helmholtz equation, J. Functional Anal. 36 (1980) 105-113.

# BOUNDARY VALUE PROBLEMS ANALYSIS AND PSEUDO-DIFFERENTIAL OPERATORS IN ACOUSTICS

Marc Durand
U.E.R. de Mathématiques
Université de Provence
13331 MARSEILLE CEDEX 13
France

## P R E F A C E

It is well known that partial differential problems can be solved by using integrals (more or less singular). Moreover, since some years, such problems are tried to be solved by integral problems defined on the boundary of the concerned domain (P.J.T. Filippi[1]). Furthermore differential problems are not sufficient to account for many physical problems. In particular boundary conditions can often be integral.

In 1730, D. Bernouilli introduced integral equations in the study of a vibrating string. The ideas where developed two centuries ago, mainly when the Green's kernels have been introduced (cf. the works of H.A. Schwarz, H. Poincaré, V. Volterra, and others). Then I. Fredholm began to construct a satisfactory theory of integral equations, and D. Hilbert developed a functional framework. Afterwards many works studied some classes of integral equations and singular integral equations (S.G. Mikhlin, N.I. Muschelishvili,...). Perhaps, in this way the last refinements are due to B.V. Chvedelidze[2]. But beyond the solutions of particular equations, a theory is needed that allows to put the problems correctly, with adapted functional spaces. This is the aim of the Pseudo-differential operators theory (PDO's). They are a class of integro-differential operators that contains differential operators and their inverses. The theory fit well with the elliptic linear problems (as the Helmholtz equation), and can be applied to more general equations (v.g. parabolic equations). Furthermore, the Fourier integral operators where introduced by L. Hörmander[3] and others to study hyperbolic problems. This powerful tool is very useful to obtain information about the nature of the solutions of linear partial differential equations (regularity, propagation of singularities..)

In the last years, J.M. Bony[4] introduced a new class of PDO's, namely the paradifferential operators, adapted to the non-linear problems, at least to some of them.

In a first chapter, we give some well known results about the fundamental equations of acoustics, and the usual methods to solve them : principles of radiation, of limiting amplitude or limiting absorption (B.R. Vainberg[5]) and integral methods (A.P. Calderon[6]).

In the second chapter, we investigate the PDO's theory. First we introduce some classical problems which allow to conceive what is needed (1). In the second paragraph, about the PDO's, we give the basis of a symbolic calculus which makes the use of the theory very easier (2.1) and we study the oscillating integrals (2.2). Thus we can introduce the Fourier integral operators (2.3) whose the PDO's are a particular case (2.4). At the end of this paragraph we give the main results of regularity about PDO's (2.5) and some applications to elliptic problems (2.6). A third paragraph (3) is aimed at the study of boundary value problems (BVP's). We must define a notion of regularity of a PDO along the boundary, namely the "transmission property" (3.1). The Calderon's projectors may then be defined correctly (3.2), they were already introduced in the first chapter, to solve classical problems. To complete the theory, a generalization of the classical Poisson kernels is needed (3.3), and we may introduce an operator algebra, which defines the good framework for the resolution of general elliptic boundary value problems (L. Boutet de Monvel[7,8]).

In the last chapter, we show how the PDO theory allows to make the problem of the scattering of an acoustical wave by a thin obstacle "well posed" (M.Durand[9]). The problem is solved via a double-layer potential. The very singular integral equation that is obtained is well defined (although the kernel is like $\frac{1}{r^2}$ on a one-dimensional curve). The theory provides a relatively simple numerical procedure for solving the integral equation. On the other hand, it is shown that there is no jump of the solution across the edge of the screen, this holds without supplementary conditions (as the classical "edge conditions").

In such a short text, we cannot give the proofs. Our aim is to set forth a theory, which gives a good framework for solving the usual problems encountered in physics. To examine the question thoroughly, the interested reader could use the works of R.T. Seeley[10] or F. Treves[11], or the book of J.L. Lions and E.Magenes[12] for a good sight at BVP's.

# I. - BOUNDARY VALUE PROBLEMS

In this first chapter we recall some results about the problems encountered by acousticians, in order to make clearer what follows.

We consider a domain $\Omega$ with boundary $\Gamma$ in the three-dimensional space $\mathbb{R}^3$ . For the moment, we suppose that $\Gamma$ is smooth. We wish to study the Helmholtz equation in $\Omega$ . In order to precise the problem, functional spaces are needed and we must give boundary conditions as well as conditions at infinity when $\Omega$ is not bounded.

## 1 - Interior problem in a bounded domain .

Let us consider the boundary value problem

$$\begin{cases} (\Delta + k^2)\, u = f, \\ (\alpha \partial_n u + \beta u)\,|_\Gamma = g, \quad \alpha \in \mathbb{C}\ ,\ \beta \in \mathbb{C} \end{cases}$$

with $f \in C^\infty(\Omega)$ , $g \in C^\infty(\Gamma)$ , , the support of f is compact. u is seeked in $\mathscr{D}'(\Omega)$, space of distributions in $\Omega$ . Obviously this problem is equivalent to the following one

$$\begin{cases} (\Delta + k^2)\, u = \varphi \\ (\alpha \partial_n u + \beta u)\,|_\Gamma = 0 \end{cases}$$

We consider the unbounded operator A from $L^2(\Omega)$ into $L^2(\Omega)$ with domain $D(A) = \{u \in L^2(\Omega)\,;\ \Delta u \in L^2(\Omega)$ and $(\alpha \partial_n u + \beta u)\,|_\Gamma = 0\}$. On D(A) , $A = -\Delta$.

When $\alpha = 0$ , (1) is the Dirichlet problem, it is the Neumann problem if $\beta = 0$. If $\alpha/\beta$ is real, the operator A is self-adjoint. Its spectrum contains only real and positive eigenvalues. If $k^2$ is not an eigenvalue of A , for every $\varphi$ in $L^2(\Omega)$ there exists one and only one solution in the space $H^0(\Delta) = \{u \in L^2(\Omega);\ \Delta u \in L^2(\Omega)\}$ , and u is known to belong to the Sobolev space $H^1(\Omega) = \{u \in L^2(\Omega);\ \nabla u \in L^2(\Omega)\}$.

$\alpha$ and $\beta$ can be chosen as functions of x. If $\alpha/\beta$ is not real, the operator A is not self-adjoint, the eigenvalues are shown to be in a cone in $\mathbb{C}$ with vertex at the origin and axis $\mathbb{R}_+$ , except for a finite number of them.

A boundary condition with "oblique derivative" can be considered :

$$(\alpha \partial_n u + Tu)\,|_\Gamma = 0,$$

where T is a vector field tangent to $\Gamma$ . If $\alpha \neq 0$ , the problem is still elliptic. But if $\alpha$ can be zero, the ellipticity of the problem is lost, except in dimension 2 .

When $\alpha$ and $\beta$ are regular, a priori estimates are given, the solution u is as regular as the data.

2 - Exterior problem in an unbounded domain $\tilde{\Omega}$.

Now we suppose that $\Omega$ is the exterior of a regular bounded domain $\tilde{\Omega}$. We consider the problem

$$(2) \quad \begin{cases} (\Delta + k^2) u = f, \\ (\alpha \partial_n u + \beta u)|_\Gamma = 0, \end{cases}$$

$\alpha$ and $\beta$ are regular functions, $\alpha/\beta$ is real, supp f is compact. It is shown that the spectrum of A is $\mathbb{R}_+$, there is no eigenvalue. There exists an infinity of solutions decaying at infinity as $1/r^{\frac{n}{2}-1}$, but no solution decreases faster. To obtain uniqueness, it is needed choosing among the solutions which are $0(\frac{1}{r^{(n/2)-1}})$ at infinity. Three ways allow to solve this problem.

a- <u>Principle of radiation</u> - The solutionsof the Helmholtz equation are the amplitudes of forced vibrations, and we seek for u such that $v(x,t) = u(x)\, e^{-i\omega t}$ is a solution of the wave equation with a source $f(x)\, e^{-i\omega t}$. A plane wave v propagating in the x-direction is the sum of an outgoing wave $v_1 = F(t - \frac{x}{a})$ and an incoming wave $v_2 = F(t + \frac{x}{a})$, with $a = \frac{\omega}{k}$. We have $\frac{\partial v_1}{\partial r} + \frac{1}{a}\frac{\partial v_1}{\partial t} = 0$ and $\frac{\partial v_2}{\partial r} - \frac{1}{a}\frac{\partial v_2}{\partial t} = 0$. As $v = u\, e^{-i\omega t}$, we deduce that

$$\frac{\partial u_1}{\partial r} - iku_1 = 0 \quad \text{and} \quad \frac{\partial u_2}{\partial r} + iku_2 = 0.$$

When spherical waves are generated by sources distributed in a bounded part of $\mathbb{R}^3$, it is still obtained the sum of outgoing and incoming waves, associated to the following conditions

$$\frac{\partial u}{\partial r} - iku = \sigma(\frac{1}{r}) \qquad \text{(outgoing wave)},$$

$$\frac{\partial u}{\partial r} + iku = \sigma(\frac{1}{r}) \qquad \text{(incoming wave)}.$$

Then, to solve the Helmholtz equation, we shall impose the condition that the incoming wave is zero, that is the solution is a spherical wave proceeding from the source and going to infinity. Finally we seek for the solution of the following problem :

$$\begin{cases} \text{To find } u \in L^2_{loc}(\Omega) \text{ such that} \\ (\Delta + k^2)u = 0, \\ (\alpha\partial_n u + \beta u)\big|_\Gamma = g, \quad g \in L^2(\Omega), \\ \left.\begin{array}{l} u(r) = 0(\frac{1}{r}), \\ \dfrac{\partial u}{\partial r} - iku = \sigma(\frac{1}{r}) \end{array}\right\} \quad r \to \infty . \end{cases}$$

This problem is well posed. It has one and only one solution. This result can be extended to a very general class of differential operators that don't need to be elliptic, but only hypoelliptic. The radiation conditions are usefull to choose among various elementary solutions of the problem.

b - <u>Principle of limit absorption</u> - We still consider a forced vibration $fe^{-i\omega t}$, but we can suppose there is a small complex absorption. The wave equation is modified into

$$\Delta v - \frac{1}{a^2}\frac{\partial^2 v}{\partial t^2} - \beta\frac{\partial v}{\partial t} = f(x)\, e^{-i\omega t}, \quad \beta > 0,$$

and we seek for solutions $v(x,t) = u(x)\, e^{-i\omega t}$. The Helmholtz equation is changed into :

$$(\Delta + k^2 + i\beta\omega)u = f .$$

Let us put $q = q_o + iq_1 = \sqrt{k^2 + i\beta\omega}$, with $q_1 > 0$. The elementary solutions of the operator are $e^{-iqr}/r$ and $e^{iqr}/r$. We seek for a solution in $L^2(\Omega)$, as $(k^2 + i\beta\omega)$ is not in the spectrum of the operator A. The only possible solution is $e^{iqr}/r = (e^{iq_0 r}/r)\, e^{-q_1 r}$, which decreases at infinity. The limiting value of the solution when $\beta \to 0$ is associated to the elementary solution $e^{ikr}/r$, which is the amplitude of an outgoing wave :

$$\frac{e^{ikr}}{r}\, e^{-i\omega t} = \frac{e^{i(kr-\omega t)}}{r}\,, \quad \text{and} \quad \frac{\partial}{\partial r}(e^{ikr}/r) - ik(e^{ikr}/r) = o(1/r)\,.$$

Then we obtain the same solution as with the principle of radiation. This method can be extended to general operators, by adding absorption terms not so elementary as a constant $\beta$, but which are differential terms.

c - Principle of limit amplitude - In this third case, we consider the wave problem with null initial conditions :

$$\begin{cases} \Delta v - \dfrac{1}{a^2}\dfrac{\partial^2 v}{\partial t^2} = f\, e^{-i\omega t}\,, \\ v(x,0) = v_t(x,0) = 0\,, \end{cases}$$

and we add boundary conditions as in what precedes. We seek for a solution u of the Helmholtz equation as $u(x) = \lim_{t\to\infty} v(x,t)\, e^{+i\omega t}$. The solution v is seeked as an outgoing wave and the limit u is obtained by a convolution with the elementary solution $e^{ikr}/r$ which veryfies the outgoing radiation condition. This principle can be extended to general operators, but it is not so easy to see that it still gives the same results as the two preceding ways.

Remarks - The boundary conditions can be non-local, given by an integral equation.

- A difficulty is to define what is an elliptic problem.

When the boundary conditions are differential, a boundary value problem is elliptic if the operator is elliptic and if the boundary conditions "cover" the operator, i.e. there are enough and not too much conditions. When the conditions are integral, a good definition of ellipticity can be given by using the pseudo-differential operators.

- When the problem is elliptic, a non-bounded operator A with domain D(A) can be associated to the problem. This operator is of normal type (i.e. its image is closed, its kernel and cokernel have a finite dimension) and the problem is solved by a Fredholm alternative. It is usual to transform such problems into integral ones, what we shall explain now.

## 3 - Differential problems as integral problems.

To illustrate this method, we limit ourselves to the study of classical Dirichlet and Neumann problems in a bounded regular domain $\Omega$ with boundary $\Gamma$ . Let us consider :

$$(D) \begin{cases} \Delta u = 0 & \text{in } \Omega , \\ B(\gamma u) = f & \text{on } \Gamma , \end{cases}$$

where $(\gamma u) = (u_o, u_1)$ are the first two traces of u on $\Gamma$ , $B(\gamma u) = B_o u_o + B_1 u_1$ , the $B_i$ are differential operators defined on $\Gamma$ . Let G be a fundamental solution of $\Delta$ in the whole space. For $x \in \Omega$ , the Green's formula gives the relation ,

$$u(x) = -\int_\Gamma u_0(y) \frac{\partial G}{\partial n_y} (x-y)\, d\Gamma(y) + \int_\Gamma u_1(y)\, G(x-y)\, d\Gamma(y) . \tag{1}$$

By using the Plemelj's formulas,

we obtain for $x \in \Gamma$ :

$$u_0(x) = 2\int_\Gamma u_1(y)\, G(x-y)\, d\Gamma(y) - 2\int_\Gamma u_0(y) \frac{\partial}{\partial n_y} G(x-y)\, d\Gamma(y), \qquad (2)$$

that is

$$(2) \quad u_o = K_o u_o + K_1 u_1 .$$

Conversely let $(u_o, u_1)$ be a solution of the equation (2) in $\Gamma$. By the formula (1) we construct a function $u$ defined in $\Omega$ and $\complement\bar{\Omega}$, which verifies $\Delta u = 0$ in $\complement\Gamma$. The first trace $(\gamma_o u)$ of $u$ is obviously

$$\gamma_0 u(x) = \frac{u_0}{2}(x) - \int_\Gamma u_0(y) \frac{\partial}{\partial n_y} G(x-y)\, d\Gamma(y) + \int_\Gamma u_1(y)\, G(x-y)\, d\Gamma(y),$$

that is : $\gamma_o u = (1/2)u_o + (1/2)(K_o u_o + K_1 u_1) = u_o$ .

The Green's formula for $u$ is, for $x \in \Omega$ ,

$$u(x) = -\int_\Gamma u_0(y) \frac{\partial}{\partial n_y} G(x-y)\, d\Gamma(y) + \int_\Gamma (\gamma_1 u)(y)\, G(x-y)\, d\Gamma(y).$$

By subtracting (3) from (1), we obtain

$$v(x) = \int_\Gamma [u_1(y) - (\gamma_1 u)(y)]\, G(x-y)\, d\Gamma(y), \quad x \in \Omega \cup \complement\bar{\Omega},$$

where $v(x) \equiv 0$ for $x \in \Omega$ , $\Delta v = 0$ in $\complement\bar{\Omega}$ . $v$ is continuous across $\Gamma$ , then $v$ is zero at infinity, and $v \equiv 0$ out of $\Omega$ . Finally $v$ is identically zero, and $(u_1 - \gamma_1 u)$ , which is the jump of $v$ across $\Gamma$ , is zero : $\gamma_1 u = u_1$. Then we have proved the equivalence between the problem (D) and the integral one (I) defined on $\Gamma$ :

$$(I) \quad \begin{cases} (1 - K_o)\, u_o - K_1 u_1 = 0 , \\ B_o u_o + B_1 u_1 = f . \end{cases}$$

In the general case, let us consider an elliptic differential problem $(A, B_j)$ (A is an unbounded operator on $L^2(\Omega)$ and the $B_j$ are operators on the trace spaces defined on $\Gamma$ ). We suppose that order(A) = 2m, and order($B_j$) = j , j=0,1,....,m-1 . An operator $\mathscr{C}^+$ can be defined on $(L^2(\Gamma))^{2m}$

such that the problems

$$(D)\begin{cases} Au = 0 & \text{in } \Omega , \\ B_j u = g_j & \text{in } \Gamma , \end{cases} \quad \text{and} \quad (I)\begin{cases} \mathscr{C}^+(\gamma u) = \gamma u \\ B_j(\gamma u) = g_j \end{cases}$$

are equivalent ($\gamma u = (\gamma_o u, \ldots, \gamma_{2m-1} u)$ ). When ($\gamma u$) is known, the solution $u$ of (D) is given as a multiple-layer potential. The Calderon's operator $\mathscr{C}^+$ is a quasi-projector, that is $\mathscr{C}_o^+ \mathscr{C}^+ = \mathscr{C}^+ + R$ where $R$ is an operator from $\mathscr{D}'(\Gamma)$ to $C^\infty(\Gamma)$ .

The aim of these lectures is to define the class of operators which are needed to make clear and correct the definition and the solution of problems like (I).

<u>Remark</u> - In the case of irregular domains (scattering by an irregular obstacle), the Calderon's projectors $\mathscr{C}^+$ cannot be used easily, as the pseudo-differential operators are defined on smooth domains only. Nevertheless, by using Green's formula, integral operators on $\Gamma$ are introduced and the differential problem is equivalent to an integral one.

The boundary value problem can also be solved by variational methods, in wheighted Sobolev spaces that allow to specify the regularity of the solution at the corners.

## II. - PSEUDO-DIFFERENTIAL OPERATORS

### 0. - Notations

The notations are the usual ones in the books of functional analysis. Let us specify someones

$$D_k = \frac{1}{i}\frac{\partial}{\partial x_k}$$

$\Omega$ is an open regular domain in $\mathbb{R}^n$.

$C_o^\infty(\Omega) = \{u \in C^\infty(\Omega) \quad ; \text{ supp } u \text{ is a compact of } \Omega\}$.

$\mathcal{S}$ is the Schwartz space of rapidly decreasing $C^\infty$ functions, and its dual $\mathcal{S}'$ is the space of tempered distributions.

$\mathcal{D}'$ is the space of distributions, $\mathcal{E}'$ the space of distributions with compact support.

The Fourier transform of a function $u \in \mathcal{S}$ is

$$\hat{u}(\xi) = \int e^{-i <x, \xi>} u(x)\, dx ,$$

and in the sequel we note $< x, \xi > = x.\ \xi = \sum_i x_i \xi_i$ .

Instead of Pseudo-Differential Operators, we shall write PDO's.

### 1. - Introduction

1.1. - Let P(D) be a differential operator with constant coefficients defined on $\mathbb{R}^n$

$$P(D) = \sum_{|\alpha| \leqslant m} a_\alpha D^\alpha , \quad \alpha = (\alpha_1 , \ldots , \alpha_n) \text{ is a}$$

multi-index, $a_\alpha \in \mathbb{C}$ .

The expression $P(D)u$ with $u \in \mathcal{S}$, can be explicited as follows :

$$P(D)u(x) = (2\pi)^{-n} \int e^{ix.\xi} (Pu)^\wedge(\xi)\, d\xi$$

$$= (2\pi)^{-n} \int e^{ix.\xi} P(\xi)\, \hat{u}(\xi)\, d\xi$$

$$= (2\pi)^{-n} \iint e^{i(x-y).\xi} P(\xi)\, u(y)\, dy\, d\xi .$$

Here $P(\xi)$, $\xi \in \mathbb{R}^n$, is apolynomial defined (replacing $D^\alpha$ by $\xi^\alpha$ in $P(D)$) by : $P(D)(e^{ix.\xi}) = P(\xi)\, e^{ix.\xi}$.

In the sequel we shall consider operators such that

$$Au(x) = \int e^{ix.\xi} a(x,\xi)\, \hat{u}(\xi)\, d\xi = \int e^{i(x-y).\xi} a(x,\xi)\, u(y)\, dy\, d\xi .$$

This class of operators will contain differential operators and their inverses (or pseudo-inverses). We shall have to define the functional spaces on which these operators act and the classes of functions a (so-called symbols).

1.2. - Parametrices of elliptic equations

a - Consider a partial differential equation with constant coefficients

$$P(D)u = f$$

with $f \in C_o^\infty(\mathbb{R}^n)$ (with compact support).

We should like to write

$$u(x) = (2\pi)^{-n} \int e^{ix.\xi} \hat{f}(\xi) \frac{1}{P(\xi)}\, d\xi .$$

Usually the right hand side is without sense, except if $P(\xi) \neq 0$. If the set of the zeros of $P(\xi)$ is contained in a ball of center 0 and radius $\rho$ in $\mathbb{R}^n$, one can consider a function $\chi \in C^\infty(\mathbb{R}^n)$, $\chi(\xi) = 0$ if $|\xi| < \rho$, $\chi(\xi) = 1$ if $|\xi| > \rho' > \rho$,

and we note :

$$v(x) = (2\pi)^{-n} \int e^{ix.\xi} \frac{\hat{f}(\xi)}{P(\xi)} \chi(\xi)\, d\xi .$$

Then $P(D)\ v(x) = f(x) - \underline{R}\, f(x)$, with

$$\underline{R}f(x) = (2\pi)^{-n} \int e^{ix.\xi}\, \hat{f}(\xi)\, [1 - \chi(\xi)]\, d\xi .$$

It is easy to construct a tempered distribution K such that :

$$v = K * f$$

and a function $h \in \mathcal{S}$ such that

$$\underline{R}f = h * f .$$

Thus, if $\underline{K}$ is the operator $u \to K * u$, then

$$P(D)\underline{K} = I - \underline{R} \quad (I : \text{Identity mapping}).$$

The operator $\underline{K}$ is said to be a parametrix of the operator P. It is easy to show that $\underline{R}$ maps the space $\mathcal{E}'$ of distributions with compact supports into $C^\infty(\mathbb{R}^n)$.

The class of elliptic operators (i.e. $P_m(\xi) \neq 0$ when $\xi \neq 0$ if $P_m$ is the principal part of P) is the most important of the classes of operators P such that the zeros of $P_m$ are bounded. At first, the PDO's were defined to solve elliptic problems.

b - Now consider an elliptic differential operator with $C^\infty$ coefficients defined in an open domain $\Omega \subset R^n$ :

$$P(x, D) = \sum_{|\alpha| \leqslant m} a_\alpha(x)\, D^\alpha ,$$

and the associated polynomial

$$P(x, \xi) = \sum_{|\alpha| \leqslant m} a_\alpha(\xi)\, \xi^\alpha .$$

As in the previous case, we consider the equation

$$P(x,D)u = f ,$$

and we note :

$$v(x) = (2\pi)^{-n} \int e^{ix.\xi}\, k(x, \xi)\, \hat{f}(\xi)\, d\xi ,$$

where $k(x, \xi)$ should be equal to : $\dfrac{\chi(\xi)}{P(\xi)}$ .

It is easily shown that $D^\alpha(e^{ix.\xi}u) = e^{ix.\xi}(D+\xi)^\alpha u$, then

$$P(x, D)\, v(x) = (2\pi)^{-n} \int e^{ix.\xi} P(x, D+\xi)\, k(x,\xi)\, \hat{f}(\xi)\, d\xi .$$

In order to find u, we have to solve the equation (with a parameter $\xi$):

$$P(x, D+\xi)k(x,\xi) = 1.$$

It is noted that

$$P(x, D+\xi) = P_m(x,\xi) + \sum_{i=1}^{m} P_i(x,\xi,D)$$

where $P_i(x, \xi, D)$ is a differential operator of order i whose coefficients are homogeneous polynomials in $\xi$ of degree (m - i). Then if we note

$$k(x,\xi) = \sum_{i=0}^{\infty} k_i(x,\xi)$$

where $k_i(x, \xi)$ is a polynomial in $\xi$ of degree $d_i \to -\infty$, the following successive equations are obtained :

$$P_m(x,\xi)\, k_o(x,\xi) = 1$$

$$P_m(x,\xi)\, k_i(x,\xi) = -\sum_{j=0}^{i-1} P_{i-j}(x,\xi,D)\, k_j(x,\xi), \quad i > 0 .$$

In order to make the series in k convergent, we use truncature functions $\chi_i(\xi)$ which are identically zero in balls of radii $\rho_i$, $\rho_i \to +\infty$. We note $\ell(x,\xi) = \sum_i \chi_i(\xi)\, k_i(x,\xi)$ and we obtain :

$$P(x, D+\xi)\, \ell(x,\xi) = 1 - r(x,\xi).$$

The operator $\underline{R}$ defined by :

$$\underline{R}f(x) = (2\pi)^{-n} \int e^{ix.\xi} r(x,\xi)\, \hat{f}(\xi)\, d\xi$$

is a mapping from $\mathcal{E}'(\Omega)$ into $C^\infty(\Omega)$. We have obtained the relation :

$$P\,\underline{K} = I - \underline{R},$$

$\underline{K}$ is a parametrix of P, that is a pseudo-inverse.

1.3. - Wave equation. Consider the initial problem in $\mathbb{R} \times \mathbb{R}^n$ :

$$\begin{cases} \Delta u - \dfrac{\partial^2 u}{\partial t^2} = 0 \\ u(0,x) = 0 \\ \dfrac{\partial u}{\partial t}(0,x) = f , \qquad f \in C_0^\infty(\mathbb{R}^n). \end{cases}$$

By a Fourier transformation in x ,

$$\hat{u}_x(t,\xi) = A(\xi)\, e^{i|\xi|t} + B(\xi)\, e^{-i|\xi|t}$$

is obtained, with $A + B = 0$ and $i|\xi|(A - B) = \hat{f}$.

The solution is then given by the relation :

$$u(t,x) = (2\pi)^{-n} \Big[ \int e^{i(x.\xi + t|\xi|)} \frac{1}{2i|\xi|} \hat{f}(\xi)\, d\xi - \int e^{i(x.\xi - t|\xi|} \frac{1}{2i|\xi|} \hat{f}(\xi)\, d\xi \Big].$$

It is well known that the singularities are given at the critical points of the phase function. Here, these points are $x = \pm\, t \dfrac{\xi}{|\xi|}$, which specify the "light cone" in the (x, t)space : $|x|^2 = t^2$.

When t is small, every solution of the Cauchy problem with an hyperbolic operator is a sum of expressions such as :

$$A\, f(x) = \int e^{iS(x,\xi)}\, a(x,\xi)\, \hat{f}(\xi)\, d\xi ,$$

where $S(x, \xi)$ is the solution of the characteristic equation with initial data $S(x, 0) = \langle x, \xi \rangle$ .

The general form of the solution is

$$A\, f(x) = \iint e^{i\phi(x,y,\xi)}\, a(x,y,\xi)\, f(y)\, dy\, d\xi ,$$

where : $\phi(x,y,\xi) = S(x,\xi) - \langle y, \xi \rangle$ .

## 2. - Pseudo-Differential Operators (PDO's)

2.0. - Introduction. In what precedes, the solution of differential problems were given by oscillatory integrals $\int e^{i\varphi(x,\xi)}\, a(x,\xi)\, v(\xi)\, d\xi$.

In the simplest case, the phase function will be linear, roughly speaking as $< x, \xi >$. It is the case of PDO's.

The amplitude function, as it was shown in the study of an elliptic problem with smooth coefficients (1.2.b), is given by an infinite series whose convergence is not quite clear.

In a first section, we shall specify the class of such amplitudes with some details about their asymptotic development. Then, we shall define the admissible phase functions.

The second section gives the conditions of convergence of the oscillatory integrals.

Then , the Fourier integral operators can be introduced in the third section. They allow to consider very general problems. For elliptic problems (as the Helmhotz equation) it is sufficient to study the PDO's (with a linear phase function) which are introduced in section four. Their main properties are stated in the fifth section.

2.1. - Amplitudes and symbols - Phase functions -

Definition 1. Let m, $\rho$ , $\delta$ be real numbers, $0 \leqslant \delta < \rho \leqslant 1$. $S^m_{\rho,\delta}(\Omega \times R^N)$ is the set of all functions $a \in C^\infty(\Omega \times R^N)$ such that for every compact set $K \subset \Omega$ and all multi-indices $\alpha$ , $\beta$ , the estimate

$$|D^\beta_x \, D^\alpha_\xi \, a(x, \xi)| \leqslant C_{\alpha,\beta,K}(1 + |\xi|)^{m-\rho|\alpha| + \delta|\beta|}, \; x \in K, \xi \in R^N ,$$

is valid for some constant $C_{\alpha,\beta,K}$ .

$S^m_{\rho,\delta}$ is the class of amplitudes of order m and type ( $\rho$, $\delta$ ). When the dimension of $\Omega$ is N (when PDO's are considered), these amplitudes are called symbols.

We note :

$$S^{\infty}_{\rho,\delta} = \bigcup_m S^m_{\rho,\delta} \; ; \quad S^{-\infty}_{\rho,\delta} = \bigcap_m S^m_{\rho,\delta} \; .$$

Remarks :

- Positively homogeneous functions of degree m with respect to $\xi$ for large $|\xi|$ , $C^{\infty}$ in x and $\xi$ , belong to $S^m_{1,0}$. Such functions are sufficient to study elliptic operators.

- The parametrices of parabolic operators (as the heat operator) have amplitudes in the class $S^m_{1/2,1/2}$ that we do not consider here. Nevertheless many hypoelliptic operators have parametrices with amplitudes in $S^m_{\rho,\delta}$ , $\delta < \rho$ .

- In the sequel we shall say that a function is homogeneous if it is positively homogeneous for large $|\xi|$ . This(inexact) terminology is usual in the papers about PDO's.

To define the amplitudes (or symbols) as converging series, we need some description of the class $S^m_{\rho,\delta}$ . First, we must specify the topological structure of this class to be able to use convergence properties :

Proposition 1. $S^m_{\rho,\delta}$ $(\Omega \times \mathbb{R}^N)$ is a Frechet space with the topology defined by taking as semi-norms the best constants $C_{\alpha,\beta,K}$ which can be used in the definition of amplitudes . These spaces increase when $\delta$ or m increases or when $\rho$ decreases. If a $\in S^m_{\rho,\delta}$ , then

$$a^{\alpha}_{\beta} = (i\,D_{\xi})^{\alpha}\,(i\,D_x)^{\beta}\; a \in S^{m-\rho|\alpha|+\delta|\beta|}$$

and if b $\in S^{m'}_{\rho,\delta}$ , then $a\,b \in S^{m+m'}_{\rho,\delta}$ .

The amplitudes can be approximated by regular ones (what is useful in the calculus of oscillatory integrals), namely :

Proposition 2. Let $a \in S^{m}_{\rho,\delta}(\Omega \times \mathbb{R}^N)$ and $\chi \in S(\mathbb{R}^N)$, with $\chi(0) = 1$. Let us note $a_\varepsilon(x,\xi) = \chi(\varepsilon\xi)\, a(x,\xi)$. Then $a_\varepsilon \in S^{-\infty}_{\rho,\delta}$, and $a_\varepsilon \to a$ in $S^{m'}_{\rho,\delta}$ if $m' > m$.

It can be noticed that the operators with amplitude in $S^{-\infty}_{\rho,\delta}$ will be regularizing as the operators R in sections 1.2a and b.

Now we can define asymptotic expansions of amplitudes.

Definition 2. Let a and $(a_k)_{k=0,1,\dots}$ amplitudes belonging to $S^m_{\rho,\delta}$ and to $S^{m_k}_{\rho,\delta}$, $m_k \to -\infty$. We say that a is the asymptotic sum of the $a_k$, and write :

$$a \sim \sum_k a_k$$

if

$$a - \sum_0^{n-1} a_k \in S^{m_n}_{\rho,\delta} .$$

Let us notice that the series is not convergent in the usual sense, but it does not matter. The notion of convergence in definition 2 is sufficient for the sequel.

Proposition 3. Let $a_k \in S^{m_k}_{\rho,\delta}(\Omega \times \mathbb{R}^N)$, $m_k \to -\infty$, $m_{k+1} < m_k$. Then there exists $a \in S^{m_0}_{\rho,\delta}$ such that

$$a \sim \sum_k a_k .$$

a is uniquely defined modulo $S^{-\infty}$.

By a function of truncature as $\chi$ in 1.2.b, a series $(a'_k)$ can be defined, which converges (in an usual sense) to a, and $a_k - a'_k \in S^{-\infty}_{\rho,\delta}$

for every k. As a is uniquely defined modulo $S^{-\infty}_{\rho,\delta}$, it is clear that the formal convergence of definition 2 is sufficient as long as the calculus are done modulc $S^{-\infty}_{\rho,\delta}$

To end this section, let us define the phase functions.

Definition 3. A phase function on $\Omega \times \mathbb{R}^N$ is afunction $\varphi(x,\xi)$ such that :

- $\varphi$ is real, $C^\infty$ for $\xi \neq 0$, positively homogeneous of degree one with respect to $\xi$ ;

- if $\xi \neq 0$, $\varphi$ has no critical points, that is $d\varphi \neq 0$.

In the introduction, we considered two phase functions :

$\varphi(x,y,\xi) = < x - y, \xi >$ for the elliptic operators, and

$\varphi(x, y, t, \xi) = < x - y, \xi > \pm t|\xi|$ for the wave operator. In the first case $(x, y) \in \mathbb{R}^{2n}$, $\xi \in \mathbb{R}^n$. In the second one, $(x,y,t) \in \mathbb{R}^{2n+1}$, $\xi \in \mathbb{R}^n$. In the two cases $d\varphi \neq 0$ for $\xi \neq 0$, and the surfaces

$$\tilde{C}_\varphi = \{(X,\xi) ; \xi \neq 0 , \nabla_\xi \varphi = 0\} \ (X = (x,y) \text{ or } X = (x,y,t))$$

have projections in $\Omega$ which are respectively $\{(x,y) ; x = y\}$ and $\{(x,y,t) ; |x - y|^2 = t^2\}$. The surface $\tilde{C}_\varphi$ contains the bicharacteristics of the associated operator. We shall say that $\varphi$ is a "regular phase function" if $d(\frac{\partial\varphi}{\partial\xi_i})$ are linearly independant on $\tilde{C}_\varphi$, that is $\tilde{C}_\varphi$ is a surface of codimension N.

2.2. - Oscillatory integrals. Let $a \in S^m_{\rho,\delta}$ $(\Omega \times \mathbb{R}^N)$. To simplify the notations, we shall omit to write $\rho$, $\delta$ when there is no ambiguity. We want to define a distribution $T_a$ by the relation :

$$< T_a, u > = \iint e^{i\varphi(x,\xi)} a(x,\xi) u(x)\, dx\, d\xi ,$$

$u \in C_o^\infty(\Omega)$ , $\varphi$ being a phase function. This integral is convergent if $m + N < 0$. We shall extend it to every $a \in S^\infty$ , , by using partial integrations :

Proposition 4. It $\varphi$ is a phase function, there exists a differential operator $L = \Sigma\, a_j \frac{\partial}{\partial \xi_j} + \Sigma\, b_j \frac{\partial}{\partial x_j} + c$ , with $a_j \in S^o$ , $b_j$ and c in $S^{-1}$, such that

$$^tL\, e^{i\varphi} = e^{i\varphi} .$$

Then if $a \in S^{-\infty}$, the integral (which is convergent)

$$I_\varphi(au) = \iint e^{i\varphi(x,\xi)} a(x,\xi) u(x)\, dx\, d\xi$$

can be written

$$I_\varphi(au) = \iint e^{i\varphi(x,\xi)} L^k[a(x,\xi) u(x)]\, dx\, d\xi .$$

Now, if $a \in S^m_{\rho,\delta}$ (we know that there exists $a_\varepsilon \to a$, $a_\varepsilon \in S^{-\infty}_{\rho,\delta}$) , $L^k a \in S^{m-kt}_{\rho,\delta}$ , where $t = \min(\rho, 1-\delta)$ and the integral is convergent when k is sufficiently large.

Proof of the proposition . When $\xi \neq 0$, we know that

$$M(x,\xi) = \sum_{j=1}^{N} |\xi|^2 \left|\frac{\partial\varphi}{\partial\xi_j}\right|^2 + \sum_{j=1}^{n} \left|\frac{\partial\varphi}{\partial x_j}\right|^2 \neq 0 .$$

Let $h \in C_o^\infty(\mathbb{R}^N)$, $h \equiv 1$ in a neighbourhood of 0, and let us define (h is a function of $\xi$ ):

$$a'_j = \frac{i}{M}(1-h)\, |\xi|^2 \frac{\partial\varphi}{\partial\xi_j} \in S^o ,$$

$$b'_j = -\frac{i}{M}(1-h) \frac{\partial\varphi}{\partial x_j} \in S^{-1} .$$

Then, $[\sum_j a'_j \frac{\partial}{\partial\xi_j} + \sum_j b'_j \frac{\partial}{\partial x_j} + h]\, e^{i\varphi} = e^{i\varphi}$ .

Let $a_j = -a'_j, b_j = -b'_j, \; c = h - \sum_j \frac{\partial a'_j}{\partial \xi_j} - \sum_j \frac{\partial b'_j}{\partial x_j} \in S^{-1}$.

The corresponding operator L verifies $^tL e^{i\varphi} = e^{i\varphi}$. q.e.d.

We can deduce a first result :

Corollary 1. If $\varphi$ is a phase function, the integral $I_\varphi(au)$ can be extended to every $a \in S^\infty_{\rho,\delta}$ and $u \in C^\infty_o(\Omega)$. $I_\varphi(au)$ is a continuous function of $a \in S^m_{\rho,\delta}$, for every fixed m. The linear form $A : u \to I_\varphi(au)$ is a distribution of order less than k if $m - k\rho < -N$ and $m - k(1-\delta) < -N$.

Remarks : If $d_\xi \varphi \neq 0$ for $\xi \neq 0$ (i.e. $C_\varphi = \emptyset$) we could choose $M(x, \xi) = \sum_j |\xi|^2 \left|\frac{\partial \varphi}{\partial \xi_j}\right|^2 \neq 0$. Then $L = \sum a_j \frac{\partial}{\partial \xi_j} + c$, and

$$I_\varphi(au) = \int \left[ \int e^{i\varphi(x,\xi)} L^k a(x,\xi)\, d\xi \right] u(x)\, dx .$$

The function $I_\varphi(a) = \int e^{i\varphi(x,\xi)} L^k a(x,\xi)\, d\xi$ is $C^\infty$ in x, then the distribution A of the corollary is a $C^\infty$ function. More precisely if $\varphi$ is regular, one can prove that $I_\varphi(a)$ is $C^\infty$ if a and all its derivatives are zero on $\tilde{C}_\varphi$. The singularities of the oscillatory integrals are on $\tilde{C}_\varphi$.

2.3. - Fourier integral operators. The oscillatory integrals allow to define very general operators which contain differential operators and their parametrices.

Let $\Omega_1$ and $\Omega_2$ be open sets in $\mathbb{R}^{n_1}$ and $\mathbb{R}^{n_2}$, $\varphi$ a phase function defined in $\Omega_1 \times \Omega_2 \times \mathbb{R}^N$, $a \in S^m_{\rho,\delta}(\Omega_1 \times \Omega_2 \times \mathbb{R}^N)$.

Let us consider the operator

$$u \to Au(x) = \iint e^{i\varphi(x,y,\xi)} a(x,y,\xi)\, u(y)\, dy\, d\xi, \tag{1}$$

with $u \in C_o^\infty(\Omega_2)$, $x \in \Omega_1$, which can be written :

$$< Au, v > = \int\int\int e^{i\varphi(x,y,\xi)} \; a \,.\, v(x)\, u(y)\, dx\, dy\, d\xi \,, \tag{2}$$

with $u \in C_o^\infty(\Omega_2)$, $v \in C_o^\infty(\Omega_1)$.

The expression (2) is needed if Au is a distribution only. We shall say that a phase function $\varphi$ is an "operator phase function" if for each fixed x (respectively y), it has no critical point $(y, \xi)$ (respectively $(x, \xi)$) with $\xi \neq 0$. In that case (2) is not needed.

If $\varphi$ is an "operator phase function", the following theorem defines the Fourier integral operators :

THEOREM 1.

a) - The oscillatory integral (1) is well defined. If $\max(m - k\rho, m - k(1 - \delta)) < -N$, A is a continuous linear map from $C_o^k(\Omega_2)$ into $\mathscr{D}'^k(\Omega_1)$ with a distribution kernel K(A) given by the relation :

$$< K(A), u > = \int\int e^{i\varphi(x,y,\xi)} \; a(x,y,\xi)\, u(x,y)\, dx\, dy\, d\xi$$

with $u \in C_o^\infty(\Omega_1 \times \Omega_2)$

b) - When $\max(m - k\rho, m - k(1 - \delta)) < -N-j$, A is a continuous linear map from $C_o^k(\Omega_2)$ to $C^j(\Omega_1)$ and from $\mathscr{E}'^j(\Omega_2)$ to $\mathscr{D}'^k(\Omega_1)$.

c) - Let $R_\varphi = \{ (x,y) \in \Omega_1 \times \Omega_2 ;\ \varphi'_\xi \neq 0 \ \forall \xi \neq 0 \}$. The oscillatory integral

$$K_A(x, y) = \int e^{i\varphi(x,y,\xi)} \; a(x,y,\xi)\, d\xi \,, \quad (x,y) \in R_\varphi \,,$$

defines a function in $C^\infty(R_\varphi)$ which is equal to the distribution K(A). When $R_\varphi = \Omega_1 \times \Omega_2$, A is a continuous map from $\mathscr{E}'(\Omega_2)$ to $C^\infty(\Omega_1)$.

We remark that $C_\varphi = \mathrm{pr}_{\Omega_1 \times \Omega_2} \tilde{C}_\varphi$ is the complement of $R_\varphi$ . $C_\varphi$ defines a relation between $\Omega_1$ and $\Omega_2$ as follows :

If $K \subset \Omega_2$ ,

$$C_\varphi K = \{ x \in \Omega_1 ; (x, y) \in C_\varphi \text{ for some } y \in K \} .$$

Then, it is easy to prove the following result :

THEOREM 2. If $u \in \mathcal{E}'(\Omega_2)$, sing supp $Au \subset C_\varphi$ sing supp u.

By definition sing supp u is the closure of the points where u is not $C^\infty$ , as supp u is the closure of the points where u is not zero.

Remarks : We have already seen that in the case of PDO's (that is $\varphi(x,y\ \xi) = \langle x-y, \xi \rangle$), $C_\varphi$ is the diagonal in $\Omega_1 \times \Omega_2$ . The theorem 2 shows that for a PDO A,

sing supp A u $\subset$ sing supp u,

it is said that PDO's are pseudo-local operators (as differential operators are local, that is supp $Pu \subset$ supp u).

For the wave operator ( $\varphi(x,y,t,\xi) = \langle x-y, \xi \rangle \pm t\,|\xi|$ , $n_1 - 1 = n_2 = N$, and $C_\varphi$ is the light cone.

2.4. - Pseudo differential operators. For elliptic or hypoelliptic operators, it is sufficient to restrict ourselves to the PDO's, in a first time. The theory is not so rich, but it is very easier. We consider the operator $u \to Au$ defined by the oscillatory integral

$$Au(x) = \iint e^{i(x-y).\xi}\, a(x,y,\xi)\, u(y)\, dy\, d\xi ,$$

with $u \in C_0^\infty(\Omega)$, , $\varphi(x,y,\xi) = \langle x-y, \xi \rangle$ , $n_1 = n_2 = N = n$, $\Omega$ is an open domain in $\mathbb{R}^n$ and $a \in S^m_{\rho,\delta}(\Omega \times \Omega \times \mathbb{R}^n)$ :

We remark that A is defined on spaces of functions u with compact supports, but Au has not a compact support.

Then, it is not easy to define B(Au) directly, that is the composition of PDO's is not easily defined). In order to avoid this sort of difficulties (which are not essential), we use the following definition :

Definition 4. The PDO A is "properly supported" if the set $\{(x, y) \in \operatorname{supp} K(A) \; ; \; x \in L \text{ or } y \in L\}$ is compact for every compact $L \subset \Omega$ (i.e. both projections of supp K(A) in $\Omega$ are proper).

We recall that K(A) was defined in theorem 1 as the kernel of the operator A.

It is shown that every PDO is the sum of a properly supported operator and of a PDO with $C^\infty$ kernel (which maps $\mathcal{E}'$ in $C^\infty$). Moreover if u has compact support and A is properly supported, the Au also has compact support. As we shall do with operators modulo operators with $C^\infty$ kernels, henceforth we can consider the properly supported operators only.

The definition of PDO's via amplitudes makes some results easier. Nevertheless, this notion is not well adapted because the same operator can have many amplitudes. In the first section, operators A such as

$$Au(x) = \int e^{i\langle x,\xi\rangle} \alpha(x, \xi) \, \hat{u}(\xi) \, d\xi$$

were considered, which can be written :

$$Au(x) = \iint e^{i(x-y).\xi} \alpha(x, \xi) \, u(y) \, dy \, d\xi .$$

The function $\alpha$ is a symbol of A, and we shall see that each operator is defined by one symbol only. To explicit the relation between amplitudes and symbols, we remark that :

$$A(e^{i\,x.\eta}) = \sigma_A(x,\eta)\, e^{ix.\eta}$$

where $\sigma_A(x,\eta) = \iint a(x, x+y, \eta+\xi)\, e^{-iy.\xi}\, dy\, d\xi$.

In the introduction, we noticed that the symbol of a differential operator is a polynomial in $\xi$, $P(x, \xi)$ such that $P(e^{ix.\xi}) = P(\xi) \times e^{ix.\xi}$. It is natural to hope that $\sigma_A(x,\eta)$ is the symbol of A. A tedious calculus gives the following formal development of $\sigma_A$ :

$$\sigma_A(x,y) \sim (2\pi)^n \sum_{\alpha} (i\,D_\eta)^\alpha\, D_y^\alpha\, a(x,y,\eta)\, \frac{1}{\alpha!}\, \Big|_{y=x},$$

and obviously :

$$Au(x) = (2\pi)^{-n} \int e^{ix.\eta}\, \sigma_A(x,\eta)\, \hat{u}(\eta)\, d\eta .$$

Then, $\sigma_A(x,y) = (2\pi)^n\, \alpha(x,\eta)$.

$L^m_{\rho,\delta}$ is the class of PDO's with amplitude a in $S^m_{\rho,\delta}$ $(\Omega \times \Omega \times R^n)$.

Then, the following result can be proved :

<u>Proposition 5</u>. Let $A \in L^m_{\rho,\delta}(\Omega)$ be properly supported. Then A can be written in one and only one way as

$$Au(x) = (2\pi)^{-n} \int e^{ix.\xi}\, \sigma_A(x,\xi)\, \hat{u}(\xi)\, d\xi$$

and $\sigma_A \in S^m_{\rho,\delta}$ $(\Omega \times R^n)$. $\sigma_A$ is said the symbol of A.

$L^{-\infty}_{\rho,\delta}(\Omega)$ is the class of PDO's with amplitude in $S^{-\infty}$, they are infinitely regularizing operators which map $\mathcal{E}'(\Omega)$ into $C^\infty(\Omega)$. The map $A \to \sigma_A$ is an isomorphism of $L^m_{\rho,\delta}(\Omega)/L^{-\infty}_{\rho,\delta}(\Omega)$ onto

$$\frac{S^m_{\rho,\delta}(\Omega \times R^n)}{S^{-\infty}_{\rho,\delta}(\Omega \times R^n)}$$

that is the correspondence (modulo regularizing operators, with symbols in $S^{-\infty}$) between PDO's and symbols is one-to-one.

The transpose ${}^tA$ of a properly supported PDO A is still a PDO, with symbol

$$\sigma_{{}^tA}(x,\xi) \sim \sum_{\alpha} [(i D_\xi)^\alpha D_x^\alpha \sigma_A(x,-\xi)] \frac{1}{\alpha!}$$

If A and B are PDO's, A being properly supported, then BA is a PDO whose the symbol is :

$$\sigma_{BA}(x,\xi) \sim \sum_{\alpha} [(i D_\xi)^\alpha \sigma_B(x,\xi)] D_x^\alpha \sigma_A(x,\xi) \frac{1}{\alpha!}$$

Remarks - If $\sigma_A \in S^m_{\rho,\delta}$ , $D^\alpha_\xi D^\alpha_x \sigma_A \in S^{m+(\delta-\rho)|\alpha|}_{\rho,\delta}$ (see proposition 1 in section 2.1). It is obvious that the successive terms of the formal series defining $\sigma_{{}^tA}$ or $\sigma_{BA}$ have decreasing orders, as $\delta - \rho < 0$. When $\delta = \rho$ ,the theory is not so easy and needs many refinements.

- PDO's are a very general class of Fourier integral operators. In a precise manner, if a phase function $\varphi$ defined in $\Omega \times \Omega \times \mathbb{R}^n$ is linear in $\xi$ and if $C_\varphi = \{(x,x);\ x \in \Omega\}$ , then the corresponding Fourier integral operators with amplitudes in $S^m_{\rho,\delta}$ are PDO's belonging to $L^m_{\rho,\delta}(\Omega)$.

- All the definitions are invariant by diffeomorphism. Fourier integral operators and PDO's can be defined on varieties by using partitions of unity and charts. The symbols are functions defined on the cotangent space $T^*(\Omega) \setminus \{0\}$. The phase functions $\varphi$ can be defined globally (although the expression $< x-y, \xi >$ has no meaning on a variety).

2.5 - Continuity of PDO's. - The PDO's map $C^\infty_0(\Omega)$ into $C^\infty(\Omega)$ , $\mathcal{E}'(\Omega)$ into $\mathcal{D}'(\Omega)$. In this section we shall define functional spaces which allow to say how continuous the PDO's are. These Sobolev spaces are very well adapted to the theory because they specify the regularity of functions by using their Fourier transforms.

Definition 5 - Let s be a real number. $H^s(\mathbb{R}^n)$ is the space of tempered distributions u whose Fourier transform $\hat{u}$ verify the estimate :

$$\int |\hat{u}(\xi)|^2 (1 + |\xi|^2)^s \, d\xi = \|u\|_s^2 < \infty .$$

$\|u\|_s$ defines a norm in $H^s$. These spaces are Hilbert spaces with the scalar product $(u,v)_s = \int \hat{u}(\xi)\, \bar{\hat{v}}(\xi)(1 + |\xi|^2)^s \, d\xi$. It is easy to see that

$H^0(\mathbb{R}^n) = L^2(\mathbb{R}^n)$ ,

$$H^m(\mathbb{R}^n) = \{u \in L^2(\mathbb{R}^n);\ D^\alpha u \in L^2(\mathbb{R}^n)\ \forall |\alpha| \leqslant m\}.$$

Moreover $H^{-s}(\mathbb{R}^n)$ is the dual of $H^s(\mathbb{R}^n)$ via the duality product :

$$< u, v >_{H^s \times H^{-s}} = \int u.v\, dx .$$

Definition 6 - If $\Omega$ is a regular domain in $\mathbb{R}^n$ with boundary $\Gamma$ ( $\Gamma$ is a closed regular curve, $\Omega$ is on one side of $\Gamma$ ), $H^s(\Omega)$ , $s \geqslant 0$ , is the space of functions $u$ defined in $\Omega$ that can be extended to a function $\tilde{u} \in H^s(\mathbb{R}^n)$ .

Remarks - The condition on the regularity of $\Omega$ can be weakened, but $\Gamma$ must not be too bad - for example it may not be the trajectory of a brownian motion.

- It is easy to show that $H^2(\mathbb{R}^n) = \{u \in L^2 ;\ \Delta u \in L^2\}$. But if $\Omega \neq \mathbb{R}^n, H^2(\Omega)$ is strictly contained in $H(\Delta, \Omega) = \{u \in L^2(\Omega);\ \Delta u \in L^2(\Omega)\}$.

- $H^s(\Omega)$ is provided with the most natural norm :

$$\| u \|_s = \inf_{\tilde{u}} \| \tilde{u} \|_s$$

where the inf is taken on all the extensions $\tilde{u}$ of $u$ in $H^s(\mathbb{R}^n)$ .

- $C_o^\infty(\Omega)$ is dense in $H^s(\Omega)$ if $s \leqslant 1/2$ . For $s > 1/2$ , we denote $\overset{\circ}{H}{}^s(\Omega)$ the closure of $C_o^\infty(\Omega)$ in $H^s(\Omega)$ . This space can be considered as the space of functions $u$ in $H^s(\Omega)$ whose the traces of order $[s]$ (enter part of $s$) on $\Gamma$ are zero. The dual of $\overset{\circ}{H}{}^s(\Omega)$) is a space of distributions, as $C_o^\infty(\Omega)$ is dense in $\overset{\circ}{H}{}^s(\Omega)$ . .

Definition 7 - Let $s$ be a positive real number. $H^{-s}(\Omega)$ is the dual space of $\overset{\circ}{H}{}^s(\Omega)$ (if $s \leqslant 1/2$ , $\overset{\circ}{H}{}^s = H^s$) .

The Sobolev spaces with non-integer $s$ are usefull to define the traces on the boundary, thanks to the following result :

Theorem of traces - If $\gamma$ is the trace operator from $C^\infty(\overline{\mathbb{R}^n_+})$ to $C^\infty(\mathbb{R}^{n-1})$ defined by $\gamma u = (\gamma_o u, \ldots, \gamma_k u)$, where $(\gamma_k u)(x') = \frac{\partial^k}{\partial x_n^k} u(x', 0)$, and $x = (x_1, \ldots, x_n) = (x', x_n)$, then for every $s > (k + 1/2)$, $\gamma$ can be extended to a continuous surjective map from $H^s(\mathbb{R}^n_+)$ to $\prod_{i=0}^{k} H^{s-i-1/2}(\mathbb{R}^{n-1})$.

Now we can got the main results about the continuity of PDO's :

THEOREM 3 - Let A a PDO in $L^0_{\rho,\delta}(\Omega)$ whose the symbol is $\sigma_A$. Let us suppose that there exists a constant M such that for every compact K in $\Omega$, $\lim_{\xi\to\infty} [\sup_{x\in K} |\sigma_A(x,\xi)|] < M$,

then there exists a formally self-adjoint PDO R with $C^\infty$ kernel, such that for every $u \in L^2_c(\Omega)$ (supp u is compact in $\Omega$):

$$\|Au\|_0^2 \leqslant M^2 \|u\|_0^2 + (Ru, u).$$

Corollary 2 - With the same hypothesis as in the theorem, if moreover the kernel $K_A$ of the operator A has compact support, there exists a PDO $A_1$ such that $A - A_1 \in L^{-\infty}_{\rho,\delta}(\Omega)$, and $\forall u \in L^2_c(\Omega)$,

$$\|A_1 u\|_0 \leqslant M \|u\|_0 .$$

Corollary 3 - If $A \in L^m_{\rho,\delta}(\Omega)$, , A is a continuous operator from $H^s_c(\Omega)$ to $H^{s-m}_{loc}(\Omega)$.

We say that $u \in H^s_{loc}(\Omega)$ if $\varphi u \in H^s(\Omega)$ for every function $\varphi \in C^\infty_o(\Omega)$.

Corollary 4 - If $A \in L^0_{\rho,\delta}(\Omega)$ and if $\lim_{\xi\to\infty} \sup_{x\in K} \sigma_A(x,\xi) = 0$, then the PDO A is a compact linear map from $H^s_c(\Omega)$ into $H^s_{loc}(\Omega)$.

2.6 Elliptic PDO's. - Before studying the boundary value problems, let us consider the elliptic operators. The main idea is that, modulo regularizing operators, the elliptic operators are invertible. This allows to prove the regularity of elliptic problems and to reduce their resolution into a Fredholm alternative. To simplify, we suppose here that $\rho = 1$ and $\delta = 0$.

<u>Definition</u> 7 - A symbol $a \in S^m(\Omega)$ is elliptic if $|a| \gtrsim |\xi|^m$, i.e. for any compact set $K$ there exists a constant $C$ such that if $x \in K$ and $\xi$ is large, $a(x,\xi) \geq C|\xi|^m$. A PDO $A \in L^m$ is elliptic if its symbol $\sigma_A$ is elliptic.

<u>Remark</u> - If $a$ is elliptic, there exists $b \in S^{-m}$ such that $(ab-1) \in S^{-\infty}_{\rho,\delta}$

<u>Proposition</u> 6 - Let $A$ be an elliptic PDO of degree $m$. Then there exists a properly supported PDO $B$ such that $AB \sim BA \sim \mathrm{Id}$ ($A \sim B$ means that $A - B \in L^{-\infty}$).

<u>Proof</u> - We consider $\sigma_A^{-1}(x,\xi)$ for large $\xi$, and let $b_1(x,\xi) \sim \sigma_A^{-1}(x,\xi)$. It is the symbol of an operator $B_1 \in L^{-m}$ and $-R = A \circ B_1 - 1 \in L^{-1}(\Omega)$. Then $R^k \in L^{-k}(\Omega)$ and we consider a properly supported operator $B \sim \sum_{k=0}^{\infty} B_1 \circ R^k$. Then :

$$A \circ B \sim \sum_{k \geq 0} (A \circ B_1) \circ R^k = (1+R) \sum_{k \geq 0} R^k \sim \mathrm{Id}.$$

In the same way a left pseudo-inverse $B'$ can be constructed. Moreover

$$B' \circ A \circ B \sim B' \text{ , as } A \circ B \sim \mathrm{Id} \text{ , } B' \circ A \circ B \sim B \text{ , as } B' \circ A \sim \mathrm{Id},$$

then $B' \sim B$. <u>q.e.d.</u>

Such an operator $B$ is called a "parametrix" of the PDO A. Now we can give the main result of regularity :

THEOREM 4 - If $f \in \mathscr{D}'(\Omega)$ and if $A$ is a properly supported elliptic PDO, then $\text{sing supp } f \subset \text{sing supp } (Af)$, and there exists $u \in \mathscr{D}'(\Omega)$ such that

$Au = f \pmod{C^\infty}$ , that is $Au - f \in C^\infty(\Omega)$.

Remark - The first assertion means that f is as regular as Af. It is well known, for example, that if $\Delta u$ is regular, then u is regular.

- These results can be refined. If $Au \in H^s(\Omega)$ , then $u \in H^{s+m}(\Omega)$. When the operator is hypoelliptic and non-elliptic, the same results of regularity as in the theorem are valid, but if $Au \in H^s$, then $u \in H^{s+m'}$ with an $m' < m$. The proof is trivial by using a parametrix.

Now we can show how an elliptic problem on a compact variety can be changed into a Fredholm alternative. Let V be a smooth compact variety without boundary (for example a sphere). Let A an elliptic PDO on V, B a parametrix of A : $BA - I = R$ , R is a smoothing operator, mapping $\mathcal{D}'(V)$ into $C^\infty(V)$ . Then R is a compact operator from $C^\infty(V)$ into $C^\infty(V)$, and the kernel of A ($\{u; Au=0\}$) is a finite dimensional subspace of $C^\infty(V)$. In the same manner it is shown that ImA is closed, with finite codimension (this last assertion uses the adjoint operator $A^*$). The number

$$\chi(A) = \dim \ker(A) - \dim \operatorname{coker}(A)$$

is the index of A , it is attached to the symbol of A.

3 - Boundary value problems (BVP's) .

The aim of this chapter is to show how boundary value differential problems can be transformed into integral equations on the boundary. We shall need to define Poisson operators and trace operators as pseudodifferential operators. In the first chapter we have already recalled how a boundary value problem can be transformed into an integral one on the boundary in the case of the Laplacian. This will lead us in what follows.

$\Omega$ denotes a smooth domain with boundary $\Gamma$.

3.1 - Transmission property - Let $u \in C^\infty(\bar{\Omega})$ and $\tilde{u}$ an extension of $u$. If $P$ is a differential operator, $P\tilde{u}|_\Omega = Pu$. That is the values of $Pu$ inside $\Omega$ does not depend of the values of $\tilde{u}$ outside $\Omega$. For PDO's, this is false. To be able to solve BVP's with PDO's, we shall need "regular PDO's" i.e. operators with the following transmission property :

Definition 1 - The PDO $A$ defined in a neighborhood of $\bar{\Omega}$ has the "transmission property" along $\Gamma$ (or is "regular") if $P(u^o)|_\Omega \in C^\infty(\bar{\Omega})$ when $u \in C_o^\infty(\bar{\Omega})$, if $u^o$ denotes the extension of $u$ by zero in $\mathbb{R}^n \setminus \bar{\Omega}$.

The differential operator are regular PDO's. The following result can be proved :

Proposition 1 - If $A \sim \Sigma a_j$ is a PDO defined in a neighborhood of $\bar{\Omega}$, $m_j = \text{order}(a_j) \to -\infty$, then $A$ **is regular iff for every** $j$ the expression $\left[a_j(x, \xi', \xi_n) - e^{-im_j} a_j(x, \xi', -\xi_n)\right]$ and all its derivatives in $x_n$ and $\xi$ are zero in the variety $(x_n = 0, \xi' = 0, \xi_n > 0)$.

This means that the symbol is like a polynom around $x_n = 0, \xi' = 0$. One can show that this condition is equivalent to the following one :

The operator $u \to P(u^o)|_\Omega$ from $C_o^\infty(\bar{\Omega})$ into $C^\infty(\Omega)$ can be extended as an operator from $H_c^s(\bar{\Omega})$ into $H_{loc}^{s-s_o}(\bar{\Omega})$ for every $s > -1/2$.

3.2 Calderon's projectors - Here we develop the ideas that we introduced in the first chapter. We consider the BVP

$$\begin{cases} Au = 0 \text{ in } \Omega , \\ B_j u = g_j \text{ in } \Gamma . \end{cases}$$

$A$ is an elliptic differential operator of order $2m$. Let $u \in C_o^\infty(\bar{\Omega})$ and $u^o$ the extension of $u$ by zero to the whole of $\mathbb{R}^n$. The jump formula for every differential operator of degree $\ell$ can be written :

$$P(u^o) = (Pu)^o + \tilde{P}(\gamma u) \text{ , where , if order } (P_j) \leqslant j-1 \text{ and } D' = (D_1, \ldots, D_{n-1}) ,$$

$$\tilde{P}(\gamma u) = \frac{1}{i} \sum_{j=0}^{\ell-1} \sum_{k=0}^{\ell-1-j} P_{j+k+1}(D') D_n^j (u_\ell \otimes \delta)$$

with $u_k(x') = D_n^k u(x',0)$ and $D_n^j(u_k \otimes \delta) = u_k(x') \otimes \delta_{x_n}^j$. Then if $T$ is a parametrix of $A$, for $u \in C_o^\infty(\bar{\Omega})$, $\tilde{A}(\gamma u)$ is obviously in $C^\infty(\bar{\Omega})$ and $T\tilde{A}(\gamma u)$ also. We consider the successive traces $\gamma_k(T\tilde{A}(\gamma u))$ which will be shown to be well defined. If the coefficients of $A$ are constants, $T$ can be choosen as an elementary solution of $A$, then

$$TA(u^o) = u^o = T(Au)^o + T\tilde{A}(\gamma u) = T\tilde{A}(\gamma u)$$

and the traces of $u$ on $\Gamma$ are that of $T\tilde{A}(\gamma u)$. If we note

$\mathscr{C}^+(\gamma u) = (\gamma_k T\tilde{A}(\gamma u))_{k=0,\ldots,2m-1}$, then the BVP is equivalent to :

$$\mathscr{C}^+(\gamma u) = \gamma u ,$$

$$B_j(\gamma u) = g_j .$$

We note that if $\psi = (\psi_0, \ldots, \psi_{2m-1}) \in C_o^\infty(\Gamma)$, then when $\Omega = \mathbb{R}_+^n$,

$$(\mathscr{C}^+ \psi)_k = \lim_{x_n \to 0} D_{x_n}^k T\tilde{A}(\psi \otimes \delta_{x_n}) .$$

When $2m$ functions $\psi_i$ are given on the boundary, usually one can construct a function in $\Omega$ that verifies $Au = 0$. But it is not sure that the successive traces of $u$ are the functions $\psi_i$. This is the case if the $\psi_i$ are in the image of $\mathscr{C}^+$, that is $\mathscr{C}^+(\psi) = \psi$.

To develop the theory, a precise definition of Poisson operators (which map the boundary values into functions defined in $\Omega$ ) is needed.

3.3 - Poisson operators - In two examples we'll define Poisson kernels before giving a general definition. We choose $\Omega = \mathbb{R}^n_+$ , $\Gamma = \mathbb{R}^{n-1}$ .

Problem I - To find $u \in H^2(\Omega)$ such that

$$\begin{cases} \Delta u = 0 , \\ \gamma_0 u = \varphi \in H^{3/2}(\Gamma) . \end{cases}$$

Obviously the solution is

$$u(x', x_n) = K\varphi(x', x_n) = \frac{1}{(2\pi)^{n-1}} \int e^{ix'\xi'} e^{-x_n|\xi'|} \widehat{\varphi}(\xi')\, d\xi' .$$

The operator $K$ is said a "Poisson operator", it has a kernel, namely the Poisson kernel, known as $\frac{\partial}{\partial n} G_x(y)$ where $G$ is the elementary solution.

Problem II - To find an operator $K_j$ such that if $\psi$ is a function on $\Gamma$ , the function $u = K_j \psi \in C^\infty(\Omega)$ and verifies $\gamma_j u = \psi$ .

We'll use a truncature function $\alpha \in C_0^\infty(\overline{\mathbb{R}}_+)$ , $\alpha(0) = 1$ , and we put

$$\widehat{K_j\psi}(\xi', x_n) = \frac{1}{j!} x_n^j \alpha(x_n(1 + |\xi'|^2))\, \widehat{\psi}(\xi') = k_j(x, \xi')\, \widehat{\psi}(\xi') .$$

Then :
$$K_j \psi(x) = \frac{1}{(2\pi)^{n-1}} \int e^{ix'\xi'} k_j(x, \xi')\, \widehat{\psi}(\xi')\, d\xi' .$$

If $\widehat{\psi}(\xi')$ exists, obviously $\gamma_j(K_j\psi) = \psi$ , and $\psi$ is $C^\infty$ in $\Omega$ (when $x_n > 0$, $k_j(x, \xi')$ has a bounded support in $\xi'$).

We can now define the Poisson symbols $k_j$ such that the associated operators map $C_0^\infty(\Gamma)$ into $C^\infty(\Omega)$ , $\mathcal{E}'(\Gamma)$ into $\mathcal{D}'(\Omega)$ and $H_c^s(\Gamma)$ into $H_{loc}^{s-m+1/2}(\overline{\Omega})$.

Definition 2 - A Poisson symbol of degree $m$ is a function $k(x, \xi')$ in $C^\infty(\overline{\mathbb{R}}^n_+ \times \mathbb{R}^{n-1})$ such that for every integer $p$ , multi-indices $\alpha$ and $\beta$ , for every compact $K \subset \overline{\Omega}$ , there exists a constant $C$ such that :

$$\sup_{x \in K} |x_n^p D_x^\beta D_{\xi'}^\alpha k(x, \xi')| \leq C(1 + |\xi'|)^{m-|\alpha|-p+\beta_n}$$

$\mathcal{K}^m$ is the class of Poisson kernels of degree $m$.

Definition 3 - A smoothing Poisson operator is an operator $K$ defined by a kernel $k(x,y') \in C^\infty(\overline{\mathbb{R}}^n_+ \times \mathbb{R}^{n-1})$ :

$$K\varphi(x) = \int_\Gamma k(x, y')\ \varphi(y')\ d\Gamma(y') .$$

Such an operator maps $\mathcal{E}'(\Gamma)$ into $C^\infty(\bar{\Omega})$. The operator associated to the symbol $k(x, \xi')$ is noted $k(x,D)$.

Definition 4 - A Poisson operator of degree $m$ is the sum of an operator $k(x,D)$ associated to a symbol $k \in \mathcal{K}^m$, and a smoothing Poisson operator.

Remarks - As for the PDO's, it is shown that these definitions are invariant by diffeomorphisms. A smoothing Poisson operator has a null symbol. A Poisson symbol can be modified for small $\xi'$, because if $k(x, \xi')$ has a compact support in $\xi'$, the operator $k(x,D)$ is smoothing.

- If $k \in \mathcal{K}^m$, $k \sim \sum_{j=1}^{\infty} k_j$ with $k_j \in \mathcal{K}^{m_j}$, $m_j \to -\infty$.

Let us have a look at the both preceding examples :

Problem I - $k(x, \xi') = e^{-x_n|\xi'|}$. Then, by using that $t^k e^{-t} < \infty$ ,

$$\left| x_n^p\ D_{x_n}^{\beta}\ D_{\xi'}^{\alpha}\ e^{-x_n|\xi'|} \right| \leqslant C(1 + |\xi'|)^{-|\alpha| - p + |\beta|}$$

thus $k \in \mathcal{K}^o$.

Problem II - It is shown that $k \in \mathcal{K}^{-j}$ and $k \sim \sum_0^\infty k_q$, with

$$k_q = \sum_{p=0}^{q} \frac{1}{j!}\ x_n^j\ (x_n|\xi'|)^p\ |\xi'|^{-2q}\ a_{p,q}\ D^p \alpha(x_n|\xi'|) ,$$

where the $a_{p,q}$ are constant.

Remark - In these examples the symbols $k$ (problem I) and $k_q$ (problem II) are homogeneous in $(1/x_n, \xi')$. Such symbols are sufficient in the usual cases.

The Poisson operators are "good" operators. They have all the usual properties that we can expect :

- If $k \in \mathcal{K}^m$, it maps $H^s_c(\Gamma)$ into $H^{s-m+1/2}_{loc}(\overline{\Omega})$.

- If $k \in \mathcal{K}^m$, if $q \in S^{m'}_{\rho,\delta}(\mathbb{R}^{n-1} \times \mathbb{R}^{n-1})$, then $q(x',\xi')k(x,\xi')$ is in $\mathcal{K}^{m+m'}$, and $K \circ Q$ is a Poisson operator (if $Q$ is properly supported).

- If $P$ is a differential operator in a neighborhood of $\overline{\Omega}$, and $K$ a Poisson operator, $P \circ K$ is a Poisson operator.

We must now precise some results about traces and transmission properties.

Proposition 2 - If $K$ is a Poisson operator, with symbol $k(x,\xi')$, then $\gamma_0 \circ K$ is a PDO on $\Gamma$ with symbol $\sigma(x',\xi') = k(x',0,\xi')$.

If $Q$ is a PDO in $\mathbb{R}^n_+$ with symbol $q = \Sigma q_j$, each $q_j$ being a rational fraction, for every $u \in C^\infty_0(\mathbb{R}^{n-1})$, for every $\mu$ and $\nu$ in $\mathbb{N}$, one can define

$$Q^{\mu,\nu} u = \lim_{x_n \to 0_+} D^\mu_n Q(u \otimes \delta^\nu_{x_n}) \in C^\infty(\mathbb{R}^{n-1})$$

and $Q^{\mu,\nu} u$ is an OPD on $\mathbb{R}^{n-1}$ with symbol

$$\sigma(x',\xi') = \frac{1}{2\pi} \sum_j \int_{\Gamma_{\xi'}} (D_n + \xi_n)^\mu q_j(x',0,\xi) \xi^\nu_n \, d\xi_n$$

where $\Gamma_{\xi'}$ is a circle in the complex half-plane $\mathrm{Im}\,\xi_n > 0$, that encloses the poles of $q_j(x',0,\xi',\xi_n)$.

Proposition 3 - The PDO $P$ is regular iff for every integer $p$, the operator $K_p : u \to P(u \otimes \delta^{(p)})$ is a Poisson operator.

Proposition 4 - If $P$ is a regular PDO and $K$ a Poisson operator, if $P_\Omega$ is defined by $P_\Omega u = P(u^o)|_\Omega$, then $P_\Omega \circ K$ is also a Poisson operator.

At last, the following result is very usefull :

Proposition 5 - The parametrix of a regular elliptic PDO is regular.

We have defined two classes of operators : the PDO's and the Poisson operators. When the PDO's are regular, they can be composed with Poisson operators. Thanks to the second proposition, we may define the traces of Ku , K being a Poisson operator. All that is the foundation of a mathematical theory well adapted to the solution of elliptic problems.

Remark - We can also define an algebra of operators which contains all these operators. To do it, we must still precise two notions :

-A "trace operator" is an operator from $C_0^\infty(\bar{\Omega})$ into $C^\infty(\Gamma)$:

$$Tu = \Sigma\, Q_i\,[(P_i u)|_\Gamma] \quad ,$$

where the $P_i$ are regular PDO's in $\Omega$ and the $Q_i$ are PDO's on $\Gamma$ . The class of trace operators contains the usual traces and the transposes of Poisson operators.

-A "singular Green operator" is, roughly speaking, a sum $\Sigma\, K_i T_i$ , where the $T_i$ are trace operators of degree zero, and the $K_i$ are Poisson operators.

Then if P is a regular PDO in $\Omega$ , G a singular Green operator, K a Poisson operator, T a trace operator, and Q a PDO in $\Gamma$ , the following matrix operator

$$A = \begin{pmatrix} P_\Omega + G & K \\ T & Q \end{pmatrix}$$

is an operator well defined on $C_0^\infty(\bar{\Omega}) \oplus C_0^\infty(\Gamma)$ with values in $C^\infty(\bar{\Omega}) \oplus C^\infty(\Gamma)$ . These operators generate an algebra and can be extended to Sobolev spaces.

A is said elliptic if there exists an inverse of A (modulo smoothing operators) in the algebra.

In the first chapter and in the last section 3.2, we introduced the Calderon's projectors. Thanks to the theory of PDO's, the reduction of an elliptic boundary value problem to an integral one is well defined. It is known that sometimes (for example for the Neumann problem with double-layer potentials) the singularity of the kernel of the integral operator is a source of great difficulties. In the context of PDO's, there is no matter. All the encountered integral operators are well defined as PDO's or Poisson operators,trace operators, singular Green operators. Their definition is independant of the coordinates. The problem of numerical calculus is not solved here. On the other hand, the theory that we developed allows to choose the functional spaces correctly and to obtain "well-posed" problems.

Unfortunately, when $\Gamma$ is not smooth, the PDO's are not defined. They still give a good look at what happens, but they may not be applied directly. We'll only give an example.

## III. - SCATTERING OF AN ACOUSTICAL WAVE BY A THIN OBSTACLE

Let S the source of an acoustical wave scattered by a thin screen which is perfectly reflecting. The screen is put over a perfectly reflecting ground. The boundary conditions are then Neumann conditions. Thanks to a symmetry with respect to the ground, the differential problem is put in the whole space. We consider the case of the space $\mathbb{R}^2$ only, to simplify the problem. If $u$ is the speed potential and $f$ the source vibration, we must consider the following problem :

$$\begin{cases} (\Delta + k^2)\, u = f \quad \text{in} \quad \Omega = \mathbb{R}^2 \setminus \Gamma \\ \partial_n u\big|_\Gamma = 0\,, \\ \text{Sommerfeld conditions at infinity.} \end{cases}$$

$\Gamma$ is a segment of a piecewise $C^1$-curve in $\mathbb{R}^2$, supp $f \cap \Gamma = \emptyset$. The Sommerfeld conditions are the outgoing wave conditions given by one among the three principles (radiation, limiting amplitude or limiting absorption) introduced in the first chapter.

The problem is changed into the following one :

$$
(D) \begin{cases}
\text{For every distribution } \chi \text{ in } H^{-1/2}(\Gamma), \text{ to find } u \text{ such that} \\
\varphi u \ H^1(\Omega) \text{ for every function } \varphi \in C_o^\infty(\mathbb{R}^2)), \text{ such that :} \\
(\Delta + k^2)\, u = 0 \quad \text{in} \quad \Omega, \\
\quad \partial_n u \big|_\Gamma = \chi, \\
\text{Sommerfeld conditions.}
\end{cases}
$$

The functional spaces have been chosen to obtain a "well posed" problem. This problem is not very easy, because the domain $\Omega$ is irregular. Such domains are forbidden in the usual studies about partial differential equations (cf. J.L. Lions and E. Magenes [12]), mainly because the domain is on the two sides of its boundary $\Gamma$.

The traces of $u$ and of the normal derivatives $\partial_n u$ on $\Gamma$ are given by the Plemelj's formulas :

If $u(x) = -\int_\Gamma \mu(y)\, \partial_{n_y} G_x(y)\, d\Gamma(y)$, $x \in \Omega$,

then:

$$\operatorname{Tr} u(x) = \frac{\mu}{2}(x) - \int_\Gamma \mu(y)\, \partial_{n_y} G_x(y)\, d\Gamma(y),$$

$$\operatorname{Tr} (\partial_n u)(x) = -\lim_{\tilde{x} \to x} \int_\Gamma \partial_{n_x} \partial_{n_y} G_{\tilde{x}}(y)\, \mu(y)\, d\Gamma(y),$$

where $G_x(y) = G(|x - y|) = H_o^{(1)}(k|x - y|)$ is the Green function of the operator $(\Delta + k^2)$ that verifies the Sommerfeld's conditions at infinity. Then as

$$u(x) = -\int_\Gamma \mu(y)\, \partial_{n_y} G_x(y)\, d\Gamma(y) \ ,$$

where $\mu$ is the jump of $u$ across the screen $\Gamma$, the problem (D) generates the following integral equation :

$$\lim_{\tilde{x} \to x} \int_\Gamma \partial_{n_x} \partial_{n_y} G_{\tilde{x}}(y)\, \mu(y)\, d\Gamma(y) = -\chi(x) \ .$$

Such an integral operator is not well defined, as the kernel singularity is like $\frac{1}{r^2}$ and $\dim(\Gamma) = 1$. We should introduce the Hadamard's finite parts, but these finite parts are not defined when the exponent of the singularity is an entire number, what is our case!

On the other hand, the thin screen can be replaced by a sequence of thick screens $\Gamma_\varepsilon$, which are regular and enclose domains $\tilde{\Omega}_\varepsilon$ whose the exterior is $\Omega_\varepsilon$. The associated differential problems are well defined :

$$(D_\varepsilon) \begin{cases} \text{For every distribution } \chi^\varepsilon \text{ in } H^{-1/2}(\Gamma_\varepsilon) \text{, to find } u^\varepsilon \text{ such that} \\ \varphi u^\varepsilon \in H^1(\Omega_\varepsilon) \text{ for every function } \varphi \in C_0^\infty(\mathbb{R}^2) \text{, such that :} \\ (\Delta + k^2)\, u^\varepsilon = 0 \quad \text{in } \Omega_\varepsilon , \\ \qquad \partial_n u^\varepsilon \big|_{\Gamma_\varepsilon} = \chi^\varepsilon , \\ \text{Sommerfeld conditions.} \end{cases}$$

Thanks to the PDO's theory, the Poisson operators and the Calderon's projectors, the problems $(D_\varepsilon)$ are easily shown to be equivalent to the integral problems :

$$(K_\varepsilon) \lim_{\tilde{x}\to x} \int_{\Gamma_\epsilon} \partial_{n_x} \partial_{n_y} G_{\tilde{x}}(y)\, \mu^\varepsilon(y)\, d\Gamma_\varepsilon(y) = K_\varepsilon \mu^\varepsilon(x) = -\chi(x) .$$

The integral operator is well defined, it is a PDO of degree 1 (locally its principal symbol is $\sigma_{K_\varepsilon} = |\xi|$ ).

Afterwards $\varepsilon$ tends to zero. This part of the work is relatively delicate. The solutions of the problems $(D_\varepsilon)$ must tend to the solution of the problem (D). Problems $(D_\varepsilon)$ are "well posed" (i.e. existence, uniqueness, continuous dependance of the data) and this property must remain true for the limit problem (D). The key of the proof is an a priori estimate :

If $u^{\varepsilon,z}$ is the solution of the problem $(D_{\varepsilon,z})$ ( problem $(D_\varepsilon)$ with $k^2 = z$, $z \notin \mathbb{R}$) ) and $\chi^\varepsilon$ the data of the boundary condition, then for every ball $B_R$ centered at the origin and with radius $R$ that contains the curves $\Gamma_\varepsilon$ , there exists a constant $C$ independent of $\varepsilon$ and $z$ such that :

$$\|u^{\varepsilon,z}\| + \|\nabla u^{\varepsilon,z}\| + \|\Delta u^{\varepsilon,z}\| \leqslant C\, \|\chi^\varepsilon\|_{H^{-1/2}(\Gamma_\varepsilon)} ,$$

wherethe norms in the left member are the usual norms in the space $L^2(\Omega_\varepsilon \cap B_R)$. This uniform estimate is proved by using the limiting absorption principle. This estimate allows $\varepsilon$ to go to zero.

On the other hand, problems $(K_\varepsilon)$ approximate a problem (K) which must preserve the properties of problems $(K_\varepsilon)$ . To show the convergence of problems $(K_\varepsilon)$, the regular curves $(\Gamma_\varepsilon)$ are mapped onto the unit circle $\mathbb{T}$ . The transformed problems are defined on the set $\mathbb{T}$ . Then convergence results are obtained. The problems $(K_\varepsilon)$ approximate a problem (K) defined on the curve $\gamma$ which represents the curve $\Gamma$ traveled on the two sides.

The operator K is not a classical PDO because the curve $\gamma$ is not regular (cf.the end of chapter II).

Problems $(D_\varepsilon)$ and $(K_\varepsilon)$ are equivalent, that is the solution of $(D_\varepsilon)$ generates that of $(K_\varepsilon)$ and the converse is true. Problems (D) and (K) must preserve this property.

At last the problem (K) is changed into the problem (I) by an easy calculation. We obtain the following result :

THEOREM - The differential problem (D) is equivalent to the following integral one:

$$
(\mathrm{I}) \quad \begin{cases} \text{For every } \chi \in H^{-1/2}(\Gamma), \text{ to find } \mu \in \mathring{H}^{1/2}(\Gamma) \text{ such that} \\[2ex] \lim\limits_{\tilde{x}\to x} \int_\Gamma \partial_{n_x} \partial_{n_y} G_{\tilde{x}}(y)\, \mu(y)\, d\Gamma(y) = I\mu = -\chi . \end{cases}
$$

When $\Gamma = (0,1)$, $\mathring{H}^{1/2}(\Gamma)$ is the space of functions $u \in L^2(0,1)$ such that, if $\tilde{u}$ is an extension of $u$ by zero to $\mathbb{R}$, then $\tilde{u} \in H^{1/2}$. This space is strictly included in $H^{1/2}(0,1)$. When $\Gamma$ is any curve, $H^{1/2}(\Gamma)$ is defined in the same manner.

It is shown that the integral problem (I) has one and only one solution in $\mathring{H}^{1/2}(\Gamma)$.

Remarks - These results are not so easy than we could hope. The curve $\gamma$ is too irregular. When the obstacle is regular and thick, the PDO's theory gives the results in a quite simple manner : the integral problems are well defined and the solutions are obtained in the good functional spaces.

When the obstacle is thin, a complete study needs working with functional analysis. Thanks to the PDO's, we can well understood the problem; it is possible to define it in a precise manner. We can say that the PDO's lead us to the solution. To obtain a very precise formulation, in a mathematical manner, the PDO's theory is necessary but not sufficient, a framework of functional analysis is also needed.

BIBLIOGRAPHY


1. FILIPPI P.J.T., Layer potentials and acoustic diffraction, *J.S.V.*, 54 (4), 1-29, 1977.

2. CHVEDELIDZE, B.V., Linear discontinuous boundary value problems in function theory, singular integral equations and some of its applications (in Russian), *Tr. Tbilisk. Matem. In. Ia An Gruz, SSR* , 23, 3-158, 1956.

3. HÖRMANDER, L., Fourier integral operators I, *Acta Math.*, 127, 79-183, 1971.

4. BONY, J.M., Calcul symbolique et propagation des singularités pour les équations aux dérivées partielles non linéaires,*Ann.Sci.Ec.Norm. Sup.*, 4th sery, 14, 209-246, 1981.

5. VAINBERG, B.R.,Principles of radiation, limit absorption and limit amplitude in the general theory of partial differential equations, *Russ. Math. Surveys*, 21 (3), 115-193, 1966.

6. CALDERON, A.P., Communication at the "Joint Symposium of the Russian and American mathematical Society", Novisibirsk, 1963.

7. BOUTET DE MONVEL, L. Comportement d'un opérateur pseudo-différentiel sur une variété à bord, *J. Anal. Math*, 17, 241-304, 1966.

8. BOUTET DE MONVEL, L. Boundary problems for pseudo-differential operators, *Acta Mathematica*, 126, 11-51, 1971.

9. DURAND, M., Layer potentials and boundary value problems for the Helmholtz equation in the complement of a thin obstacle, to appear in *Math. Meth. Appl. Sci.*

10. SEELEY, R.T. Topics in pseudo-differential operators, in *Pseudo-Differential Operators*, CIME, Edizioni Cremonese, Roma, 1969.

11. TREVES, F. *Introduction to pseudo-differential and Fourier integral operators*, vol. 1, Plenum Press, New-York, 1980.

12. LIONS, J.L. and MAGENES, E. *Problèmes aux limites non homogènes et applications*, vol I, Dunod, Paris, 1968.

# Parametrices, singularities, and high frequency asymptotics in the theory of sound waves

Hans-Dieter Alber
Department of Mathematics
University of Bonn
Wegelerstr. 10
5300 Bonn 1, West-Germany

## 1. The acoustic equation

Let $\Omega \subseteq R^3$ be an unbounded domain with bounded complement $B = R^3 \setminus \bar{\Omega}$ and boundary $\partial\Omega \in C^\infty$. Let $a,b \in C^\infty(\bar{\Omega})$ with $C_1 \leq a(x), b(x) \leq C_2$ for $x \in \Omega$ and for suitable constants $C_1, C_2 > 0$. In the next section I want to study the singularities of solutions $v(x,t)$ of the following Dirichlet and Neumann problems

$$\partial_t^2 v - b\nabla(\frac{1}{a}\nabla v) = f(x,t) \text{ in } \Omega \times R^+$$

$$v|_{\partial\Omega \times R^+} = 0$$

$$v(x,o) = v_o(x), \ \partial_t v(x,o) = v_1(x)$$

and

$$\partial_t^2 v - b\nabla(\frac{1}{a}\nabla v) = f(x,t) \text{ in } \Omega \times R^+$$

$$\partial_n v(x,t) = 0 \qquad (x,t) \in \partial\Omega \times R^+$$

$$v(x,o) = v_o(x), \; \partial_t v(x,o) = v_1(x)$$

where f, $v_o, v_1$ are given functions, and where $\partial_n$ denotes the derivative of v with respect to the normal n at the boundary $\partial\Omega \times R^+$. To do this, I need results from the solution theory, and in particular, from the regularity theory of these problems. These results are well known and can be proved, for example, by semi-group theory. Therefore, in this introductory section I only give some definitions and state the results needed later on.

The differential operator $b\nabla(\frac{1}{a}\nabla)$ is symmetric on the space $C_o^\infty(\Omega)$ with the scalar product

$$(\frac{1}{b}u,v)_\Omega = \int_\Omega \frac{1}{b(x)} u(x)\, \bar{v}(x)dx.$$

By $A_D : D(A_D) \subseteq L_2(\Omega) \to L_2(\Omega)$ and $A_N : D(A_N) \subseteq L_2(\Omega) \to L_2(\Omega)$ I denote the selfadjoint extensions of this differential operator corresponding to the Dirichlet boundary value problem and to the Neumann boundary value problem. In the following definitions and statements I drop the subscripts D and N, if these definitions, statements, and results hold both in the Dirichlet case and in the Neumann case.

For $T > 0$ let $Z_T = \Omega \times (0,T)$. Let $f : Z_T \to C$ and $t \in (0,T)$. For brevity I denote by f(t) the function $x \to f(x,t) : \Omega \to C$. $H_m(\Omega)$, $H_m(Z_T)$ are the usual Sobolev spaces. To state the following theorem in a simple way, I introduce the following notation:

Let $f \in L_2(Z_T)$ with $\partial_t^j f \in L_2(Z_T)$, $j=1,\dots,m$. For $1 \leq j \leq m$ I define the operator

$$F_j : D(F_j) \subseteq H_1(\Omega) \times L_2(\Omega) \to H_1(\Omega) \times L_2(\Omega)$$

with

$$D(F_j) = D(A) \times H_1(\Omega)$$

by

$$F_j \begin{pmatrix} v_1 \\ v_2 \end{pmatrix} = \begin{pmatrix} 0 & 1 \\ A & 0 \end{pmatrix} \begin{pmatrix} v_1 \\ v_2 \end{pmatrix} + \begin{pmatrix} 0 \\ \partial_t^{j-1} f(o) \end{pmatrix}.$$

Now I have the following result:

<u>Theorem 1</u>: Let $m \geq 1$, let $f \in H_{m-1}(Z_T)$ and in addition let $\partial_t^m f \in L_2(Z_T)$. For the initial conditions $u_o, u_1$ assume that

$$\begin{pmatrix} u_o \\ u_1 \end{pmatrix} \in D(F_m \circ \dots \circ F_1).$$

Then there exists a unique solution $u \in H_{m+1}(Z_T)$ of

$$\begin{aligned} &(\partial_t^2 - A)u = f \\ &u(0) = u_o,\ u_t(0) = u_1 \end{aligned} \tag{1}$$

with $u(t) \in D(A)$, for all $t \in (0,T)$.

The following energy inequality shows that a domain of dependence exists for this solutions. This domain of dependence is determined by characteristic manifolds to the hyperbolic operator $\partial_t^2 - b\nabla(\frac{1}{a}\nabla)$. Let $\chi : \Omega \to R^+$ be a continuous and piecewise continuously differentiable function, which satisfies piecewise the eikonal equation

$$|\nabla\chi|^2 = \frac{a}{b}.$$

The manifold $\{t = \chi(x)\}$ is then characteristic.

Let $K = \{(x,t) \in \Omega \times R^+ | t < \chi(x)\}$, and for $t_o > 0$ let $K_{t_o} = \{x \in \Omega \mid (x,t_o) \in K\}$. I call the set $K \cap \{0 < t < t_o\}$ the backward propagation cone associated with $K_{t_o}$. It is clear that a forward propagation cone associated with $K_{t_o}$ can be defined similarly. The following energy inequality holds:

Theorem 2: Let u be the solution of (1). Then

$$||u(t)||_{E,K_t} \leq ||u(0)||_{E,K_o} + \int_o^t ||b^{1/2} f(\tau)||_{K_\tau} d\tau,$$

where

$$||v||^2_{E,K_t} = (\frac{1}{b} v_t, v_t)_{K_t} + (\frac{1}{a} \nabla v, \nabla v)_{K_t}.$$

The equation (1) also can be solved if f is a distribution. In this case also the solution u is a distribution. First I have to introduce a suitable space of distributions. Let $S_D = S_D(\Omega \times R)$ be the space of functions $\phi \in C^\infty(\bar{\Omega} \times R)$ with $[b\nabla(\frac{1}{a}\nabla)]^m \phi|_{\partial\Omega \times R} = 0$ for all $m = 0,1,2,\ldots$, and with

$$p_{\alpha m}(\phi) = \sup_{z\in\Omega\times R} (1 + |z|)^m D^\alpha\phi(z)| < \infty$$

for all multi-indices $\alpha \in N_o^4$ and $m \geq 0$.

The space $S_N$ is defined in the same way with the Neumann boundary condition $\partial_n[b\nabla(\frac{1}{a}\nabla)]^m \phi(x,t) = 0$, $x \in \partial\Omega$, instead of the Dirichlet condition.

Supply $S$ with the topology induced by the semi-norms $p_{\alpha m}$ and let $S'$ denote the dual of $S$. For $u \in S'$ I define $\partial_t u$ and $Au$ by

$$(\partial_t u,\phi) = -(u,\partial_t \phi)$$

$$(Au,\phi) = (u,A\phi).$$

Let $\tau \in R$ and let $\in S'$ with $f = 0$ for $t < \tau$.

$u \in S'$ is called distribution solution of (1), if

$$\begin{aligned}(\partial_t^2 - A)u &= f \\ u &= 0 \ , \ t < \tau\end{aligned} \tag{2}$$

is satisfied. Note that a classical solution v of (1) also defines a distribution solution $v' \in S'$ in the sense of (2), if $v'$ is defined by

$$(v',\phi) = \int_{-\infty}^{\infty} \int_{\Omega} \frac{1}{b(x)} v(x,t)\phi(x,t)dx\,dt,$$

for $\phi \in S$.

Theorem 3: For every $f \in S'$ with $f = 0$ for $t < \tau$ there exists a uniquely defined solution $u \in S'$ of (2).

## 2. Propagation of singularities

Let $x_o \in \Omega$ and $R \in S'$ be the solution of

$$\begin{aligned}(\partial_t^2 - A)\ R &= \delta(t)\ \delta(x-x_o) \\ R &= 0 \ , \ \text{for } t < 0.\end{aligned}$$

According to theorem 3 this solution exists and is uniquely defined. I call R the Riemann function to the Dirichlet or Neumann problem of the hyperbolic differential operator $\partial_t^2 - b\nabla(\frac{1}{a}\nabla)$. I want to determine the singularities of R. If $a = b = 1$, then R is the Riemann function to the wave equation, and it is well known that

$$R(x,x_o,t) = \frac{\delta(t-|x-x_o|)}{4\pi|x-x_o|}$$

for small t, as long as the support of R does not intersect the boundary $\partial\Omega \times \mathbb{R}$. In this case R can also be represented by the Fourier integral

$R(x,x_o,t) =$

$$= (2\pi)^{-3} \int_o^\infty \int_{|\omega|=1} (e^{i\zeta(t-\omega(x-x_o))} - e^{i\zeta(-t-\omega(x-x_o))}) \frac{\zeta}{2i} \, d\omega \, d\zeta, \quad t > 0,$$

where this integral has to be explained siutably as a distribution by applying it to a test function. Note that the expression under the integral sign is a solution of the wave equation which depends on parameters $\omega,\zeta$. Having this in mind I want to do the same thing for general a,b. I want to represent R in the form

$$R(x,t) = \int_o^\infty \int_{|\omega|=1} e^{i\zeta\chi^+(x,t,\omega)} \sum_{m=-1}^{\infty} \zeta^{-m} z_m^+(x,t,\omega,\zeta) d\omega \, d\zeta$$
$$+ \int_o^\infty \int_{|\omega|=1} e^{i\zeta\chi^-(x,t,\omega)} \sum_{m=-1}^{\infty} \zeta^{-m} z_m^-(x,t,w,\zeta) d\omega \, d\zeta, \tag{3}$$

where the functions $\chi^\pm$, $z_m^\pm$ have to be determined suitably, such that the expressions under the integral signs, which I denote by $v^\pm(x,t,\omega,\zeta)$, are solutions to the equation $(\partial_t^2 - A)\, v = 0$, for $t \geq 0$, and such that

$$\begin{aligned} &v^+(x,o,\omega,\zeta) + v^-(x,o,\omega,\zeta) = 0 \\ &\partial_t v^+(x,o,\omega,\zeta) + \partial_t v^-(x,o,\omega,\zeta) = (2\pi)^{-3} \zeta^2 e^{-i\zeta\omega(x-x_o)}, \end{aligned} \tag{4}$$

at least for all x in a neighbourhood of $x_o$. A simple calculation shows that

$$(\partial_t^2 - A)\, v^{\pm} =$$

$$= e^{i\zeta\chi} \sum_{m=-1}^{\infty} [-(\chi_t^2 - \frac{b}{a}\,|\nabla\chi|^2)\zeta^{2-m} z_m$$

$$+ [2(\chi_t\, \partial_t\, z_m - \frac{b}{a}\,\nabla\chi\cdot\nabla z_m) + ((\partial_t^2 - A)\chi)z_m]i\zeta^{1-m} \tag{5}$$

$$+ (\partial_t^2 - A)z_m\, \zeta^{-m}].$$

I rearrange this series according to the powers of $\zeta$. $v^{\pm}$ is a solution to $(\partial_t^2 - A)v = 0$ if the new series vanishes term by term. The term of highest order with respect to $\zeta$ vanishes, if

$$\chi_t^{\pm 2} - \frac{b}{a}\,|\nabla\chi^{\pm}|^2 = 0, \tag{6}$$

the other terms vanish if

$$2(\chi_t^{\pm}\, \partial_t\, z_m^{\pm} - \frac{b}{a}\,\nabla\chi^{\pm}\nabla z_m^{\pm}) + ((\partial_t^2 - A)\chi^{\pm})z_m^{\pm} = i(\partial_t^2 - A)z_{m-1}^{\pm} \tag{7}$$

for $m \geq -1$, where $z_{-2} = 0$.

(6) is a first order partial differential equation for $\chi^{\pm}$, the characteristic equation for the hyperbolic operator $\partial_t^2 - b\nabla(\frac{1}{a}\,\nabla)$. If $\chi^{\pm}$ has been determined from this equation, it can be inserted (7), which then is a system of first order partial differential equations for the functions $z_m^{\pm}$, which can be solved recursively. To satisfy (4) I require

$$\chi^{\pm}(x,o,\omega) = -\,\omega\cdot(x-x_o). \tag{8}$$

(6) then implies that

$$\chi_t^+(x,o,\omega) = -\chi_t^-(x,o,\omega) = \left(\frac{b(x)}{a(x)}\right)^{1/2} . \tag{9}$$

It then follows

$$\partial_t v^+(x,o,\omega,\zeta) + \partial_t v^-(x,o,\omega,\zeta) =$$

$$= e^{-i\zeta\omega\cdot(x-x_0)}[i\zeta^2(\chi_t^+ z_{-1}^+ + \chi_t^- z_{-1}^-)$$

$$+ \sum_{m=-1}^{\infty} \zeta^{-m}(\partial_t z_m^+ + \partial_t z_m^- + i\chi_t^+ z_{m+1}^+ + i\chi_t^- z_{m+1}^-)] .$$

Thus (4) is satisfied for all x in a neighbourhood of $x_o$ and for all $\zeta \geq 1$ if

$$z_m^+(x,o,\omega,\zeta) = -z_m^-(x,o,\omega,\zeta), \tag{10}$$

and

$$\begin{aligned} &\chi_t^+ z_{-1}^+(x,o,\omega,\zeta) + \chi_t^- z_{-1}^-(x,o,\omega,\zeta) = -i H_1(x)H_2(\zeta) \\ &i\chi_t^+ z_{m+1}^+ + i\chi_t^- z_{m+1}^- + \partial_t z_m^+ + \partial_t z_m^- = 0, \text{ for } t = 0, \end{aligned} \tag{11}$$

where the function $H_1 \in C_o^\infty(\Omega)$ is equal to $(2\pi)^{-3}$ in a neighbourhood of $x_o$, and where $H_2 \in C^\infty(R)$ with

$$H_2(s) = \begin{cases} 0 , & s < 1/2 \\ 1 , & s > 1 \end{cases} .$$

(10) and (11) are sufficient to determine $z_m^\pm$ at t= 0 completely. (8), (9), and (11) thus yield the initial conditions at t = 0 for the solutions of the first order equations (6) and (7). By (10), (11) and (7) the functions $z_m(x,t,\omega,\zeta)$ vanish identically if $\zeta < 1/2$, since $H_2(\zeta)$ has been inserted. Thus the singularities of the integrands

in (3) have been removed. For $\zeta > 1$ the functions $z_m(x,t,\omega,\zeta)$ are completely independent of $\zeta$.

According to the usual theory of first order partial differential equations the solution $\overset{+}{\chi}$ of (6) is constructed in the following way. For $x' \in \Omega$ I have $\nabla\overset{+}{\chi}(x',o,\omega) = -\omega$ and $\overset{+}{\chi}_t(x',o,\omega) = \left(\frac{b(x')}{a(x')}\right)^{1/2}$. Now consider the characteristic curve $s \to (x(s),\ t(s),\ \xi(s),\ \eta(s)) \in \Omega \times R \times R^3 \times R$ of (6) through the point $(x(o),\ t(o),\ \xi(o),\ \eta(o)) = \left(x',o,-\omega,\ \left(\frac{b(x')}{a(x')}\right)^{1/2}\right)$. These characteristic curves are the integral curves of the Hamiltonian vector field to the function $p(x,t,\xi,\eta) = \eta^2 - \frac{b(x)}{a(x)}|\xi|^2$, or equivalently, they are solutions of the following system of ordinary differential equations

$$\begin{aligned} \frac{dx}{ds} &= \nabla_\xi p = -2\xi\,\frac{b(x)}{a(x)} \\ \frac{dt}{ds} &= p_\eta = 2\eta \\ \frac{d\xi}{ds} &= -\nabla_x p = |\xi|^2 \nabla \frac{b(x)}{a(x)} \\ \frac{d\eta}{ds} &= -p_t = 0. \end{aligned} \tag{12}$$

It follows that

$$\eta(s) = \left(\frac{b(x')}{a(x')}\right)^{1/2}, \quad t(s) = 2\left(\frac{b(x')}{a(x')}\right)^{1/2} s.$$

Along these characteristic curves I have $p(x,t,\xi,\eta) = \text{const}$, hence $p(x,t,\xi,\eta) = 0$, due to the choice of $\eta(o)$, from which it follows that

$$|\xi|^2 = \frac{a(x)}{b(x)}\,\frac{b(x')}{a(x')}, \quad \left|\frac{dx}{ds}\right|^2 = 4\,\frac{b(x)}{a(x)}\,\frac{b(x')}{a(x')}.$$

These curves are also called null-bicharacteristics of the operator $\partial_t^2 - b\nabla(\frac{1}{a}\nabla)$. They exist for $|s|$ sufficiently small. The value of the function $\chi^+$ can be calculated from

$$\frac{d\chi^+}{ds} = \xi\,\nabla_\xi p + \eta\, p_\eta = 2p = 0,$$

hence $\chi^+$ is constant along the bicharacteristics, and in particular, equal to zero on the bicharacteristics passing over $(x_o,o)$. Also, $\nabla\chi^+ = \xi$, $\chi_t^+ = \eta$, hence

$$\chi_t^+ = \left(\frac{b(x')}{a(x')}\right)^{1/2} . \tag{13}$$

In the same way the function $\chi^-$ is constructed by considering the bi-characteristic through the point $(x(o), t(o), \xi(o), \eta(o)) = (x',o,-\omega, -(\frac{b(x')}{a(x')})^{1/2})$. It follows from (12) that this bicharacteristic coincides with the bicharacteristic passing through $(x',o,\omega, (\frac{b(x')}{a(x')})^{1/2})$, which is used in the construction of $\chi^+$, if the parameter s is replaced by -s.

Finally, it remains to discuss the functions $z_m^\pm$. For $t < 0$ I define $z_m^\pm(x,t,\omega,\zeta) = 0$. For $t > 0$ these functions are determined by the equations (7). These are linear equations for the functions $z_m^\pm$, hence for $x' \in \Omega$ it suffices to consider the characteristic curve $s \to (x(s), t(s))$ passing through $(x(o), t(o)) = (x',o)$, which is a solution of

$$\frac{dx}{ds} = -\,2\,\frac{b}{a}\,\nabla\chi^\pm$$

$$\frac{dt}{ds} = 2\,\chi_t^\pm .$$

As $\nabla\chi^{\pm} = \zeta$, $\chi_t^{\pm} = \eta$, it follows from (12) that these characteristic curves of (7) coincide with the projection of the characteristic curves of (6) to $\Omega \times R$, and therefore the functions $z_m^{\pm}$ can be determined in the same region as $\chi^{\pm}$. It follows from the initial conditions (10), (11) and from the equations (7), that for $t = 0$ all functions $z_m^{\pm}$ vanish for $x$ outside $\operatorname{supp} H_1$. Therefore these functions vanish outside the region, whose boundary is given by the characteristics to the equation (7) passing through the boundary of $\operatorname{supp} H_1$, hence, vanish outside the forward propagation cone to the operator $\partial_t^2 - b\,\nabla(\frac{1}{a}\nabla)$ associated with $\operatorname{supp} H_1$. In particular, $\operatorname{supp} z_m^{\pm}$ does not intersect the boundary $\partial\Omega \times R$ for sufficiently small $t > 0$.

Thus $\chi^{\pm}$ and $z_m^{\pm}$ have been determined. However, the series $\Sigma\, \zeta^{-m} z_m^{\pm}$ determined in this way does not converge, in general. Hence, R can not be represented by the integrals in (3), and the ansatz has to be modified. A simple modification is to use the truncated series to represent R in the form

$$
\begin{aligned}
R(x,t) = &\int_0^{\infty} \int_{|\omega|=1} e^{i\zeta\chi^{+}(x,t,\omega)} \sum_{m=-1}^{\ell} \zeta^{-m} z_m^{+}(x,t,\omega,\zeta)\,d\omega\, d\zeta \\
&+ \int_0^{\infty} \int_{|\omega|=1} e^{i\zeta\chi^{-}(x,t,\omega)} \sum_{m=-1}^{\ell} \zeta^{-m} z_m^{-}(x,t,\omega,\zeta)\,d\omega\, d\zeta + r_\ell(x,t).
\end{aligned}
\tag{14}
$$

The functions $\chi^{\pm}$, $z_m^{\pm}$ are determined as above. It is clear, however, that the truncated series $v_\ell^{\pm} = e^{i\zeta\chi^{\pm}} \sum_{m=-1}^{\ell} \zeta^{-m} z_m^{\pm}$ does not satisfy the equation $(\partial_t^2 - A)v = 0$. Instead, it follows from (5), (6), and (7) that

$$(\partial_t^2 - A)v_\ell^\pm = e^{i\zeta\chi^\pm}\,\zeta^{-\ell}(\partial_t^2 - A)z_\ell^\pm. \tag{15a}$$

Also, the following initial conditions are satisfied

$$\begin{aligned} &v_\ell^+(x,o,\omega,\zeta) + v_\ell^-(x,o,\omega,\zeta) = 0 \\ &\partial_t\, v_\ell^+(x,o,\omega,\zeta) + \partial_t\, v_\ell^-(x,o,\omega,\zeta) = e^{-i\zeta\omega\cdot(x-x_o)} \\ &\qquad [\zeta^2\, H_1(x)\, H_2(\zeta) + \zeta^{-\ell}\,\partial_t(z_\ell^+ + z_\ell^-)]. \end{aligned} \tag{15b}$$

Thus a remainder term $r_\ell$ has to be introduced. It will be seen that this is a smooth function, and consequently the singularities of R are represented by the integrals in (14), which therefore define a Parametrix for $\partial_t^2 - b\,\nabla\,\frac{1}{a}\,\nabla$.

It still remains to give a meaning to the integrals in (14), since the integrands are increasing with respect to $\zeta$. As noted at the beginning of this section, the Riemann function for the wave equation is a distribution. This is true also in the general case, and therefore the integrals in (14) must be interpreted as distributions in $S'$, which are defined as follows: Note first that $\chi_t^+ = \left(\frac{b(x')}{a(x')}\right)^{1/2} \geq C > 0$, by (13). Thus, $e^{i\zeta\chi^+} = \zeta^{-1}\, Le^{i\zeta\chi^+}$ if $L = -\,i(\chi_t^+)^{-1}\,\partial_t$, and therefore, by formal partial integration,

$$\int_{-\infty}^{\infty}\int_{\Omega}\Big[\int_{o}^{\infty}\int_{|\omega|=1} v_\ell^+\, d\omega\, d\zeta\Big]\phi(x,t)\,dx\,dt =$$

$$= \int_{o}^{\infty}\int_{\Omega}\int_{o}^{\infty}\int_{|\omega|=1} e^{i\zeta\chi^+}\sum_{m=-1}^{\ell}\zeta^{-m-k}\, L^{*k}(z_m^+\,\phi)\,d\omega\, d\zeta\, dx\, dt \tag{16}$$

$$+ \sum_{j=1}^{k} [\int_{o}^{\infty} \int_{|\omega|=1} \int_{\Omega} e^{-i\zeta\omega\cdot(x-x_o)}$$

$$[i(\chi_t^+)^{-1} \sum_{m=-1}^{\ell} \zeta^{-m-j} L^{*j-1}(z_m^+ \phi)]_{t=o} \, dx \, d\omega \, d\zeta].$$

The test function $\phi \in S$ is chosen such that supp $\phi$ is contained in the region where $\chi^+$ and $z_m^+$ are defined. This imposes a restriction on the extension of supp $\phi$ only in the direction of the positive t-axis, since by definition $z_m(x,t,\omega,\zeta) = 0$ for $t < 0$.

The second term on the right hand side of (16) is a sum of usual Fourier integrals, which exist, if evaluated in the order indicated in this formula. To see this, note that $\phi$, $z_m^+$, and $\chi^+$ are infinetely differentiable functions, and that $\partial_t^j z_m^+(\cdot,o,\omega,\zeta) \in C_o^\infty(\Omega)$. Hence, the integration with respect to x can be extended to all of $R^3$, and the usual partial integration technique can be used for the proof of existence. The first integral on the right hand side of (16) exists, if k is chosen sufficiently large. To see this, note that $\chi^+$ is independent of $\zeta$ and that $z_m^+$ is independent of $\zeta$ if $\zeta > 1$. For k sufficiently large the value of this integral is independend of the particular choice of k. The right hand side of (16) thus defines a distribution, which is formally denoted by the first integral in (14). The second integral in (14) is defined in the same way.

<u>Theorem 4:</u> There exists $t_o > 0$ such that the Riemann function R is given by (14) for $t < t_o$, where $r_\ell \in H_{\ell-1}(Z_{t_o})$.

Proof: There exists $t_o > 0$ such that the functions $\chi^{\pm}$, $z_m^{\pm}$ are defined for $t < t_o$, and such that the support of the distribution $P_\ell \in S'$ defined by the sum of the integrals in (14), which is contained in the forward propagation cone associated with supp $H_1$, does not intersect $\partial\Omega \times R$. By definition $P_\ell$ and therefore also $(\partial_t^2 - A)P_\ell \in S'$ vanishes for $t < 0$, hence the solution $r_\ell \in S'$ of

$$(\partial_t^2 - A)r_\ell = -(\partial_t^2 - A)P_\ell + \delta(t)\delta(x-x_o)$$

$$r_\ell = 0, \; t < 0,$$

exists, by theorem 3. Clearly, $P_\ell + r_\ell$ is equal to the uniquely defined Riemann function. It remains to show that $r_\ell$ is smooth. Let $\phi \in S$ with supp $\phi \leq \{t < t_o\}$. By partial integration it then follows that

$$\int_{\Omega\times R} (v_\ell^+ + v_\ell^-)(\partial_t^2 - A)\phi \, dx\, dt = \tag{17}$$

$$\int_{\Omega\times R^+} (\partial_t^2 - A)(v_\ell^+ + v_\ell^-)\phi \, dx\, dt + \int_\Omega e^{-i\zeta\omega\cdot(x-x_o)}$$

$$[\zeta^2 H_1(x)H_2(\zeta) + \zeta^{-\ell} \partial_t(z_\ell^+ + z_\ell^-)]\phi_{|t=o} \, dx$$

$$= \int_{\Omega\times R^+} [e^{i\zeta\chi^+}(\partial_t^2 - A)z_\ell^+ + e^{i\zeta\chi^-}(\partial_t^2 - A)z_\ell^-]\zeta^{-\ell} \phi \, dx\, dt$$

$$+ \int_\Omega e^{-i\zeta\omega\cdot(x-x_o)}[\zeta^2 H_1(x)H_2(\zeta) + \zeta^{-\ell} \partial_t(z_\ell^+ + z_\ell^-)]\phi(x,o)dx,$$

where I used (15a) and (15b). As will be seen in a moment, this implies

$$\int_{\Omega\times R} \int_o^\infty \int_{|\omega|=1} (v_\ell^+ + v_\ell^-)d\omega \, d\zeta(\partial_t^2 - A)\phi \, dx\, dt = \tag{18}$$

$$= \int_{\Omega\times R^+} \int_0^\infty \int_{|\omega|=1} [e^{i\zeta\chi^+}(\partial_t^2 - A)z_\ell^+ + e^{i\zeta\chi^-}(\partial_t^2 - A)z_\ell^-]\zeta^{-\ell} \, d\omega \, d\zeta \, \phi(x,t)dx \, dt$$

$$+ \int_0^\infty \int_{|\omega|=1} \int_\Omega e^{-i\zeta\omega\cdot(x-x_o)}[\zeta^2 H_1 H_2 + \zeta^{-\ell} \partial_t(z_\ell^+ + z_\ell^-)]\phi(x,o)dx \, d\omega \, d\zeta.$$

The last integral is equal to $\phi(x_o,o) + \int_\Omega u_1(x)\phi(x,o)dx$, where

$$u_1(x) = \int_0^1 \int_{|\omega|=1} e^{-i\zeta\omega\cdot(x-x_o)}[\zeta^2(H_2(\zeta)-1)H_1(x) + \zeta^{-\ell} \partial_t(z_\ell^+ + z_\ell^-)]d\omega \, d\zeta$$

$$\in C_o^{\ell-2}(\Omega),$$

by the definition of $H_2$ and of $z_\ell^\pm$. The second last integral is equal to $\int_{\Omega\times R} u_2(x,t)\phi(x,t)dx \, dt$, where

$$u_2(x,t) = \begin{cases} 0 & , \ t < 0 \\ \int_0^\infty \int_{|\omega|=1} [e^{i\zeta\chi^+}(\partial_t^2 - A)z_\ell^+ + e^{i\zeta\chi^-}(\partial_t^2 - A)z_\ell^-]\zeta^{-\ell} \, d\omega \, d\zeta, & t \geq 0, \end{cases}$$

which implies that $u_2$ is $\ell$-2 times continuously differentiable with respect to x and t for $t > 0$, since $\chi^\pm$ and $z_\ell^\pm$ are independent of $\zeta$ for $\zeta > 1$. Also, $u_2$ vanishes in a neighbourhood of $\partial\Omega \times R$ for $t < t_o$. Thus $-(\partial_t^2 - A)P_\ell + \delta(t)\delta(x-x_o) = - u_2 - \delta(t)u_1(x)$, which implies that $r_\ell$ is equal to the classical solution of

$$(\partial_t^2 - A)r_\ell = - u_2 \text{ in } \Omega \times R^+$$

$$r_\ell(x,o) = 0, \ \partial_t r_\ell(x,o) = - u_1(x) \in C_o^{\ell-2}(\Omega)$$

and consequently $r_\ell \in H_{\ell-1}(Z_{t_o})$, since $u_1 \in C_o^{\ell-2}(\Omega)$, $u_2 \in C_o^{\ell-2}(\Omega \times R^+)$, and $u_2$ vanishes in a neighbourhood of $\partial\Omega \times R$ for $t < t_o$, which shows that the assumptions of theorem 1 are satisfied.

The formula (18) is correct, since the right hand side of equation (16) with $\phi$ replaced by $(\partial_t^2 - A)\phi$ is equal to

$$\lim_{s\to\infty} \int_{-\infty}^{\infty} \int_{\Omega} \int_{o}^{s} \int_{|\omega|=1} v_\ell^+ (\partial_t^2 - A)\phi \, d\omega \, d\zeta \, dx \, dt,$$

and in this expression partial integration with respect to x and t is possible and leads to the result given in (18).

The proof is complete.

It can be shown that the singular support of the distribution defined by the first integral in (14) is contained in the set

$\{(x,t) \in \Omega \times R \mid \chi^+(x,t,\omega) = 0 \text{ and } \nabla_\omega \chi^+(x,t,\omega) = 0 \text{ for some } \omega\}$.

The construction of $\chi^+$ implies that this can be only the case if $(x,t)$ lies on one of the bicharacteristics passing through $(x_o,o)$, which shows that singularities of solutions to $(\partial_t^2 - A)v = 0$ propagate along bicharacteristics.

The preceding construction breaks down if bicharacteristics intersect, and it also breaks down, if the propagation cone associated with support $H_1$ intersects the boundary. The reason is, that in formula (17) after partial integration boundary terms appear, since the function $v_\ell^+ + v_\ell^-$ does not satisfy the boundary condition. But it is possible to modify the functions $v_\ell^\pm$ such that the boundary condition is satisfied, and thus treat the reflection of singularities.

Consider a bicharacteristic of $\partial_t^2 - b\nabla \frac{1}{a} \nabla$ whose projection $\gamma$ to $\Omega \times R$ starts at $(x_o,o)$ and intersects the boundary in a point $(x_b,t_b)$, but which is not tangential to the boundary in this point. We want to study R in a neighbourhood V of $(x_b,t_b)$ in $\Omega \times R$. Let $(x,t) \in V$. By construction, the value $r_\ell(x,t)$ of the remainder term in (14) only depends on the values of the distribution defined by the integrals in (14) at points in the intersection of the backward propagation cone with vertex $(x,t)$ and the forward propagation cone associated with supp $H_1$. For, the support of the distributions defined by the integrals in (14) is contained in this forward propagation cone, and this will also be the case after the functions $v_\ell^\pm$ have been modified suitably. If we make suitable assumptions on the form of these propagation cones, for example, that all forward an backward propagation cones involved in this construction are strictly convex, then this intersection is contained in a neighbourhood of the bicharacteristic $\gamma$, which can be made arbitrarily small if the neighbourhood V and supp $H_1$ are made small. It suffices to modify and extend $v_\ell^\pm$ such that the new functions $\hat{v}_\ell^\pm$ are defined in this neighbourhood and satisfy Dirichlet or Neumann boundary conditions in the intersection $V_1$ of this neighbourhood with $\partial\Omega \times R$. $V_1$ can be made so small, that all bicharacteristics whose projection starts at supp $H_1$ and intersects the boundary in $V_1$ meet the boundary not tangentially. I define

$$\overset{\wedge +}{v}_\ell(x,t,\omega,\zeta) = e^{i\zeta\chi^+(x,t,\omega)} \sum_{m=-1}^{\ell} \zeta^{-m} z_m^+(x,t,\omega,\zeta)$$

$$+ e^{i\zeta\chi(x,t,\omega)} \sum_{m=-1}^{\ell} \zeta^{-m} z_m(x,t,\omega,\zeta),$$

where $\chi^+$, $z_m^+$ are defined as above, and $\chi$, $z_m$ are solutions of

$$(\chi_t)^2 - \frac{b}{a} |\nabla\chi|^2 = 0$$

$$\chi(x,t) = \chi^+(x,t), \quad (x,t) \in V_1$$

and

$$2(\chi_t \, \partial_t \, z_m - \frac{b}{a} \nabla\chi\cdot\nabla z_m) + ((\partial_t^2 - A)\chi) z_m = i(\partial_t^2 - A) z_{m-1}.$$

In the Dirichlet case $z_m$ must satisfy

$$z_m(x,t,\omega,\zeta) = - z_m^+(x,t,\omega,\zeta) \ , \quad (x,t) \in V_1$$

and in the Neumann case

$$z_{-1} = - z_{-1}^+$$

$$i\partial_n\chi^+(z_{m+1} + z_{m+1}^+) + \partial_n \, z_m + \partial_n \, z_m^+ = 0 \ , \quad -1 \leq m \leq \ell-1$$

if $(x,t) \in V_1$. Therefore, in the Neumann case $\overset{\wedge +}{v}_\ell$ does not satisfy the boundary condition exactly. Rather,

$$\partial_n \, \overset{\wedge +}{v}_\ell = \zeta^{-\ell} \, e^{i\zeta\chi^+} \, \partial_n(z_\ell + z_\ell^+)$$

is satisfied. Since this is a term which is homogeneous of order $-\ell$, it gives a smooth contribution to the remainder $r_\ell$. The function $\chi$ is determined as follows. For $(x',t) \in V_1$ let $\xi' = \nabla\chi^+(x,',t',\omega)$,

$\eta' = \chi_t^+(x',t',\omega)$. Then the value of the bicharacteristic used to construct $\chi^+(x,t,\omega)$ passing over $(x',t')$ is $(x',t', \xi', \eta')$ in this point. Let $\xi'_\tau$ be the component of $\xi'$ tangential to $\partial\Omega$, and let $\xi'_n$ be the component of $\xi'$ normal to $\partial\Omega$. Now consider the bicharacteristic $s \to (x(s), t(s), \xi(s), \eta(s))$ with initial values $(x(o), t(o), \xi(o), \eta(o)) = (x', t', \xi'', \eta')$, the "reflected" bicharacteristic, where $\xi''_\tau = \xi'_\tau$ and $\xi''_n = - \xi'_n$. This bicharacteristic is used as above to constructed the function $\chi(x,t,\omega)$. Note that this choice of the bicharacteristic is consistent with the initial values $\chi = \chi^+$ required for $\chi$. In the same way the function $v_\ell^-$ is constructed.

## 3. High frequency asymptotics

In the following I have to specify my investigations to the wave equation $(\partial_t^2 - \Delta)v = 0$ and to the Helmholtz equation $(\Delta+k^2)v = 0$ in the case of the Dirichlet boundary condition. I shall determine the asymptotic behaviour of the Green function to $(\Delta+k^2)u = 0$ at the boundary $\partial\Omega$. First I have to make some assumptions, which to some extend are necessary for the result, and to some extend are probably due to technical reasons. Let the obstacle $B = R^3 \setminus \bar{\Omega}$ be bounded, and let $x_o \in \Omega$.

(A 1) Assume that for every ball S containing B there exist constants $C, \alpha > 0$ such that for every solution $u(x,t)$ of the Dirichlet Problem for the wave equation with $\operatorname{supp} u(\cdot,o) \subseteq S$ the inequality

$$||u(t)||_{E,S} \leq C\, e^{-\alpha t} ||u(o)||_{E,S}$$

holds.

(A 2) None of the bicharacteristic rays to the wave operator $\partial_t^2 - \Delta$ passing over the point $(x_o, o) \in \Omega \times R$ meets the boundary $\partial\Omega \times R$ tangentially to infinite order.

It has been shown by Morawetz, Ralston, and Strauss [8], that the first assumption is satisfied if the obstacle B is non trapping. An obstacle is called non-trapping, if the length of all generalized bicharacteristic rays in $(S\setminus B) \times R$ is uniformly bounded. By generalized bicharacteristics I mean the bicharacteristics to the wave equation, which are reflected at the boundary according to the laws of geometrical optics, as discussed in the preceding section. The second assumption is necessary since we need the following result on propagation and reflection of singularities of solutions to the wave equation. This result can be deduced from the construction in the preceding section for singularities on rays which meet the boundary transversal. Remember that I only considered transversal rays. For tangential rays this construction does not work, and one has to use other methods. The most general result is the following result of Melrose and Sjöstrand [7], which I state somewhat loosly:

Theorem 5: Singularities of solutions to the wave equation propagate along generalized bicharacteristics with speed 1, if the bicharacteristics are not tangential to the boundary to infinite order.

The word singularities here means the points, where the solution is not inifinitely differentiable. Up to now it has not been determined exactly, what happens to singularities on bicharacteristics, which meet the boundary to infinite order.

Next, observe that for distributions $u(x,t) \in S_D'$ the partial Fourier transform $\hat{u}(x,k) \in S_D'$ with respect to t can be defined in the usual way: For $u \in S_D'$ and $\phi \in S_D$ I define

$$(\hat{u},\phi) = (u,\hat{\phi}),$$

where

$$\hat{\phi}(x,t) = \int_{-\infty}^{\infty} \phi(x,k)e^{+ikt}\, dk.$$

This definition makes sense, since the Fourier transformation with respect to t is a continuous, bijective mapping on $S_D$, which can be shown as usual. I have the following result:

<u>Theorem 6</u>: Assume that (A 1) is satisfied, and let $R(x,y,t) \in S_D'$ be the solution of

$$(\partial_t^2 - \Delta)R = \delta(t)\delta(x-y)$$

$$R = 0\ , \quad \text{for } t < 0.$$

(R is the Riemann function to the wave equation.) Then the Fourier transform $\hat{R} \in S_D'$ is a well defined function $(x,k) \to \hat{R}(x,y,k)$, and coincides with the Green function $G(x,y,k)$ of the Helmholtz equation $(\Delta+k^2)u = 0$, which is defined by

$$(\Delta+k^2)G = -\ \delta(x-y)$$

$$G(x,y,k) = 0, \ x \in \partial\Omega$$

$$\partial_r G = ikG + o(r^{-1}), \ r = |x| \to \infty.$$

Let me comment on this theorem. The fact, that $\hat{R}$ is represented by a function is a regularity result. For the proof of this result I need the assumption (A 1). However, $\hat{R} = G$ can also be shown in the general case, if the distribution $G \in S'_D$ is defined suitably, see [2]. Clearly, it is generally true that $\hat{R}$ is a solution of $(\Delta+k^2)\hat{R} = -\delta(x-y)$ in the sense of $S'_D$, for I have

$$((\Delta+k^2)\hat{R},\phi) = (R,[(\Delta+k^2)\phi]^\wedge) = (R,(\Delta-\partial_t^2)\hat{\phi}) = (-\delta(t)\delta(\cdot-y),\hat{\phi})$$

$$= -\int_{-\infty}^{\infty} \phi(y,k)\,dk = (-\delta(\cdot-y),\phi).$$

Using theorem 6 I shall now calculate the asymptotic behaviour of the normal derivative $\partial_n G$ at the boundary $\partial\Omega$ for large k from the representation of R constructed in the last section. It is necessary to know this normal derivative if one wants to solve the Dirichlet problem. Let (A 1) be satisfied and let $x_o \in \partial\Omega$ be a point such that (A 2) is satisfied. Consider a bicharacteristic to the wave operator passing over $(x_o,o)$ which meets the boundary $\partial\Omega \times \mathbb{R}$ transversal in a point $(x_b,t_b)$. In a neighbourhood of this point in $\bar{\Omega} \times \mathbb{R}$ the Riemann function $R(x,x_o,t)$ can be represented by the construction in the preceding section. I can assume that this neighbourhood has the form $U \times (t_b-\varepsilon,t_b+\varepsilon)$, where $\varepsilon > 0$ is choosen sufficiently small and where U is a suitable neighbourhood of $x_b$ in $\bar{\Omega}$. To calculate the Fourier transform of R with respect to t in the point $x_b$, however, it is necessary to know R in $x_b$ for all times. Therefore I make the additional assumption, that no other generalized bicharacteristic ray starting in $x_o$ passes over $x_b$. Theorem 5 then yields, that R is an infinitely differentiable function in $x_b$ for all $t > t_b$. Having this, it is an easy consequence of the assumption (A 1), that the function $(x,t) \to R(x,x_o,t)$ and all derivatives with respect to t and x vanish exponentially for $t \to \infty$, uniformly in a neighbourhood of $x_b$ in $\bar{\Omega}$. Now I choose two functions $\psi_1, \psi_2 \in C^\infty(R)$ with

$$\psi_1(s) = \begin{cases} 0, & s < \frac{1}{3}\,\mathrm{dist}\,(x_o,x_b) \\ 1, & \frac{2}{3}\,\mathrm{dist}\,(x_o,x_b) < s < t_b + \varepsilon/2 \\ 0, & t_b + \varepsilon < s \end{cases}$$

and with

$$\psi_2(s) = \begin{cases} 0, & s < \frac{2}{3}\,\mathrm{dist}\,(x_o,x_b) \\ 1 - \psi_1(s), & s > \frac{2}{3}\,\mathrm{dist}\,(x_o,x_b). \end{cases}$$

Since R vanishes in a neighbourhood of $x_b$ for $t < \frac{2}{3}\,\mathrm{dist}\,(x_o,x_b)$, I thus obtain for x in a neighbourhood of $x_b$

$$\begin{aligned} R(x,x_o,t) = {} & \Big[\int_o^\infty \int_{|\omega|=1} e^{i\zeta(t-\omega\cdot(x-x_o))} \sum_{m=-1}^{\ell} \zeta^{-m} z_m^{+} + e^{i\zeta\hat{\chi}^{+}} \sum_{m=-1}^{\ell} \zeta^{-m} \hat{z}_m^{+}\, d\omega\, d\zeta \qquad (19) \\ & + \int_o^\infty \int_{|\omega|=1} e^{i\zeta(t-\omega\cdot(x-x_o))} \sum_{m=-1}^{\ell} \zeta^{-m} z_m^{-} + e^{i\zeta\hat{\chi}^{-}} \sum_{m=-1}^{\ell} \zeta^{-m} \hat{z}_m^{-}\, d\omega\, d\zeta\Big]\psi_1(t) \\ & + \psi_1(t) r_\ell(x,t) + \psi_2(t) R(x,x_o,t). \end{aligned}$$

Here I used the fact that for the wave equation $\chi^{\pm}(x,t,\omega) = \pm t-\omega\cdot(x-x_o)$. The function $\psi_2(t)R(x,x_o,t)$ is infinitely differentiable and together with all derivatives decays exponentially. It thus follows that the Fourier transform with respect to t of this functions in the sense of $S_D'$ is equal to the usual Fourier transform, and that the normal derivative of the Fourier transform is obtained by Fourier transforming $\partial_n \psi_2(t)R(x,x_o,t)$. But according to well known results the Fourier transform of this function decays faster than any power of t for $t \to \infty$. All the same remarks apply to the function $\psi_1(t)r_\ell(x,t)$, with the difference, that $r_\ell$ is contained in $H_{\ell-1}(Z_{t_b+\varepsilon})$, and therefore, by the Sobolev imbedding theorem, the function $\partial_n \psi_1(t)r_\ell(x,t)$ is only $\ell-4$ times continuously differentiable with respect to t. The Fourier transform with respect to t of this function therefore behaves like $O(k^{-\ell+4})$, $k \to \infty$. These results show that the asymptotic behaviour of the Green function $\partial_n G(x_b,x_o,k)$ for $k \to \infty$ is given by the asymptotic behaviour of the Fourier transform of the integrals in (19). So I have to calculate these Fourier transforms. In the calculation one has to use the following facts: It follows from (12) that the two bicharacteristic rays of the wave operator passing through the point (x',o) are given by $x = \pm \omega t + x'$. Hence (7) implies that $z_{-1}^{\pm}(x,t,\omega,\zeta) = H_1(x \mp \omega t)H_2(\zeta)$ and therefore that these functions are constant for $\zeta > 1$ in a neighbourhood of all bicharacteristic rays passing through $(x_o,o)$. Thus (7) also implies that $z_m = 0$ for all $m \geq 0$ in this neighbourhood. Finally, one has that

$$\hat{\chi}^{\pm}(x,t,\omega) = \pm t - \phi(x),$$

where $\phi$ is a solution of

$$|\phi(x)|^2 = 1$$

and satisfies

$$\phi(x) = |x-x_o|,$$

for all x in a neighbourhood of $x_b$ in $\partial\Omega$. The calculation of the Fourier transform is elementary, but somewhat technical. So I only state the following result:

Theorem 7: Let the assumptions (A 1) and (A 2) be satisfied for $\partial\Omega$ and for $x_o \in \Omega$.

(i) Let $M \subseteq \partial\Omega$ be the set of all points x with the following properties:

1) x can be connected to $x_o$ by a straight line in $\Omega$, which intersects $\partial\Omega$ in x transversal.

2) No other ray starting at $x_o$ and reflected at $\partial\Omega$ according to the laws of geometrical optics passes through x.

Then

$$\partial_n G(x,x_o,k) = \frac{-ik\, n\cdot(x-x_o)}{2\pi|x-x_o|^2} e^{ik|x-x_o|} + O(1) \quad , \quad k \to \infty$$

uniformly for all x in every compact subset of M.

(ii) Let $x \in \partial\Omega$ be a point such that no (direct or reflected) ray starting at $x_o$ passes over x. Then

$$\partial_n G(x,x_o,k) = O(k^{-\infty}) \quad , \quad k \to \infty.$$

References

1. Alber, H.D., Justification of geometrical optics for non-convex obstacles,
*J. Math. Anal. Appl.* 80, 372, 1981.

2. Alber, H.D., Zur Hochfrequenzasymptotik der Lösungen der Schwingungsgleichung - Verhalten auf Tangentialstrahlen,
*Habilitationsschrift Universität Bonn,* 1982.

3. Chazarin, J., Construction de la paramétrix du probléme mixte hyperbolique pour l' equation des ordes,
*C.R. Acad. Sci. Paris* 276, 1213, 1973.

4. Duistermaat, J.J., Hörmander, L., Fourier integral operators II,
*Acta Math.* 128, 183, 1972.

5. Hörmander, L., Fourier integral operators,
*Acta Math.* 127, 79, 1971.

6. Majda, A., High frequency asymptotics for the scattering matrix and the inverse problem of acoustical scattering,
*Comm. Pure Appl. Math.* 29, 261, 1976.

7. Melrose, R., Sjöstrand, J., Singularities of boundary value problems I,
*Comm. Pure Appl. Math.* 31, 593, 1978.

8. Morawetz, C.S., Ralston, J.V., Strauss, W.A., Decay of solutions of the wave equation outside nontrapping obstacles,
*Comm. Pure Appl. Math.* 30, 447, 1977.

9. Taylor, M.E., Grazing rays and reflection of singularities of solutions to wave equations,
*Comm. Pure Appl. Math.* 29, 1, 1976.

# SOLUTION PROCEDURES FOR INTERFACE PROBLEMS IN ACOUSTICS AND ELECTROMAGNETICS

E. STEPHAN

Fachbereich Mathematik
Technische Hochschule Darmstadt
D-6100 Darmstadt, Fed. Rep. Germany

## INTRODUCTION

The main aim of this paper is to derive solution procedures for 2 and 3-dimensional interface problems governing the scattering of sound by a homogeneous medium and the scattering of time harmonic electromagnetic fields in air by metallic obstacles. Two ideas are developed. Ths first is a boundary integral procedure for the Helmholtz interface problem and for the eddy current problem (in three dimensions). The second idea is an asymptotic procedure for the scattering of time harmonic electromagnetic fields which applies for large conductivity and reflects the skin effect in metals.

In section 1 we treat the exterior Neumann problem and the interface problem for the Helmholtz equation by reducing them to pseudodifferential equations on the analytic boundary surface $S$ of the scatterer.

By means of the fundamental solution of the corresponding Helmholtz equation we present both the method of simple or double layer potentials and the direct method via Green's formula. The equations which appear in our integral equation method involve pseudodifferential operators on S . Precise existence and regularity results for these pseudodifferential equations are obtained guaranteeing the corresponding results for the exterior Neumann or interface problem.

In section 2 we present a new boundary integral procedure for eddy current problems from [14]. There, the key is the introduction of a new simple layer potential method for the boundary value problem corresponding to perfect conductors. The perfect conductor problem involves solving Maxwell's equations in the region exterior to the obstacle with tangential component of the electric field zero on the obstacle surface S . Whereas all known integral equation procedures for the perfect conductor problem lead to integral equations of second kind, our method in [15] leads to a system of first kind equations and gives a simple procedure for the calculation of the tangential component of the magnetic field on S . This enables us to formulate an integral equation procedure for the interface problem where different sets of Maxwell's equations must be solved in the obstacle and outside while the tangential components of both electric and magnetic fields are continuous across S . At the end of section 2 we illustrate our results on the eddy current problem by two-dimensional examples, namely a cylinder in a transverse-magnetic field and in a transverse-electric field.

In section 3 we present for large conductivity an asymptotic procedure for the eddy current problem, derived in [14]. Asymptotic expansions inside and outside the metallic obstacle are given in terms of a parameter $\beta$ measuring the conductivity. The inner expansion represents the skin effect in metal for monochromatic fields. The leading term outside is the solution of the infinite conductivity problem ($\beta = \infty$) and successive terms are calculated by recursion. The asymptotic procedure is based on the above new simple layer potential procedure for the perfect conductor problem, and the asymptotic procedure gives an approximate solution by solving a sequence of problems analogous to the one for perfect conductors. For transverse-magnetic electromagnetic fields in air incident on a metallic cylinder the asymptotic procedure converges due to [16] and if one stops the recursion process after $N$ terms then the error is $O(\beta^{-N-1})$ .

In section 4 we generalize our boundary integrals for two-dimensional interface problems to the case of a curvilinear or polygonal transmission curve. As examples we treat (i) the eddy current problem for an infinitely long cylinder in a transverse-magnetic field, and (ii) forward scattering with low frequency at obstacles with corners. According to [1], [3] here the essential tool for the analysis of the boundary integral equations is the Mellin transform instead of the Fourier transform in the case of smooth surfaces (interfaces) as in the above sections.

Finally, it should be mentioned that both methods, the boundary integral and the asymptotic procedure, admit of numerical implementation techniques like finite elements. For the boundary integral equations corresponding to the perfect conductor problem (in section 2) it is shown in [17] that Galerkin's method with finite elements on the obstacle surface converges with quasi-optimal order. For the asymptotic procedure with large conductivity the arizing integral equations are computed successfully by collocation in [6]. In forward scattering the use of special "corner" elements as additional trial functions increases the convergence rates of the Galerkin procedure [22]. For more details on the numerical treatment of problems like above we refer to [24].

## 1. BOUNDARY INTEGRAL METHODS FOR HELMHOLTZ INTERFACE PROBLEMS

Scattering of sound

If a sound wave meets an obstacle, it is partially reflected from it and partially transmitted through it. Let us consider a steady-state sound wave that is set up in a homogeneous medium $\Omega$ characterized by a density $\rho$, a damping coefficient $\alpha$ and sound velocity $c$ in which there is a homogeneous body $\Omega'$ of density $\rho_i$, damping coefficient $\beta$ and sound velocity $c_i$. We shall characterize the sound wave by the pressure $v$ and the angular frequency $\omega$ of the acoustic vibrations. Let the medium occupy all space $\mathbb{R}^3$ with the exception of the bounded domain $\Omega'$ occupied by the obstacle. We denote by $v_o, v_i, v_e$ the pressure of the incident, refracted and scattered wave, respectively, satisfying the homogeneous Helmholtz equations

$$\Delta v_i + k_i^2 v_i = 0 \quad , \quad k_i^2 = \frac{\omega(\omega+i\beta)}{c_i^2} \quad \text{in} \quad \Omega' \tag{1.1}$$
$$\Delta v_e + k^2 v_e = 0 \quad , \quad k^2 = \frac{\omega(\omega+i\alpha)}{c^2} \quad \text{in} \quad \Omega = \mathbb{R}^3 - \bar{\Omega}'$$

Both the total acoustic field $v = v_e + v_o$ and the incident field $v_o$ satisfy the homogeneous Helmholtz equation in the exterior domain $\Omega$. At infinity the scattered wave $v_e$ fulfills the Sommerfeld radiation condition

$$\lim_{r\to\infty} r\left(\frac{\partial v_e}{\partial r} - ikv_e\right) = 0 \quad , \quad \lim_{r\to\infty} v_e = 0 \tag{1.2}$$

Finally, on the boundary $S$ of the obstacle, the pressure and the velocity of vibrations in the body and the medium must coincide, yielding the conjugacy conditions

$$v_i = v_e + v_o \ , \ \frac{1}{\rho_i}\frac{\partial v_i}{\partial \underset{\sim}{n}} = \frac{1}{\rho}\left(\frac{\partial v_e}{\partial \underset{\sim}{n}} + \frac{\partial v_o}{\partial \underset{\sim}{n}}\right) \ , \quad \text{on } S \tag{1.3}$$

where $\frac{\partial}{\partial \underset{\sim}{n}}$ denotes differentiation with respect to the outer normal $\underset{\sim}{n}$ to $S$. Thus the <u>scattering of sound</u> is described by the interface problem (1.1) - (1.3).

For higher damping the constant $\beta$ is usually large leading to the total reflection of a plane wave at an absolutely rigid immovable obstacle. Formally this means solving only the Helmholtz equations $(1.1)_2$ in $\Omega$ for the scattered field and requireing that the normal derivative of the total acoustic field vanishes on $S$, that is

$$\begin{aligned} &\Delta v_e + k^2 v_e = 0 \quad \text{in} \quad \Omega = \mathbb{R}^3 \setminus \overline{\Omega'} \\ &\frac{\partial v_e}{\partial n} = -\frac{\partial v_o}{\partial n} \quad \text{on} \quad S \end{aligned} \tag{1.4}$$

where $v_e$ satisfies (1.2) at infinity.

In the following we assume for simplicity that $S$ is a closed analytic surface which divides $\mathbb{R}^3$ into simply connected domains, an interior $\Omega'$ (bounded) and an exterior $\Omega$ (unbounded).

In order to avoid additional difficulties we need the technical assumption:

$k^2 \neq 0$ is <u>not</u> an eigenvalue of the interior Dirichlet problem (1.5)

The uniqueness of the solution of the interface problem (1.1) - (1.3) and of the exterior Neumann problem (1.4) is wellknown ([12], [13], [25]). For brevity we give here only the uniqueness proof for the interface problem.

THEOREM 1.1: *Let* $k, k_i \in \mathbb{C}\setminus\{0\}$ *with* $0 \leq \arg k, \arg k_i < \pi$ *and let* $\mu = \frac{1}{\rho}$, $\mu_i = \frac{1}{\rho_i} \in \mathbb{C}\setminus\{0\}$ *be such that*

$$\kappa = \frac{\mu_i \overline{k_i}^2}{\mu \overline{k}^2} = \frac{\rho \overline{k_i}^2}{\rho_i \overline{k}^2} \in \mathbb{R}$$

*where* $\kappa \geq 0$ $(<0)$ *if* $\operatorname{Re} k \circ \operatorname{Re} k_i \geq 0$ $(<0)$. *Then the only solution of the homogeneous transmission problem* (1.1) - (1.3) *is a trivial solution* $v_e = v_i = 0$.

*Proof:* Let $\Omega_R = \{x \in \Omega : |x| < R\}$ contain $S$ in its interior and denote by $\Gamma_R$ the boundary of $\Omega_R$ with outer normal $\underset{\sim}{n}$. Applying Green's theorem over $\Omega'$ and $\Omega_R$ and using (1.1), (1.3) with $v_o = \frac{\partial v_o}{\partial \underset{\sim}{n}} = 0$ we obtain

$$\mu \int_{\Gamma_R} v_e \frac{\partial \overline{v}_e}{\partial n_{\sim}} ds = \mu \int_{\Omega_R} |\text{grad } v_e|^2 dx + \mu_i \int_{\Omega'} |\text{grad } v_i|^2 dx - \tag{1.6}$$

$$- \mu \overline{k}^2 \int_{\Omega_R} |v_e|^2 dx - \mu_i \overline{k_i}^2 \int_{\Omega'} |v_i|^2 dx$$

Dividing (1.6) by $\mu \overline{k}^2$ and taking imaginary parts we have

$$\text{Im}\left(\frac{1}{\overline{k}^2} \int_{\Gamma_R} v_e \frac{\partial \overline{v}_e}{\partial \underset{\sim}{n}} ds\right) = \text{Im}\left(\frac{1}{\overline{k}^2} \int_{\Omega_R} |\text{grad } v_e|^2 dx\right) + \kappa \, \text{Im}\left(\frac{1}{\overline{k_i}^2} \int_{\Omega'} |\text{grad } v_i|^2 dx\right) \tag{1.7}$$

For $\text{Im } k > 0$ it follows from [13], since $v$ satisfies the radiation condition, that the left side of (1.7) tends to zero for $R \to \infty$. Hence (1.7) reduces to

$$\frac{\text{Re } k \cdot \text{Im } k}{|k|^4} \int_{\Omega} |\text{grad } v_e|^2 dx + \kappa \frac{\text{Re } k_i \cdot \text{Im } k_i}{|k_i|^4} \int_{\Omega'} |\text{grad } v_i|^2 dx = 0 \, . \tag{1.8}$$

For $\text{Re } k \neq 0$ it follows from the sign of $\kappa$ that

$$\int_{\Omega} |\text{grad } v_e|^2 dx = 0$$

yielding $v_e = 0$ in $\Omega$. Therefore transmission conditions (1.3) give $v_i = \frac{\partial v_i}{\partial \underset{\sim}{n}} = 0$ on $S$ and thus Green's theorem over $\Omega'$ gives

$$- \int_{\Omega'} |\operatorname{grad} v_i|^2 \, dx + \bar{k}_i^2 \int_{\Omega'} |v_i|^2 dx + \int_S v_i \frac{\partial \overline{v_i}}{\partial n} \, ds = 0$$

implying $v_i = 0$ in $\Omega'$ .

If $\operatorname{Re} k = 0$ , $\operatorname{Re} k_i \neq 0$ , then we from (1.8)

$$\int_{\Omega'} |\operatorname{grad} v_i|^2 \, dx = 0$$

hence $\operatorname{grad} v_i = 0$ in $\Omega'$ . The transmission conditions give now $\frac{\partial v_e}{\partial \underset{\sim}{n}} = 0$ on $S$ . Therefore by the uniqueness theorem for the exterior Neumann problem we have $v_e = 0$ in $\Omega$ . Again using the transmission conditions we deduce $v_i = 0$ in $\Omega'$ .

If $\operatorname{Re} k = \operatorname{Re} k_i = 0$, then $\bar{k}^2$ and $\bar{k}_i^2$ are both negative and therefore from (1.6) we obtain with $\kappa \geq 0$

$$\int_{\Omega} |v_e|^2 dx = \int_{\Omega'} |v_i|^2 dx = 0$$

which implies that $v_e = v_i = 0$ .

If $\operatorname{Im} k = 0$ we have from (1.7)

$$\operatorname{Im} \int_{\Gamma_R} v_e \frac{\partial \bar{v}_e}{\partial \underset{\sim}{n}} \, ds \geq 0$$

The radiation condition yields, as $R \to \infty$ ,

$$k \int_{\Gamma_R} |v_e|^2 ds + \operatorname{Im} \int_{\Gamma_R} v_e \frac{\partial \bar{v}_e}{\partial \underset{\sim}{n}} \, ds = o(1) \ .$$

Since both terms on the left side are non negative we have

$$\int_{\Gamma_R} |v_e|^2 ds = o(1) \qquad \text{as} \qquad R \to \infty \ .$$

Hence, Rellich's Theorem [20] implies that $v_e = 0$ in $\Omega$. The proof is now completed by means of arguments similar to the ones above.

In the following we first give a boundary integral equation method based on simple layers for solving both the interface problem (1.1) - (1.3) and the exterior Neumann problem (1.4). Then we give the corresponding double layer procedure. To this end we introduce the simple layer $V_\gamma$ with continuous density $\psi$ on the surface $S$ by

$$V_\gamma(\psi)(x) = \int_S \psi(y)\, \phi_\gamma(|x-y|)\, dS_y \ , \quad x \in \mathbb{R}^3 \tag{1.9}$$

Here

$$\phi_\gamma(|x-y|) = \frac{e^{i\gamma|x-y|}}{4\pi|x-y|} \tag{1.10}$$

is the fundamental solution of the Helmholtz equation $\Delta w = -\gamma^2 w$ satisfying the Sommerfeld radiation condition for $\operatorname{Re}\gamma \neq 0$. There hold the following well-known properties of the simple layer potential ([4], [26]).

LEMMA 1.2: *For any complex* $\gamma$, $0 \le \arg\gamma \le \pi/2$ *and any continuous* $\psi$ *on* $S$:

(i) $V_\gamma(\psi)$ *is continuous in* $\mathbb{R}^3$

(ii) $\Delta V_\gamma(\psi) = -\gamma^2 V_\gamma(\psi)$ *in* $\Omega \cup \Omega'$

(iii) $V_\gamma(\psi)(x) = O(|x|^{-1} e^{i\gamma|x|})$ *as* $|x| \to \infty$

(iv) $(\frac{\partial}{\partial \underset{\sim}{n}} V_\gamma(\psi) \frac{+}{(x)}) = \mp \frac{1}{2}\psi(x) + \int_S K_\gamma(x,y)\psi(y)\, dS_y$ *on* $S$

*where the function* $K_\gamma$ *is* $O(|x-y|^{-1})$ *as* $y \to x$ *and* $\pm$ *denotes the limit to* $S$ *from* $\Omega$ *and* $\Omega'$, *respectively.*

In order to describe the mapping properties of $V_\gamma$ and $K_\gamma$ as pseudodifferential operators acting in Sobolev spaces we first diescuss some geometric ideas (see [14]). We introduce coordinate systems on $S$. These consist of a finite number of coordinate patches $S_1 .., S_N$ covering $S$. For each patch there is a region $\Gamma_k \subset \mathbb{R}^2$ and a map $X_k$ such that $x = \underset{\sim}{X}_k(u)$, $u = (u_1, u_2) \in \mathbb{R}^2$, covers $S_k$. The mappings are compatible on overlapping regions. To say that $S$ is a regular analytic surface means that the individual maps from $\Gamma_k$ to $\Gamma_e$ on overlaps are analytic and that $\underset{\sim}{X}_{k,u_1}$ and $\underset{\sim}{X}_{k,u_2}$ are linearly independent.

We use the $\underset{\sim}{X}_k$ to generate local coordinate systems in $\mathbb{R}^3$ and set

$$\underset{\sim}{e}_1(u) = X_{u_1}, \quad \underset{\sim}{e}_2(u) = X_{u_2}, \quad \underset{\sim}{e}_3(u) = \underset{\sim}{e}_1(u) \times \underset{\sim}{e}_2(u) \tag{1.11}$$

Then the equations

$$x = \underset{\sim}{X}(u) + u_3\, \underset{\sim}{e}_3(u), \quad u \in \Gamma, \quad |u_3| < \delta$$

will define a coordinate system for a region $U_k \subset \mathbb{R}^3$ with $u_3 = 0$ corresponding to $S_k$. We will assume that $u_3 > 0$ corresponds to $\Omega$.

For simplification we further assume that the coordinate systems are orthonomal, that is,

$$\underset{\sim}{e}_i(u) \cdot \underset{\sim}{e}_j(u) = \delta_{ij} .$$

Following the ideas of [21] we introduce a partition of unity $\sum \xi_k \equiv 1$ subordinate to the $S_k$ and define $V_\gamma(\psi)$ by

$$V_\gamma(\psi)(x) = \sum_k \int_{\Gamma_k} \psi\,(\underset{\sim}{X}_k(u))\xi_k(u)\ \phi_\gamma\ (|x-X_k(u)|)\ du \tag{1.12}$$

Here the orthonomality of the coordinate system implies that the surface element is unity. For $x \in S$, (1.12) gives

$$V_\gamma(\psi)(x) = \sum_j \sum_k \xi_j \int_{\Gamma_k} \psi(\underset{\sim}{X}_k(u))\ \xi_k(X_k(u))\ \phi_\gamma(|X_j-X_k(u)|)\ du \tag{1.13}$$

Formula (1.13) is the basis for the idea of pseudodifferential operators on $S$. If $\psi \in C_o^\infty(S_k)$ for some patch $S_k$ then $V_\gamma(\psi)$ will be in $C^\infty(S_k)$. The idea is to extend that definition to $\psi$'s which need not to be $C^\infty$ but lie in some Sobolev space on $S$. It is clear from (1.13) that one need concentrate only on the quantities $\chi V_\gamma(\psi)$ where $\chi$ and $\psi$ have support in the same patch $S_k$.

Let $\chi,\ \psi \in C_o^\infty(S_k)$. Then we have

$$\chi V_\gamma(\psi) = \chi(X(U)) \int_{\Gamma_k} \psi(X(u))\ \phi_\gamma(|X(U)-X(u)|)\ du = \int_{\mathbb{R}^2} \tilde{\psi}(u)\ K_\gamma(U,u-U)du \tag{1.14}$$

with the kernel

$$K_\gamma(U,u-U) = \chi(X(U))\ \phi_\gamma(|X(U)-X(u)|)\ .$$

Let us introduce the Fourier transform $\psi^\wedge$ of $\tilde{\psi}$ by

$$\psi^\wedge(\xi) = (2\pi)^{-1} \int_{\mathbb{R}^2} \tilde{\psi}(u)e^{-i\xi\cdot u}du \tag{1.15}$$

Then we can

$$\chi V_\gamma(\psi) = \int_{\mathbb{R}^2} e^{i\xi\cdot x}\ \psi^\wedge(\xi)\ a_\gamma(U,\xi)\ d\xi \tag{1.16}$$

with

$$a_\gamma(U,\xi) = (2\pi)^{-1}\ \chi(X(u)) \int_{\mathbb{R}^2} e^{-i\xi\cdot\eta}\ K_\gamma(U,\eta)\ d\eta.$$

Now, $a_\gamma(U,\xi)$ is called the <u>symbol</u> of $V_\gamma$ .

Suppose that $K_\gamma(U,\eta)$ has an asymptotic expansion of the form

$$K_\gamma(U,\eta) \sim \sum_{n=r}^{\infty} K_\gamma^n(U,\eta) \tag{1.17}$$

where $K_\gamma^n$ is homogeneous of degree $n$ in $\eta$ . Then $a_\gamma$ , the (distributional) Fourier transform of $K_\gamma$ , has the form

$$a_\gamma(U,\xi) \sim \sum_{n=r}^{\infty} a_\gamma^n(U,\xi)$$

where $a_\gamma^n$ is homogeneous of degree $-n-2$ in $\xi$ . If (1.17) holds then $V_\gamma$ obtained by (1.16) is called a pseudodifferential operator of <u>order</u> $r$ and $a_\gamma^r(U,\xi)$ is its <u>principal</u> symbol. $V_\gamma$ is called <u>elliptic</u> if $a_\gamma^r(U,\xi) \neq 0$ for $\xi \neq 0$ .

Before we cite some results from [21] on pseudodifferential operators on $S$ we recall the definition of Sobolev spaces on compact manifolds $S$ . Via diffeomorphisms $\chi$ mapping any domain $U \subset S$ onto open sets $U_\chi$ in $\mathbb{R}^2$ the Sobolev space $H^r(S)$ is the completion of $C^\infty(S)$ , the space of infinitely differentiable functions on $S$ , in the norm

$$\|\chi\psi\|_r^2 = \int_{\mathbb{R}^2} (1+|\xi|^2)^r \, |\widehat{\chi\psi}(\xi)|^2 d\xi \quad , \quad \psi \in C_0^\infty(S) \, , \tag{1.18}$$

defined by a partition of unity subordinate to a covering of $S$ by domains of charts [9].

LEMMA 1.3: *Let* $A$ *be a pseudodifferential operator of order* $r$ *on* $S$ . *Then*

(i) $A$ *is a continuous map from* $H^t(S)$ *into* $H^{t-r}(S)$ *for any* $t$

(ii) *If* $A$ *is elliptic the map* $A : H^t(S) \to H^{t-r}(S)$ *is Fredholm*

(iii) *If* $A$ *is elliptic then* $\psi \in H^t(S)$ *and* $A\psi \in H^s(S)$ *implies* $\psi \in H^{s+r}(S)$ *and there is a constant* $C_{t,s}$ *such that*

$$\|\psi\|_{s+r} \le C_{t,s} \left( \|A\psi\|_s + \|\psi\|_t \right)$$

Now we apply the above ideas to $V_\gamma$ and show first that the expansion (1.17) holds. Since $S$ is assumed to be analytic it follows that the functions $X$ are analytic and that

$$|\underset{\sim}{X}(U) - \underset{\sim}{X}(u)| = \sum_{\nu=1}^{\infty} M_\nu(U,u-U)$$

where $M_\nu$ is homogeneous of degree $\nu$ in $u-U$. Moreover the orthonormality of the coordinate system yields

$$M_1(U,u-U) = |u-U|$$

Next we find from (1.10) that

$$\phi_\gamma(r) = r^{-1} \sum_{j=0}^{\infty} \frac{\delta^j}{i!} r^j \quad , \quad \delta \in C \ , \ r = |x-y| \tag{1.19}$$

Thus we obtain

$$\phi_\gamma(|X(U) - X(u)|) = |u-U|^{-1} + \sum_{\nu=0}^{\infty} k_\gamma^\nu(U,u-U) \tag{1.20}$$

with $k_\gamma^\nu$ homogeneous of degree $\nu$ in $u-U$. Substituting (1.20) into (1.14) yields (1.17) with $r = -1$ and

$$K_\gamma^{-1}(U,\eta) = \chi(X(U)) \quad |\eta|^{-1} \tag{1.21}$$

Hence application of Fourier transform (1.15) $(\eta \to \xi)$ gives the principal symbol

$$a_\gamma^{-1}(U,\xi) = \quad \chi(X(U)) \frac{1}{2} \quad |\xi|^{-1} \tag{1.22}$$

of our pseudo-differential operator $V_\gamma$ . From (1.19) follows

$$\phi_\gamma(r) = \phi_i(r) + (i\gamma+1) + \Phi_\gamma(r) , \quad \Phi_\gamma(r) = \sum_{k=1}^{\infty} \frac{\delta_k}{k!} r^k \tag{1.23}$$

yielding the following perturbation result.

LEMMA 1.4 [14]: (i) $V_\gamma = V_i + \tilde{W}_\gamma$ *where* $\tilde{W}_\gamma$ *is a continuous map from* $H^t(S)$ *into* $H^{t+3}(S)$ .

(ii) $V_i$ *and* $V_{k_i}$ *map bijectively* $H^r(S)$ *onto* $H^{r+1}(S)$ *for any* $r$ .

*Proof:* Due to (1.23) the assertion (i) follows from the decomposition

$$V_\gamma(\psi) = V_i(\psi) + \Gamma_\gamma(\psi) + W_\gamma(\psi) \tag{1.24}$$

with

$$\Gamma_\gamma(\psi) = (i\gamma+1)\frac{1}{4\pi}\int_S \psi \, dS_y, \quad W_\gamma(\psi)(x) = \frac{1}{4\pi}\int_S \psi(y) \, \Phi_\gamma(|x-y|) \, dS_y$$

Since $W$ is a pseudodifferential operator of order $-3$ and $\Gamma_\gamma$ takes $H^r(S)$ into $H^t(S)$ for any $t$ .

(ii) From (1.23), (1.16) and (1.22) we see that $V_i$ is an elliptic pseudodifferential operator of order $-1$ . Thus by Lemma 1.3 $V_i$ is a Fredholm operator from $H^r(S)$ into $H^{r+1}(S)$ for any $r$ . Moreover $V_i$ is self-adjoint from $H^{-1/2}(S)$ to $H^{1/2}(S)$ (= dual space of $H^{-1/2}(S)$) since for any $\psi, \chi \in C_o^\infty(S)$ there holds

$$\int_S \psi(x)\, V_i(\chi)(x)\, dS_x = \int_S \chi(x)\, V_i(\psi)(x)\, dS_x$$

because $\phi_i$ depends only on $|x-y|$. Therefore $V_i$ is bijective from $H^{-1/2}(S)$ onto $H^{1/2}(S)$ if $V_i(\psi) = 0$ implies $\psi = 0$. Then by Lemma 1.3 (iii) the assertion holds for any $r$.

The injectivity of $V_i$ follows by standard arguments: Suppose $V_i(\psi) = 0$ for $\psi \in H^{-1/2}(S)$. Then by Lemmy 1.3 (iii) we have $\psi \in H^r(S)$ for any $r$ and hence $\psi$ is continuous. Thus due to Lemma 1.2 the potential $v(x) = \int_S \psi(y)\, \phi_i(|x-y|)\, dS_y$ is continuous in $\mathbb{R}^3$ satisfying $\Delta v = v$ in $\Omega \cup \Omega'$, $v = O(|x|^{-1} e^{-|x|})$ as $|x| \to \infty$ and $v \equiv 0$ on $S$. Application of Green's theorem over $\Omega_R = \Omega' \cup \{x, |x| < R\}$ gives

$$0 = \int_{\Omega_R} (\Delta v - v) v\, dx = -\int_{\Omega_R} (|\operatorname{grad} v|^2 + |v|^2)\, dx + \int_{\Gamma_R} v \frac{\partial v}{\partial \underset{\sim}{n}}\, ds$$

Thus

$$\int_{\Omega_R} (|\operatorname{grad} v|^2 + |v|^2)\, dx = \int_{\Gamma_R} v \frac{\partial v}{\partial \underset{\sim}{n}} R^2\, d\omega$$

and the integral on the right side vanishes as $R \to \infty$, because $v$ and $\frac{\partial v}{\partial n}$ are both $O(\frac{e^{-R}}{R})$ as $R \to \infty$. Hence $\|v\|_{\tilde{H}^1(\mathbb{R}^3)} \equiv 0$ implies $v \equiv 0$ in $\mathbb{R}^3$ and $\frac{\partial v}{\partial \underset{\sim}{n}} = 0$ on $S$. Now the jump relations (Lemma 1.2 (iv)) give $\psi = (\frac{\partial v}{\partial n})^- - (\frac{\partial v}{\partial n})^+ = 0$. The bijectivity of $V_{k_i}$ follows from the result on $V_i$, because we can write the equation $V_{k_i}(\psi) = f$ as $\psi + V_i^{-1} \tilde{W}_{k_i} \psi = V_i^{-1} f$, a Riesz-Schauder system. A similar argument as above shows that

$\psi + V_i^{-1} \tilde{W}_{k_i} \psi = 0$ implies $\psi = 0$. □

Via Lemma 1.2 (iv) there is defined an operator $K_\gamma$ by

$$K_\gamma(\psi)(x) = \frac{-1}{4\pi} \int_S \frac{\partial}{\partial n_x} \frac{e^{i\gamma|x-y|}}{|x-y|} \psi(y)\, dS_y \tag{1.25}$$

which (for $\gamma=0$) is the adjoint to the operator of the double layer potential

$$N_\gamma(\psi)(x) = -\frac{1}{4\pi} \int_S \frac{\partial}{\partial n_y} \frac{e^{i\gamma|x-y|}}{|x-y|} \psi(y)\, dS_y \tag{1.26}$$

LEMMA 1.5: $I + 2K_k$ *is bijective from* $H^r(S)$ *onto* $H^r(S)$ . *Moreover,*

$$(I + 2K_k)^{-1} = I + R_k \tag{1.27}$$

*where* $R_k$ *is continuous from* $H^r(S)$ *into* $H^{r+1}(S)$ .

*Proof:* Lemma 1.2 (iv) shows that $K_k$ is a pseudodifferential operator of order minus one hence takes $H^r(S)$ into $H^{r+1}(s)$ . Thus $I + 2K_k$ is a Riesz-Schauder operator. To show that it is bijective it suffices to show that $(I + 2K_k)\psi = 0$ implies $\psi = 0$ . If it is bijective the form (1.27) follows from the theory in [ 21 ].

Suppose, then, that $(I + 2K_k)\psi = 0$ . As before, we can use Lemma 1.3 to conclude that $\psi$ is smooth. Now define $v$ by $v = V_k(\psi)$ . We will have $(\Delta+k^2)v = 0$ in $\Omega$ and $(I + 2K_k)\psi = 0$ on $S$ implies

$\frac{\partial v}{\partial \underset{\sim}{n}} = 0$ on $S$. Now, uniqueness of this exterior Neumann problem gives $v \equiv 0$ in $\Omega$. But we can also set $v = V_k(\psi)$ in $\Omega'$. Assuming that $k^2 \neq 0$ is not an eigenvalue of the interior Dirichlet problem we deduce $v \equiv 0$ in $\Omega'$. Then by the jump relations in Lemma 1.2 (iv) we have

$$\psi(x) = \left(\frac{\partial}{\partial \underset{\sim}{n}} V_k \psi\right)^-(x) - \left(\frac{\partial}{\partial \underset{\sim}{n}} V_k(\psi)\right)^+(x) = 0 , \quad x \in S .$$

□

Now we are in the position to solve (1.4) by a simple layer method. Namely, setting $v_e = V_k(\psi)$ the exterior Neumann problem (1.4) is transformed into a Fredholm integral equation of second kind on $S$ for the unknown layer $\psi$

$$\psi(x) + 2 \int_S K_k(x,y)\psi(y)\, dS_y = 2 \frac{\partial v_o}{\partial n}(x) \quad , \quad x \in S . \tag{1.28}$$

which we abbreviate with the notation (1.25) by

$$(I + 2K_k)\psi = 2 \frac{\partial v_o}{\partial n} \tag{1.29}$$

As a consequence of Lemma 1.2 and Lemma 1.5 there holds the following result.

THEOREM 1.6: *If* $\psi \in C^o(S)$ *is a solution of* (1.28) *then* $v_e = V_k(\psi)$ *yields a (classical) solution of* (1.4). *For any real* $r$ *there exists exactly one solution of* (1.28) *for given data* $\frac{\partial v_o}{\partial \underset{\sim}{n}} \in H^r(S)$, *and if* $r \geq 1 + \epsilon$, $\epsilon > 0$, *then* $v_e = V_k(\psi)$ *solves* (1.4).

Our simple layer prodecure for the interface problem (1.1) - (1.3) proceeds as follows. With the representation formula $v_e = V_k(\psi)$ for the exterior pressure we set for the total acoustic field

$$v = V_k(\psi) + v_o \quad \text{in } \Omega, \qquad v = V_{k_i}(\chi) \quad \text{in } \Omega' \ . \tag{1.30}$$

We obtain from the boundary conditions (1.3) a coupled system of pseudo-differential equations for the unknown layers $(\psi,\chi)$ on S:

$$\begin{aligned} &V_{k_i}(\chi) = V_k(\psi) + v_o, \\ &(I - 2K_{k_i})\chi + \nu(I + 2K_k)\psi = 2\nu \frac{\partial v_o}{\partial \underset{\sim}{n}}, \quad \nu = \frac{\rho_i}{\rho} \in R \end{aligned} \tag{1.31}$$

But by evaluating the kernel function $r^{-1} e^{i\gamma r}$ for small $r$ with $\gamma = k_i$ and k one verifies as above that both $V_{k_i}$ and $V_k$ satisfy (1.24). Hence there holds

$$V_k(\psi) = V_{k_i}(\psi) + \mathcal{W}(\psi)$$

with a pseudodifferential operator $\mathcal{W}$ of order -3. Therefore multiplication of $(1.31)_1$ with the bijective operator $V_{k_i}^{-1}$ yields

$$\chi - \psi = V_{k_i}^{-1} \mathcal{W}(\psi) + V_{k_i}^{-1}(v_o) \tag{1.32}$$

Since furthermore

$$K_k(\psi) = K_{k_i}(\psi) + L(\psi)$$

with a pseudodifferential operator $L$ of order -2 the second equation in (1.31) gives

$$\chi + \nu\psi = 2K_{k_i}(\chi - \nu\psi) - 2L(\psi) + 2\frac{\partial v_o}{\partial \underset{\sim}{n}}\nu \tag{1.33}$$

The equations (1.32) and (1.33) form a Riesz-Schauder system on $H^r(S) \times H^r(S)$, $r \in R$. Each of the operators occuring on the right sides is of order at most $-1$ and the forcing terms $V_{k_i}^{-1}(v_o)$ and $\frac{\partial v_o}{\partial \underset{\sim}{n}}$ belong to $H^r(S)$ for given $v_o \in H^{r+1}(S)$. A reversal of the steps shows that if $(\psi,\chi)$ satisfy (1.32), (1.33), then they also satisfy (1.31). But the uniqueness result for (1.1) - (1.3) (Theorem 1.1) shows that the only solution of the homogeneous equations (1.32), (1.33) vanishes identically. Hence we have the following existence result for the interface problem (1.1)-(1.3) governing the scattering of sound:

THEOREM 1.7: *Let* $v_o \in H^{r+1}(S)$ *for arbitrary* $r \in R$. *Then the equations* (1.32) (1.33) *have a unique solution with* $\chi, \psi \in H^r(S)$.

For the double layer potential operator (1.26) the following results are well-known (see [4], [26]).

LEMMA 1.8: *For any complex* $\gamma$, $o \le \arg\gamma \le \pi/2$ *and any continuous* $\psi$ *on* S:

*(i)* $\Delta N_\gamma(\psi) = -\gamma^2 N_\gamma(\psi)$ *in* $\Omega \cup \Omega'$

*(ii)* $N_\gamma(\psi)(x) = O(|x|^{-2})$ *as* $|x| \to \infty$

*(iii)* $(N_\gamma \psi(x))^{\pm} = \mp \frac{1}{2}\psi(x) + \int_S \tilde{K}_\gamma(x,y)\psi(y)dS_y$ *on* $S$

*where the function* $\tilde{K}_\gamma(x,y) = -\frac{1}{4\pi}\frac{\partial}{\partial n_y}\frac{e^{i\gamma|x-y|}}{|x-y|}$ *is* $O(|x-y|^{-1})$ *as* $y \to x$.

*(iv)* $\left(\frac{\partial}{\partial \underset{\sim}{n}} N_\gamma(\psi)(x)\right)^{\pm} = -\frac{1}{4\pi}\int_S D_\gamma(x,y)\psi(y)dS_y$ *is a hyper-singular integral operator where the function*

$D_\gamma(x,y)$ *is* $O(|x-y|^{-3})$ *as* $y \to x$.

*Proof:* We only indicate (iv). Since

$$\frac{\partial}{\partial n_y} \frac{e^{i\gamma|x-y|}}{|x-y|} = \frac{n(y)\cdot(x-y)}{|x-y|^2} e^{i\gamma|x-y|} \left(\frac{1}{|x-y|} - i\gamma\right)$$

there holds

$$\frac{\partial}{\partial n_x} \frac{\partial}{\partial n_y} \frac{e^{i\gamma|x-y|}}{|x-y|} = e^{i\gamma|x-y|} \Bigg\{ n(x)\cdot n(y) \left( \frac{1}{|x-y|^3} - \frac{i}{|x-y|^2} \right) + \tag{1.34}$$

$$+ \frac{n(x)\cdot(x-y)}{|x-y|} n(y)\cdot(x-y) \left[ -\frac{3}{|x-y|^4} + \frac{3i\gamma}{|x-y|^3} + \frac{\gamma^2}{|x-y|^2} \right] \Bigg\} .$$

Let $(x,\delta) := S \cap \{x : |y-x| < \delta\}$. Then there are constants $a > o$, $M > o$, such that for any $x \in S$ the piece $(x,a)$ of $S$ is given by the analytic representation $y_3 = F(y_1,y_2)$ with the origin $x$ in the tangential-normal coordinate system (compare (1.11)). Then there holds

$$|F(y_1,y_2)| \le M(y_1^2 + y_2^2), \quad F_{y_1}(y_1,y_2)^2 + F_{y_2}(y_1,y_2)^2 \le M(y_1^2 + y_2^2) \tag{1.35}$$

$$|n(x)-n(y)| \le M|x-y| \quad \text{for} \quad y \in (x,a).$$

This yields with $n(x) = (o,o,1)$ and $y-x = (y_1,y_2,F(y_1,y_2))$

$$|n(y)\cdot(x-y)| = |[n(y)-n(x)]\cdot(x-y) + n(x)\cdot(x-y)| \tag{1.36}$$

$$\le |n(y)-n(x)|\cdot|x-y| + |F(y_1,y_2)| \le 2M|x-y|^2 .$$

Now, the assertion (iv) follows by setting $D_\gamma(x,y) = \frac{\partial}{\partial n_x} \frac{\partial}{\partial n_y} \frac{e^{i\gamma|x-y|}}{|x-y|}$ . □

Introducing again local charts on the analytic manifold $S$ and applying Fourier transform we show now that the normal derivative of the double layer potential is again a pseudodifferential operator. First we consider the case $\gamma = o$ (following an oral communication by W.L. Wendland), that is we consider $\phi_o(r) = r^{-1}$ instead of $\phi_\gamma(r) = r^{-1} e^{i\gamma r}$.

LEMMA 1.9: *The operator*

$$D_o\psi(x) = -\frac{1}{4\pi}\int_S \frac{\partial}{\partial \underset{\sim}{n}_x}\frac{\partial}{\partial \underset{\sim}{n}_y}\frac{1}{|x-y|}\psi(y)dS_y \tag{1.37}$$

*is a pseudodifferential operator of order plus one and a continuous map from* $H^r(S)$ *into* $H^{r-1}(S)$ *for any real* $r$.

*Proof:* With the notation

$$\xi=y_1=u_1,\ \eta=y_2=u_2,\ \zeta=F(u_1,u_2),\ x_i=x_i(u_1,u_2),\ i=1,2,3$$

there holds at the origin $M = (o,o,o) = x$

$$F(o,o) = F_{u_1}(o,o) = F_{u_2}(o,o) = o \quad\text{and}\quad \underset{\sim}{n}_x = (o,o,1) .$$

Thus with the surface gradient

$$(D_\xi, D_\eta, D_\zeta)\psi := \operatorname{grad}\psi - \underset{\sim}{n}\cdot\operatorname{grad}\psi\ \underset{\sim}{n} \tag{1.38}$$

we obtain

$$4\pi\frac{\partial}{\partial \underset{\sim}{n}_x}(N_\gamma\psi)(x)\Big|_{x=M} = 4\pi\frac{\partial}{\partial x_3}(N_\gamma\psi)(x) = -\frac{\partial}{\partial x_3}\int_S\frac{D_\zeta\psi(y)}{|x-y|}dS_y$$
$$-\frac{\partial}{\partial x_1}\int_S\frac{D_\xi\psi(y)}{|x-y|}dS_y - \frac{\partial}{\partial x_2}\int_S\frac{D_\eta\psi(y)}{|x-y|}dS_y + \int_S(D_\zeta\psi)(y)\frac{\underset{\sim}{n}(y)\cdot(y-x)}{|x-y|^3}dS_y \tag{1.39}$$

With our choice of orthonormal coordinate system we have $\sqrt{1+F_{u_1}^2+F_{u_2}^2} = c$

$$c\,\underset{\sim}{n}(u_1,u_2) = (-F_{u_1},-F_{u_2},1),\quad dS_y = c\,du_1\,du_2,\quad \psi_{u_1} = \psi_\xi + \psi_\zeta\cdot F_{u_1},$$
$$\psi_{u_2} = \psi_\eta + \psi_\zeta\cdot F_{u_2},\quad c\frac{\partial\psi}{\partial \underset{\sim}{n}} = -F_{u_1}\psi_\xi - F_{u_2}\psi_\eta + \psi_\zeta .$$

Hence, we find the representations

$$c^2 D_\xi\psi = \psi_{u_1}(c^2-F_{u_1}^2) - \psi_{u_2}F_{u_1}F_{u_2},\quad c^2 D_\eta = \psi_{u_2}(c^2-F_{u_2}^2) - \psi_{u_1}F_{u_1}F_{u_2}$$

and from $\underset{\sim}{n} \cdot (D_\xi, D_\eta, D_\zeta)\psi = o$ follows $c^2 D_\zeta \psi = \psi_{u_1} F_{u_1} + \psi_{u_2} F_{u_2}$.

Now with

$$\frac{\partial}{\partial x_3} \frac{1}{|x-y|}\bigg|_M = -\frac{(x_3-\zeta)}{|x-y|^3}\bigg|_M = \frac{F}{r^3}, \quad r = \sqrt{u_1^2 + u_2^2 + F^2}$$

we rewrite (1.39) in the origin $x = M$ as

$$4\pi \frac{\partial}{\partial n_x}(N_\gamma \psi)(x) = A\psi(x) + A_1\psi(x) + A_2\psi(x) + B_1\psi(x) + C_1\psi(x) + B_2\psi(x) + C_2\psi(x) + H\psi(x),$$

where

$$A\psi(x) = -\int_S \frac{(\psi_{u_1} F_{u_1} + \psi_{u_2} F_{u_2})F}{c\, r^3} du_1 du_2, \quad A_i\psi(x) = -\frac{\partial}{\partial x_i} \int_S \frac{\psi_{u_i}}{\sqrt{u_1^2+u_2^2}} du_1 du_2$$

$$B_i\psi(x) = \frac{\partial}{\partial x_i} \int_S \psi_{u_i} \left\{\frac{1}{\sqrt{u_1^2+u_2^2}} - \frac{c}{r}\right\} du_1 du_2, \quad i = 1,2$$

$$C_1\psi(x) = \frac{\partial}{\partial x_1} \int_S \frac{\psi_{u_1} F_{u_1}^2 + F_{u_1} F_{u_2} \psi_{u_2}}{c\, r} du_1 du_2, \quad C_2\psi(x) = \frac{\partial}{\partial x_2} \int_S \frac{\psi_{u_2} F_{u_2}^2 + F_{u_1} F_{u_2} \psi_{u_1}}{c\, r} du_1 du_2$$

$$H\psi(x) = \int_S \frac{(F_{u_1}\psi_{u_1} + F_{u_2}\psi_{u_2})(F - u_1 F_{u_1} - u_2 F_{u_2})}{c^2 r^3} du_1 du_2$$

Since

$$\frac{1}{\sqrt{u_1^2+u_2^2}} - \frac{1}{\sqrt{u_1^2+u_2^2+F^2}} = \frac{1}{2}(u_1^2+u_2^2)^{-3/2} F^2 + \ldots = O(\sqrt{u_1^2+u_2^2})$$

we have altogether using (1.35), (1.36) that $A$ is a pseudodifferential operator ($\psi$do) of order $-1$; $B_1$, $B_2$ are $\psi$do's of order zero; $C_1$, $C_2$, $H$ are $\psi$do's of order $-2$. Therefore the principal symbol of $D_o = \frac{\partial}{\partial n_x} N_o$ is given by

$$\sigma(D_o)(\xi) = -\frac{\xi_1^2 + \xi_2^2}{2|\xi|} = -|\xi|\frac{1}{2} \tag{1.40}$$

Hence, together with the general ideas above, the proof is complete. □

Taking $\gamma = o$ in (1.34) yields

$$\frac{\partial}{\partial n_x}\frac{\partial}{\partial n_y}\frac{1}{|x-y|} = \frac{n(x)\cdot n(y)}{|x-y|^3} - 3\,\frac{[n(x)\cdot(x-y)][n(y)\cdot(x-y)]}{|x-y|^5}$$

and therefore

$$g(x,y) := \frac{\partial}{\partial n_x}\frac{\partial}{\partial n_y}\left\{\frac{e^{i\gamma|x-y|}}{|x-y|} - \frac{1}{|x-y|}\right\} = e^{i\gamma|x-y|}\left\{n(x)\cdot n(y)\cdot\frac{-i\gamma}{|x-y|^2} + \right.$$

$$\left. + \frac{n(x)\cdot(x-y)}{|x-y|}\, n(y)\cdot(x-y)\left[\frac{3i\gamma}{|x-y|^3} + \frac{\gamma^2}{|x-y|^2}\right]\right\} .$$

Hence the corresponding operator $T\psi(x) = \int_S g(x,y)\psi(y)dS_y$ is a pseudodifferential operator of order $-1$ and for $D_\gamma = T + D_o$, a direct application of Lemma 1.9 yields the following result.

COROLLARY 1.10: *For any complex* $\gamma$, $o \leq \arg\gamma \leq \frac{\pi}{2}$, *the operator* $D_\gamma$ *given by (1.37) with* $r^{-1}e^{i\gamma r}$ *substituted for* $\frac{1}{r}$ *is a pseudodifferential operator of order plus one and maps* $H^r(S)$ *into* $H^{r-1}(S)$ *continuously for any real* $r$.

If one uses a double layer potential ansatz $v_e = N_\gamma(\psi)$, then the exterior Neuman problem (1.4) is reduced to the pseudodifferential equation on S:

$$D_\gamma\,\psi(x) = -\frac{\partial v_o}{\partial n} \quad \text{on} \quad S . \qquad (1.41)$$

In the following we treat the interface problem (1.1) - (1.3) via the direct method. With the fundamental solution $\phi_\gamma$ in (1.10) for the Helmholtz equation we obtain by Green's identity a representation formula for the solution of the exterior problem (1.4) in the form

$$v_e(x) = \int_\Omega \{v_e(\Delta+k^2)\phi_k - \phi_k(\Delta+k^2)v_e\}dy = -\int_S (\phi_k \frac{\partial v_e}{\partial \underset{\sim}{n}} - v_e \frac{\partial \phi_k}{\partial \underset{\sim}{n}})dS_y, \quad x \in \Omega . \tag{1.42}$$

Analogously, besides (1.42), for the solution of the interface problem (1.1) - (1.3) there holds for the interior field in $\Omega'$

$$v_i(x) = -\int_S (v_i \frac{\partial \phi_{k_i}}{\partial \underset{\sim}{n}} - \phi_{k_i} \frac{\partial v_i}{\partial \underset{\sim}{n}})dS_y , \quad x \in \Omega' = \mathbb{R}^3 - \overline{\Omega}. \tag{1.43}$$

With the notation (1.9) and (1.26), equations (1.42) and (1.43) read as

$$\begin{aligned} (I + 2N_k)\, v_e(x) &= -2V_k \frac{\partial v_e}{\partial \underset{\sim}{n}}(x) , \quad x \in S \\ (I - 2N_{k_i})v_i(x) &= 2V_{k_i} \frac{\partial v_i}{\partial \underset{\sim}{n}}(x), \quad x \in S \end{aligned} \tag{1.44}$$

In order to satisfy the interface condition (1.3) we have to take the normal derivative of the expression (1.42), that is with Lemma 1.2 (iv):

$$\frac{\partial v_e}{\partial \underset{\sim}{n}}(x) = \frac{1}{2} \frac{\partial v_e}{\partial \underset{\sim}{n}}(x) - \int_S \frac{\partial \phi_k}{\partial \underset{\sim}{n}_x} \frac{\partial v_e}{\partial \underset{\sim}{n}_y} dS_y + \int_S v_e \frac{\partial^2 \phi_k}{\partial \underset{\sim}{n}_x \partial \underset{\sim}{n}_y} dS_y, \quad x \in S \tag{1.45}$$

For the interior field there holds

$$\frac{\partial v_i}{\partial \underset{\sim}{n}}(x) = \frac{1}{2} \frac{\partial v_i}{\partial \underset{\sim}{n}}(x) + \int_S \frac{\partial \phi_{k_i}}{\partial \underset{\sim}{n}_x} \frac{\partial v_i}{\partial \underset{\sim}{n}_y} dS_y - \int_S v_i \frac{\partial^2 \phi_{k_i}}{\partial \underset{\sim}{n}_x \partial \underset{\sim}{n}_y} dS_y, \quad x \in S \tag{1.46}$$

or in short,

$$(1-2K_k) \frac{\partial v_e}{\partial \underset{\sim}{n}}(x) = -D_k v_e(x), \qquad (I+2K_{k_i}) \frac{\partial v_i}{\partial \underset{\sim}{n}}(x) = D_{k_i} v_i(x), \quad x \in S \tag{1.47}$$

Subtraction of $(1.47)_1$ from $(1.47)_2$ and inserting the interface conditions (1.4) yields with $\nu = \frac{\rho_i}{\rho}$,

$$-(D_k+D_{k_i})v_i - (\frac{1}{\nu} - 1) \frac{\partial v_i}{\partial \underset{\sim}{n}} + K \frac{\partial v_i}{\partial \underset{\sim}{n}} = (-I+2K_k) \frac{\partial v_o}{\partial \underset{\sim}{n}} =: f_1 \tag{1.48}$$

whereas subtraction of $(1.44)_2$ from $(1.44)_1$ gives

$$(N_k+N_{k_i})v_i+(\frac{1}{\nu}+1)V_{k_i}\frac{\partial v_i}{\partial \underset{\sim}{n}}-\frac{1}{\nu}\,V\,\frac{\partial v_i}{\partial \underset{\sim}{n}} = V_k\frac{\partial v_o}{\partial \underset{\sim}{n}} =: f_2 \qquad (1.49)$$

with smoothing pseudodifferential operators $K = (\frac{2}{\nu}K_k + 2K_{k_i})$ of order -1 and $V = V_{k_i} - V_k$ of order -3 .

The system (1.48) and (1.49) is an Agmon-Douglis-Nirenberg system on $H^{1/2}(S)\times H^{-1/2}(S)$ in the sense of ., thus defining a continuous map $A : (v_i, \frac{\partial v_i}{\partial \underset{\sim}{n}}) \to (f_1,f_2)$ from $H^{1/2}(S)\times H^{-1/2}(S)$ into $H^{-1/2}(S)\times H^{1/2}(S)$.
Moreover the difference between (1.48), (1.49) and

$$A_o = \begin{pmatrix} -2D_o & -(\frac{1}{\nu}-1)I \\ 2N_o & (1+\frac{1}{\nu})V_o \end{pmatrix}$$

is compact. Application of two-dimensional Fourier transform shows with (1.22), (1.40) that the principal symbol $\sigma(A)(\xi)$ has the form

$$\sigma(A) = \begin{pmatrix} |\xi| & -(\frac{1}{\nu}-1) \\ 0 & (1+\frac{1}{\nu})\frac{1}{2|\xi|} \end{pmatrix} \qquad (1.50)$$

Now, we can verify that there exists a $\gamma' > o$ and a $\kappa > 4$ such that

$$\mathrm{Re}(\zeta_1,\zeta_2)\begin{pmatrix} 1 & 0 \\ 0 & \kappa \end{pmatrix}\begin{pmatrix} |\xi| & (\frac{-1}{\nu}+1) \\ 0 & (1+\frac{1}{\nu})\frac{1}{2|\xi|} \end{pmatrix}\begin{pmatrix} \bar{\zeta}_1 \\ \bar{\zeta}_2 \end{pmatrix} \geq \gamma'\,(\zeta_1\,\bar{\zeta}_1 + \zeta_2\,\bar{\zeta}_2) \qquad (1.51)$$

for all $\zeta \in C^2$ and all $\xi \in R^2$ with $|\xi| = 1$, that is $A$ is strongly elliptic in the sense of [24] . By a general result for pseudodifferential operators we conclude now that the inequality (1.51) implies coerciveness in the sense of a Garding inequality.

THEOREM 1.11: *(i) There exists a real* $\gamma > 0$ *such that* (1.52)

$$\operatorname{Re} < A(v_i, \frac{\partial v_i}{\partial \underset{\sim}{n}}),\ (v_i, \frac{\partial v_i}{\partial \underset{\sim}{n}})>_o \ \geq \gamma\{||v_i||^2_{1/2} + ||\frac{\partial v_i}{\partial \underset{\sim}{n}}||^2_{-1/2}\} - |k((v_i, \frac{\partial v_i}{\partial \underset{\sim}{n}})(v_i, \frac{\partial v_i}{\partial \underset{\sim}{n}}))|$$

*for all* $(v_i, \frac{\partial v_i}{\partial \underset{\sim}{n}}) \in H^{1/2}(S) \times \tilde{H}^{-1/2}(S)$ *with a compact bilinear form* $k(\cdot,\cdot)$ *on* $(H^{1/2}(S) \times H^{-1/2}(S))^2$. *Here* $<\ ,\ >_o$ *is the* $L^2$*-scalar product*

$$< A(v_i, \frac{\partial v_i}{\partial \underset{\sim}{n}}),\ (v_i, \frac{\partial v_i}{\partial \underset{\sim}{n}})>_o = (-(D_k + D_{k_i})v_i - (\frac{1}{\nu} - 1)\frac{\partial v_i}{\partial \underset{\sim}{n}} + K\frac{\partial v_i}{\partial \underset{\sim}{n}}\ ,\ v_i)_o$$

$$+ (\ (N_k + N_{k_i})v_i + (\frac{1}{\nu} + 1)V_k\frac{\partial v_i}{\partial \underset{\sim}{n}} - \frac{1}{\nu}\ V\frac{\partial v_i}{\partial \underset{\sim}{n}},\ \frac{\partial v_i}{\partial \underset{\sim}{n}})_o$$

*(ii)* $A$ *, given by (1.48), (1.49), is a bijective mapping*

$$A\ :\ (v_i,\ \frac{\partial v_i}{\partial \underset{\sim}{n}}) \rightarrow ((-I + 2K_k)\frac{\partial v_o}{\partial \underset{\sim}{n}},\quad V_k\frac{\partial v_o}{\partial \underset{\sim}{n}})$$

*from* $H^{r+1}(S) \times H^r(S)$ *onto* $H^r(S) \times H^{r+1}(S)$ *for any real* $r$ .

*Proof:* Only (ii) is left so show. By inequality (1.52) $A$ is a Fredholm operator of index zero and therefore bijectivity follows from injectivity. But by means of similar arguments to the ones used above the uniqueness of (1.1) - (1.3) (Theorem 1.1) guarantees the uniqueness of the system (1.48), (1.49) and therefore (ii) holds.

□

## 2. A SIMPLE LAYER POTENTIAL METHOD FOR EDDY CURRENT PROBLEMS

Again let S be a closed analytic surface which divides $\mathbb{R}^3$ into simply connected disjoint domains, an interior $\Omega'$ (bounded) and an exterior $\Omega$ (unbounded).

$\Omega$ is to represent air and $\Omega'$ a metallic conductor. Thus the exterior domain $\Omega$ is characterized by constitutive parameters $\varepsilon_o, \mu_o$ denoting permittivity and permeabilty and is assumed to habe zero conductivity. The interior domain $\Omega'$ is characterized by constants $\varepsilon, \mu, \sigma$ where the conductivity $\sigma$ may be infinite. The total electromagnetic field $(\underset{\sim}{E}, \underset{\sim}{H})$ will consist of the sum of incident $(\underset{\sim}{E}^o, \underset{\sim}{H}^o)$ and scattered $(\underset{\sim}{E}^s, \underset{\sim}{H}^s)$ terms where the incident field is assumed to originate in $\Omega$. All fields are assumed to be monochromatic with frequency $\omega$.

It is noted that the incident field satisfies the time harmonic Maxwell equations almost everywhere in $\mathbb{R}^3$ and the field quantities are infinitely differentiable except at source points in $\Omega$ and $\Omega'$. Across the interface $S = \partial\Omega = \partial\Omega'$ the tangential component of the total field must be continuous. Therefore after an appropriate scaling the <u>eddy current</u> problem is given by

$$\begin{aligned} &\text{curl } \underset{\sim}{E} = \underset{\sim}{H} \quad , \quad \text{curl } \underset{\sim}{H} = \alpha^2 E \quad \text{in } \Omega \\ &\text{curl } \underset{\sim}{E} = \underset{\sim}{H} \quad , \quad \text{curl } \underset{\sim}{H} = i\beta \underset{\sim}{E} \quad \text{in } \Omega' \qquad (P_{\alpha\beta}) \\ &\text{and } (\underset{\sim}{n}\times\underset{\sim}{E})^+ = (\underset{\sim}{n}\times\underset{\sim}{E})^- \quad , \quad (\underset{\sim}{n}\times\underset{\sim}{H})^+ = (\underset{\sim}{n}\times\underset{\sim}{H})^- \quad \text{on } S \end{aligned}$$

Here $\alpha^2 = \omega^2 \varepsilon_o \mu_o$ , $\beta = (\omega\mu\sigma - i\omega^2\mu\varepsilon)$ are dimensionless parameters and $\beta = \omega\mu\sigma > 0$ if displasement currents are neglected in metal $(\varepsilon=0)$ .

Again, the superscripts plus and minus denote limits from $\Omega$ and $\Omega'$ where $\underset{\sim}{n}$ is outward directed normal on the surface S.

At higher conductivity the constant $\beta$ is usually large and this leads to the perfect conductor approximation. Formally this means solving only the Maxwell equations $(P_{\alpha\beta})$ in $\Omega$ for the scattered field and requiring that the tangential component of the total electric field vanishes on S ($\underset{\sim}{n}\times\underset{\sim}{E} = 0$ on S), i.e.

$$\begin{aligned} &\text{curl } \underset{\sim}{E}^s = \underset{\sim}{H}^s \ , \quad \text{curl } \underset{\sim}{H}^s = \alpha^2 \underset{\sim}{E}^s && \text{in } \Omega \\ &\qquad (\underset{\sim}{n}\times\underset{\sim}{E}^s)^+ = -(\underset{\sim}{n}\times\underset{\sim}{E}^o)^+ && \text{on } S \end{aligned} \qquad (P_{\alpha\infty})$$

If in addition the scattered field satisfies the Sommerfeld radiation condition the following uniqueness theorem holds:

THEOREM 2.1: *There exists at most one solution of* $(P_{\alpha\beta})$ *for any* $\alpha > 0$ *and* $0<\beta\leq\infty$ .

The proof of uniqueness for $(P_{\alpha\infty})$ can be found in [19] and the corresponding one for $(P_{\alpha\beta})$ (with $\sigma\neq 0$ in $\Omega'$) in [11] .

In order to avoid additional difficulties we need the technical assumption

$$\begin{aligned} &\text{curl } \underset{\sim}{E} = \underset{\sim}{H} \ , \ \text{curl } \underset{\sim}{H} = \alpha^2 \underset{\sim}{E} \text{ in } \Omega', \ \underset{\sim}{n}\times\underset{\sim}{E} = 0 \text{ on S implies} \\ &\underset{\sim}{E} = \underset{\sim}{H} = \underset{\sim}{0} \quad \text{on } \Omega' \end{aligned} \qquad (2.1)$$

i.e. $\alpha^2$ is not an eigenvalue of the interior Dirichlet problem for Maxwell's equations.

In the following we give new boundary integral equation methods (derived in [15]) for solving both $(P_{\alpha\beta})$ and $(P_{\alpha\infty})$ . To this end we use the simple layer $V_\gamma$ with continuous density $\psi$ on the surface S given by (1.9). For a continuous vector field $\underset{\sim}{v}$ on $S$ we define $V_\gamma(\underset{\sim}{v})$ by (1.9) again with $\underset{\sim}{v}$ replacing $\psi$. In addition to Lemma 1.2 there hold the following well-known properties of the simple layer potential (see [14],[19]).

LEMMA 2.2: *For any complex* $\gamma, 0 \le \arg\gamma \le \pi/2$ *and any continuous* $\underset{\sim}{v}$ *on* $S$:

(i) $V_\gamma(\underset{\sim}{v})$ *satisfies properties* (i) - (iii) *in Lemma* 1.2

(ii) $(\underset{\sim}{n} \times \operatorname{curl} V_\gamma(\underset{\sim}{v})(x))^{\pm} = \pm\frac{1}{2}\underset{\sim}{v}(x) + \int_S \underset{\sim}{K}_\gamma(x,y)\underset{\sim}{v}(y)\,dS_y$ *on* $S$

*where the matrix function* $\underset{\sim}{K}_\gamma$ *is* $O(|x-y|^{-1})$ *as* $y \to x$.

Our methods in [14] are based on the Stratton-Chu representation formulas of electromagnetic fields from [23] yielding for the scattered fields $(\underset{\sim}{E},\underset{\sim}{H})$ in $(P_{\alpha\infty})$:

$$\begin{aligned} \underset{\sim}{E} &= V_\alpha(\underset{\sim}{n}\times\underset{\sim}{H}) - \operatorname{curl} V_\alpha(\underset{\sim}{n}\times\underset{\sim}{E}) + \operatorname{grad} V_\alpha(\underset{\sim}{n}\cdot\underset{\sim}{E}) \\ & \qquad\qquad\qquad\qquad\qquad\qquad \text{in } \Omega \\ \underset{\sim}{H} &= \operatorname{curl} V_\alpha(\underset{\sim}{n}\times\underset{\sim}{H}) - \operatorname{curl}\operatorname{curl} V_\alpha(\underset{\sim}{n}\times\underset{\sim}{E}) \end{aligned} \tag{2.2}$$

If $\underset{\sim}{n}\times\underset{\sim}{H}$, $\underset{\sim}{n}\times\underset{\sim}{E}$ and $\underset{\sim}{n}\cdot\underset{\sim}{E}$ were all known on $S$ then (2.2) would yield a solution of $(P_{\alpha\infty})$ - but only $\underset{\sim}{n}\times\underset{\sim}{E}$ is known. The standard treatment of $(P_{\alpha\infty})$ starts from (2.2) but sets $\underset{\sim}{n}\times\underset{\sim}{H}$ and $\underset{\sim}{n}\cdot\underset{\sim}{E}$ equal to zero and replaces $-\underset{\sim}{n}\times\underset{\sim}{E}$ by an unknown tangential field $\underset{\sim}{L}$ (see [10]):

$$\underset{\sim}{E} = \operatorname{curl} V_\alpha(\underset{\sim}{L}), \quad \underset{\sim}{H} = \operatorname{curl}\operatorname{curl} V_\alpha(L).$$

Imposition of the boundary condition in $(P_{\alpha\infty})$ then yields an integral equation of second kind for $\underset{\sim}{L}$ in the tangent space to $S$. This method is analogous to solving the Dirichlet problem for the scalar Hemholtz equation with a double layer.

Our method in [15] for $(P_{\alpha\infty})$ is analogous to solving the scalar problems with a simple layer (see [7]). We set $\underset{\sim}{n}\times\underset{\sim}{E} = \underset{\sim}{0}$ in (2.2) and replace $\underset{\sim}{n}\times\underset{\sim}{H}$ and $\underset{\sim}{n}\cdot\underset{\sim}{E}$ by unknowns $\underset{\sim}{J}$ and $M$. Thus we take

$$\underset{\sim}{E} = V_\alpha(\underset{\sim}{J}) + \text{grad}\, V_\alpha(M) \;, \quad \underset{\sim}{H} = \text{curl}\, V_\alpha(\underset{\sim}{J}) \tag{2.3}$$

Having determined $\underset{\sim}{J}$ we can use Lemma 2.2 (ii) to determine $\underset{\sim}{n}\times\underset{\sim}{H}$ on $S$.

Since

$$-\alpha^2\underset{\sim}{E} = \Delta\underset{\sim}{E} = -\text{curl}\,\text{curl}\,\underset{\sim}{E} + \text{grad}\,\text{div}\,\underset{\sim}{E}$$

with $\text{curl}\,\underset{\sim}{E} = \underset{\sim}{H}$ there holds

$$-\alpha^2\underset{\sim}{E} = -\text{curl}\,\underset{\sim}{H} \quad \text{if } \text{div}\,\underset{\sim}{E} = 0 \text{ in } \Omega \,.$$

Hence the ansatz (2.3) gives a solution of $(P_{\alpha\infty})$ if $\text{div}\,\underset{\sim}{E} = 0$ holds in $\Omega$, because in (2.3) $\underset{\sim}{H}$ is automatically divergence free. But from (2.3) and Lemma 2.2 follows

$$\Delta\,\text{div}\,\underset{\sim}{E} = -\alpha^2\text{div}\,\underset{\sim}{E} \quad \text{in } \Omega \,.$$

Therefore by uniqueness of the exterior Dirichlet problem (for the Helmholtz equation) the boundary condition $\text{div}\,\underset{\sim}{E} = 0$ on $S$ guarantees $\text{div}\,\underset{\sim}{E} \equiv 0$ in the exterior domain $\Omega$. Imposing the boundary conditions of $(P_{\alpha\infty})$ in (2.3) and the contraint $\text{div}\,\underset{\sim}{E} = 0$ on $S$ we obtain a coupled system of pseudodifferential equations on the boundary surface S for the unknown densities $\underset{\sim}{J}, M$:

$$\begin{aligned} V_\alpha(\underset{\sim}{J})_T + \text{grad}_T V_\alpha(M) &= -(\underset{\sim}{n}\times\underset{\sim}{E}^o) \\ V_\alpha(\text{div}_T\underset{\sim}{J}) - \alpha^2 V_\alpha(M) &= 0 \end{aligned} \tag{$E_{\alpha\infty}$}$$

where $V_\alpha(\underset{\sim}{J})_T$ denotes the tangential component of the vector function $V_\alpha(\underset{\sim}{J})$. Here we habe used the relations with the surface divergence $\text{div}_T$ and the surface gradient $\text{grad}_T$ on $S$ for any $\gamma\in\mathbb{C}$ with

$0 \le \arg \gamma \le \frac{\pi}{2}$ :

$$\operatorname{div} V_\gamma(\underset{\sim}{v}) = V_\gamma(\operatorname{div}_T \underset{\sim}{v}) \ , \quad \operatorname{div}_T \operatorname{grad}_T V_\gamma(M) = -\gamma^2 V_\gamma(M) \ .$$

Our procedure for $(P_{\alpha\beta})$ in [14] proceeds as follows. This time we let $\underset{\sim}{E}$ and $\underset{\sim}{H}$ denote the total fields and use again (2.3) in $\Omega$ and its analog in $\Omega'$. Thus we put,

$$\underset{\sim}{E} = \underset{\sim}{E}^o + V_\alpha(\underset{\sim}{J}) + \operatorname{grad} V_\alpha(M) \ , \quad \underset{\sim}{H} = \underset{\sim}{H}^o + \operatorname{curl} V_\alpha(\underset{\sim}{J}) \ \text{in} \ \Omega, \qquad (2.4)$$

$$\underset{\sim}{E} = V_{\sqrt{i\beta}}(\underset{\sim}{j}) + \operatorname{grad} V_{\sqrt{i\beta}}(m) \ , \quad \underset{\sim}{H} = \operatorname{curl} V_{\sqrt{i\beta}}(\underset{\sim}{j}) \quad \text{in} \ \Omega'.$$

Again the constraint $\operatorname{div} \underset{\sim}{E} \equiv 0$ in $\Omega \cup \Omega'$ together with the boundary conditions in $(P_{\alpha\beta})$ give a coupled system of pseudodifferential equations for the unknown layers $(\underset{\sim}{J}, M, \underset{\sim}{j}, m)$ on $S$ :

$$\begin{aligned}
&V_\alpha(\underset{\sim}{J})_T + \operatorname{grad}_T V_\alpha(M) - V_{\sqrt{i\beta}}(\underset{\sim}{j}) - \operatorname{grad}_T V_{\sqrt{i\beta}}(m) = -\underset{\sim}{E}_T^o \\
&V_\alpha(\operatorname{div}_T \underset{\sim}{J}) - \alpha^2 V_\alpha(M) \qquad = 0 \\
&\underset{\sim}{J} + K_\alpha(\underset{\sim}{J}) + \underset{\sim}{j}\beta - K_{\sqrt{i\beta}}(\underset{\sim}{j}) = -2(\underset{\sim}{n} \times \underset{\sim}{H}^o) \qquad\qquad (E_{\alpha\beta}) \\
&V_{\sqrt{i\beta}}(\operatorname{div}_T \underset{\sim}{j}) - i\beta \, V_{\sqrt{i\beta}}(m) = 0 \ .
\end{aligned}$$

Equation $(E_{\alpha\beta})_1$ is caused by the interface condition $(\underset{\sim}{n} \times \underset{\sim}{E})^- = (\underset{\sim}{n} \times \underset{\sim}{E})^+$ on $S$, equations $(E_{\alpha\beta})_2$, $(E_{\alpha\beta})_4$ are consequences of the constraint $\operatorname{div} \underset{\sim}{E} = 0$ in $\Omega \cup \Omega'$ whereas equation $(E_{\alpha\beta})_3$ is resulting from the second interface condition $(\underset{\sim}{n} \times \underset{\sim}{H})^+ = (\underset{\sim}{n} \times \underset{\sim}{H})^-$ on $S$ together with Lemma 2.2 (ii) if we difine

$$\underset{\sim}{K}_\gamma(\underset{\sim}{v})(x) = 2 \int_S \underset{\sim}{K}'_\gamma(x,y) \underset{\sim}{v}(y) dS_y \ .$$

Collecting the above results there holds the following representation theorem.

THEOREM 2.3[14]: (i) *If* $(\underset{\sim}{J}, M)$ *solve* $(E_{\alpha\infty})$ *with* $\underset{\sim}{J}$ *differentiable and* $M$ *continuous then* (2.3) *yields a solution of* $(P_{\alpha\infty})$ .

(ii) *If* $(\underset{\sim}{J}, M, \underset{\sim}{j}, m)$ *solve* $(E_{\alpha\beta})$ *with* $\underset{\sim}{J}, j$ *differentiable and* $M, m$ *continuous then* (2.4) *yields a solution of* $(P_{\alpha\beta})$ .

In the following we derive existence, uniqueness and regularity of the boundary integral equations $(E_{\alpha\infty})$ and $(E_{\alpha\beta})$ on $S$ . As a standard procedure we first consider the half-space case which is obtained as a first approcimation to the general case by planing out the analytic surface $S$ .

In the half-space $\Omega = \{x \in \mathbb{R}^3 \mid x_3 > 0\}$ $(\underset{\sim}{n}(x) = \underset{\sim}{e}_3$ on $S = \mathbb{R}^2)$ the system $(E_{\alpha\infty})$ is reduced on $x_3 = 0$ to

$$\phi_\alpha * \underset{\sim}{J} + \frac{\partial}{\partial x_1} \phi_\alpha * \underset{\sim}{M}e_1 + \frac{\partial}{\partial x_2} \phi_\alpha * Me_2 = -4\pi (e_3 \times \underset{\sim}{E}^o) , \tag{2.5}$$

$$\phi_\alpha * \operatorname{div} \underset{\sim}{J} - \alpha^2 \phi_\alpha * M = 0$$

where the star denotes convolution. Via Fourier transformation we find in [ 14]

$$\hat{\phi}_\alpha(\xi) = (|\xi|^2 - \alpha)^{-1/2} \tag{2.6}$$

and the explicit solution of (2.5) reads with $\underset{\sim}{E}^o_T = \underset{\sim}{e}_3 \times \underset{\sim}{E}^o$ :

$$M = 4\, V_\alpha(\operatorname{div} \underset{\sim}{E}^o_T) \ , \ \underset{\sim}{J} = -\operatorname{grad}_T(4\, V_\alpha \operatorname{div} \underset{\sim}{E}^o_T) + 2(\Delta + \alpha^2) V_\alpha(\underset{\sim}{E}^o_T) \tag{2.7}$$

Let $H^r(\mathbb{R}^2)$ denote the Sobolev space of order $r$ on $\mathbb{R}^2$ , that is the completion of $C^\infty_o(\mathbb{R}^2)$ under the norm

$$\|\psi\|^2_r = \int_{\mathbb{R}^2} (1 + |\xi|^2)^r |\hat{\psi}(\xi)|^2 d\xi \ ,$$

and let $\underset{\sim}{H}^r(\mathbb{R}^2)$ be the space of vector functions with components in $H^r(\mathbb{R}^2)$ . Now, as shown in section 1 $V_\alpha$ is a pseudodifferential operator of order minus one in the sense of [ 21] with symbol (2.6) and maps $H^r(\mathbb{R}^2)$ $(\underset{\sim}{H}^r(\mathbb{R}^2))$ continuously into $H^{r+1}(\mathbb{R}^2)(\underset{\sim}{H}^{r+1}(\mathbb{R}^2))$ .

The operators $\mathrm{div}_T$ and $\mathrm{grad}_T$ are of order plus one and take $\underset{\sim}{H}^r(\mathbb{R}^2)$, $H^r(\mathbb{R}^2)$ into $H^{r-1}(\mathbb{R}^2)$, $\underset{\sim}{H}^{r-1}(\mathbb{R}^2)$ respectively. Though the result (2.7) is only formal, it indicates the mapping properties of the system $(E_{\alpha\infty})$ in the general case.

THEOREM 2.4 [14] : *For any real* $r$ *the mapping* $A_\alpha : (\underset{\sim}{J}, M) \to (-\underset{\sim}{E}^o_T, 0)$ *defined by the system* $(E_{\alpha\infty})$ *is bijective from* $\underset{\sim}{H}^{r-1}(S) \times H^r(S)$ *onto* $\underset{\sim}{H}^r(S) \times H^{r-1}(S)$ .

The proof of Theorem 2.4 given in [14] uses partition of unity and local coordinate system on the analytic surface $S$ (used in section 1) together with standard techniques for pseudodifferential operators. The expansion

$$\phi_\alpha(r) = \frac{e^{i\alpha r}}{r} = \frac{1}{r} \sum_{j=0}^{\infty} \frac{\delta^j}{j!} r^j , \quad \delta \in \mathbb{C} , \ r = |x-y| \tag{2.8}$$

induces the decomposition (see Lemma 1.4)

$$V_\alpha \psi = V_i \psi + W_\alpha \psi \tag{2.9}$$

where the Bessel potential $V_i$ is a bijective pseudodifferential operator of order minus one and $W_\alpha$ is a smoothing operator of order $-3$ . Due to the decomposition (2.9) the bijectivity of $(E_{\alpha\infty})$ is proved in [14] by a perturbation argument, since $(E_{\alpha\infty})$ differs by compact operators from

$$\begin{aligned} V_i(\underset{\sim}{J})_T + \mathrm{grad}_T V_i(M) &= \underset{\sim}{F} , \\ V_i(\mathrm{div}_T \underset{\sim}{J}) + V_i(M) &= G . \end{aligned} \tag{2.10}$$

The system (2.10) can be reduced to a Riesz-Schauder system for $(\underset{\sim}{J}, M)$ since $V_i$, $(V_i)_T$ and $(\Delta_T - I)$ are bijective and for any $\gamma \in \mathbb{C}$ , $0 \le \arg \gamma \le \frac{\pi}{2}$ , there holds

$$\operatorname{div}_T V_\gamma(\underset{\sim}{v})_T = V_\gamma(\operatorname{div}_T \underset{\sim}{v}) + J_\gamma(\underset{\sim}{v}) \tag{2.11}$$

with a continuous map $J_\gamma$ from $\underset{\sim}{H}^r(S)$ into $H^{r+1}(S)$ , $r \in \mathbb{R}$ (see [14 ]). The obtained Riesz-Schauder system for $(\underset{\sim}{J},M)$ is uniquely solvable which follows by potential theoretic arguments from Theorem 2.1 (together with jump relations). In a second step we reduce our original system $(E_{\alpha\infty})$ to a Riesz-Schauder system making use of the bijectivity of (2.10) . This again is uniquely solvable due to Theorem 3.1 and our technical assumption (2.1) .

In a similar way we treat in [14] the system $(E_{\alpha\beta})$ for the eddy current problem. Using the bijectivity of $V_{\sqrt{i\beta}}$ from $H^r(S)$ onto $H^{r+1}(S)$ $(r \in \mathbb{R})$ and the relation (2.11) we reduce $(E_{\alpha\beta})$ to a Riesz-Schauder system on $\underset{\sim}{H}^{r-1}(S) \times H^r(S)$ :

$$\begin{aligned}
\underset{\sim}{j} &= \underset{\sim}{J} + D_1(M,\underset{\sim}{J},\underset{\sim}{j}) + g \\
\underset{\sim}{j} + \underset{\sim}{J} &= K_{\sqrt{i\beta}}(\underset{\sim}{j}-\underset{\sim}{J}) - L_1(\underset{\sim}{J}) + \underset{\sim}{H}^o_T \qquad (2.12)\\
m - M &= L_2(M,\underset{\sim}{J},\underset{\sim}{j}) + f \\
i\beta m + \alpha^2 M &= D_2(M,\underset{\sim}{J},\underset{\sim}{j}) + h
\end{aligned}$$

where $\underset{\sim}{g} \in H^{r-1}(S)$ ; $f,h \in H^r(S)$ for given data $\underset{\sim}{H}^o_T \in H^{r+1}(S)$, $\underset{\sim}{E}^o_T \in H^r(S)$ . Here $D_1, D_2 K_{\sqrt{i\beta}}$ and $L_1, L_2$ are pseudo-differential operators of order minus one and minus two, respectively. Again (2.12) turns out to be uniquely solvable guaranteeing the following existence and regularity result for $(E_{\alpha\beta})$ .

THEOREM 2.5 [14] : *For any real* $r$ *and given* $\underset{\sim}{E}_T^o \in \underset{\sim}{H}^r(S)$, $\underset{\sim}{H}_T^o \in \underset{\sim}{H}^{r+1}(S)$ *the system* $(E_{\alpha\beta})$ *has a unique solution*

$$\underset{\sim}{J}, \underset{\sim}{j} \in \underset{\sim}{H}^{r-1}(S) \; ; \quad M, m \in H^r(S) \quad .$$

Remark 2.6: Since the field quantities in $(P_{\alpha\infty})$ and $(P_{\alpha\beta})$ are infinitely differentiable except at source points in $\Omega$ and $\Omega'$ the regularity of the solution $(\underset{\sim}{J}, M)$ of $(E_{\alpha\infty})$ and $(\underset{\sim}{J}, M, \underset{\sim}{j}, m)$ of $(E_{\alpha\beta})$ hinges only on the smoothness of the boundary surface $S$ .

At the end of this section we want to discuss two special cases of our eddy current problem. In each case the region $\Omega'$ is to be a cylinder of non-ferromagnetic metal parallel to the $x_3$-axis. We will then use $\Omega'$ to denote the cross sectional area in the $x_1$-$x_2$ plane with $\Gamma$ its boundary and $\Omega$ its exterior. We will also require that all fields depend only on the variables $x_1$ and $x_2$ .

For the above special case it turns out that the quantities $\underset{\sim}{J}$ and $M$ will depend only on $(x_1, x_2)$ . Then the quantities $V_\alpha(\underset{\sim}{J})$ and $V_\alpha(M)$ can be simplified. Indeed we can perform the integration with respect to the $y_3$ variable explicitly using the formula, [18],

$$\frac{1}{4\pi} \int_{-\infty}^{+\infty} \frac{e^{i\alpha\sqrt{r^2+\varphi^2}}}{\sqrt{r^2+\varphi^2}} \, d\varphi = -\frac{1}{4} H_o^{(1)}(\alpha r) \equiv K(r) \tag{2.13}$$

where $H_o^{(1)}$ is the Hankel function of first kind and order zero. Using this formula we find that $V_\alpha(\underset{\sim}{J})$ and $V_\alpha(M)$ assume the form

$$V_\alpha(\underset{\sim}{J})(x) = \int_\Gamma \underset{\sim}{J}(y)\, K(|x-y|) ds_y$$
$$V_\alpha(M)(x) = \int_\Gamma M(y)\, K(|x-y|) ds_y \qquad (2.14)$$

where here $x = (x_1, x_2)$ , $y = (y_1, y_2)$ . Thus we are reduced to one-dimensional integrals around the curve $\Gamma$ .

The first case we consider is that of transverse-magnetic (T-M) fields. Here all fields are to have the form,

$$\underset{\sim}{E} = E(x_1, x_2)\, \underset{\sim}{e}_3 \; ; \; \underset{\sim}{H} = H^1(x_1, x_2)\, \underset{\sim}{e}_1 + H^2(x_1, x_2)\, \underset{\sim}{e}_2 \qquad (2.15)$$

For this situation our ansatz is very simple for no $M$ is needed. We want to give a physical explanation for this since it helps to explain our procedure.

The quantities $\underset{\sim}{J}$ and $M$ in our general formula correspond physically to surface currents and surface charges respectively (see [23]). Physically, the charges are necessary to keep current from flowing out of the conductor. For the case (2.15) the induced eddy currents will be directed along the cylinder and no surface charges are needed to contain them, that is $M = 0$ .

The second case we treat is that of transverse-electric (T-E) fields. Here the roles of $\underset{\sim}{E}$ and $\underset{\sim}{H}$ are reversed:

$$\underset{\sim}{H} = H(x_1, x_2)\, \underset{\sim}{e}_3 \; ; \; \underset{\sim}{E} = E^1(x_1, x_2)\, \underset{\sim}{e}_1 + E^2(x_1, x_2)\, \underset{\sim}{e}_2 \qquad (2.16)$$

In this case the transverse-electric fields produce transverse currents and our physical reasoning above indicates that surface charges must be present, $M \neq 0$ .

In order to analyze our two-dimensional problems we parameterize the curve $\Gamma$ by $x = X(\tau)$ , $0<\tau<L$ . Our formulas will be simpler if we assume that $\tau$ is actually arclength, that is,

$$|\dot{X}(\tau)| \equiv 1 \ , \ ds = d\tau \tag{2.17}$$

We will describe $\underset{\sim}{J}$ and $M$ as functions of the parameter $\tau$ . We also introduce the function $\varphi$ ,

$$\varphi(t,\tau) = \dot{X}(t) \cdot \dot{X}(\tau) \tag{2.18}$$

From (2.17) it follows immediately that

$$\varphi(t,\tau) = 1 + O((t-\tau)^2) \text{ as } \tau \to t \tag{2.19}$$

As a further notational convenience we put,

$$r(t,\tau) = |X(t)-X(\tau)| \ , \ h(t,\tau) = -\frac{i}{4} H_o^{(1)}(\alpha r(t,\tau))$$
$$k(t,\tau) = \varphi(t,\tau)\, h(t,\tau) \tag{2.20}$$

Now we describe our solution procedure in the two special cases. We put,

$$\underset{\sim}{J} = J(\tau)\, \underset{\sim}{e}_3 \ , \ M \equiv 0 \quad \text{for (T-M)} \tag{2.21}$$

$$\underset{\sim}{J} = J(\tau)\dot{X}(\tau) \ , \ M = M(\tau) \text{ for (T-E)} \tag{2.22}$$

For (2.21) we have $\operatorname{div}_T \underset{\sim}{J} \equiv 0$ and our system $(E_{\alpha\infty})$ with use of (2.14) becomes the single equation,

$$\int_o^L J(\tau)\, h(t,\tau)\, d\tau = -E_o^T(X(t)) \ , \ 0<t<L$$

In the (T-E) case, we have with (2.14)

$$V_\alpha(\underset{\sim}{J})_T = \left(\int_0^L J(\tau)\ \varphi(t,\tau)\ h(t,\tau)\ d\tau\right) \dot{X}(t)$$

$$\text{grad}_T\ V_\alpha(M) = \left(\frac{d}{dt}\int_0^L M(\tau)\ h(t,\tau)\ d\tau\right) \dot{X}(t)$$

$$V_\alpha(\text{div}\underset{\sim}{J}) = \int_0^L \dot{J}(\tau)\ h(t,\tau)\ d\tau$$

Hence the system $(E_{\alpha\infty})$ becomes, in this case,

$$\begin{aligned} &\int_0^L J(\tau)\ k(t,\tau)\ d\tau + \frac{d}{dt}\int_0^L M(\tau)\ h(t,\tau)\ d\tau = -E_T^o(X(t))\cdot\dot{X}(t) \equiv f(t) \\ &-\int_0^L \dot{J}(\tau)\ h(t,\tau)\ d\tau + \alpha^2\int_0^L M(\tau)\ h(t,\tau)\ d\tau = 0 \end{aligned} \tag{2.23}$$

where

$$h(t,\tau) = \frac{1}{2\pi}\log|t-\tau| + m + O(\tau-t)\ ,\quad m \text{ a constant} \tag{2.24}$$

$$k(t,\tau) = h(t,\tau)\ (1+O(t-\tau)^2))\ .$$

The above results illustrate in a concrete manner our statements about the general system $(E_{\alpha\infty})$. (2.24) means the $J$ operator in $(2.23)_1$ is of order minus one while $M$ operator is of order zero in $(2.23)_1$. Since $\dot{J}$ is an operator of order one, the $J$ operator in $(2.23)_2$ is of order zero. Finally the $M$ operator in $(2.23)_2$ is of order minus one.

3. THE SKIN-EFFECT APPROXIMATION

In this section we describe our asymptotic solution for the eddy current problem. We carry out the calculation for the half-space problem described in section 2. We do this mainly to make the procedure clearer but we observe that if one uses the orthonormal coordinate systems as described in section 1 then in fact our calculations are locally exact.

Let us study the half space problem:

$$\begin{aligned} &\operatorname{curl} \underset{\sim}{E} = \underset{\sim}{H}\,, \quad \operatorname{curl} \underset{\sim}{H} = \alpha^2 \underset{\sim}{E} \quad \text{in} \quad x_3 > 0 \\ &\operatorname{curl} \underset{\sim}{E} = \underset{\sim}{H}\,, \quad \operatorname{curl} \underset{\sim}{H} = i\beta^2 \underset{\sim}{E} \quad \text{in} \quad x_3 < 0\,. \end{aligned} \tag{3.1}$$

We have a prescribed incident field $\underset{\sim}{E}^o$ , $\underset{\sim}{H}^o$ and the interface conditions are,

$$\underset{\sim}{E}_T^+ = \underset{\sim}{E}_T^-\,, \quad \underset{\sim}{H}_T^+ = H_T^- \quad \text{on} \quad x_3 = 0\,. \tag{3.2}$$

We assume the asymptotic form:

$$\begin{pmatrix} \underset{\sim}{E} \\ \underset{\sim}{H} \end{pmatrix} \sim \begin{pmatrix} \underset{\sim}{E}^o \\ \underset{\sim}{H}^o \end{pmatrix} + \sum_{n=0}^{\infty} \begin{pmatrix} \underset{\sim}{E}_n \\ \underset{\sim}{H}_n \end{pmatrix} \beta^{-n} \quad \text{in} \quad x_3 > 0\,, \tag{3.3}$$

$$\begin{pmatrix} \underset{\sim}{E} \\ \underset{\sim}{H} \end{pmatrix} \sim e^{\sqrt{-i}\,\beta\, x_3} \sum_{n=0}^{\infty} \begin{pmatrix} \underset{\sim}{E}_n \\ \underset{\sim}{H}_n \end{pmatrix} \beta^{-n} \quad \text{in} \quad x_3 < 0\,. \tag{3.4}$$

The idea, then, is to substitute (3.3) and (3.4) into (3.1) and (3.2) and equate coefficients of like powers of $\beta$ .

We decompose fields $\underset{\sim}{F}$ into tangential and normal components

$$\underset{\sim}{F} = \underset{\sim}{J} + f\underset{\sim}{e}_3 , \quad \underset{\sim}{J} = J^1\underset{\sim}{e}_1 + J^2\underset{\sim}{e}_2 .$$

and for any tangential vector $\underset{\sim}{J}$ we set

$$\underset{\sim}{J}^{\perp} = -J^2\underset{\sim}{e}_1 + J^1\underset{\sim}{e}_2 \equiv \underset{\sim}{e}_3 \times \underset{\sim}{J} , \quad (\underset{\sim}{J}^{\perp})^{\perp} = -\underset{\sim}{J} .$$

With the notation,

$$\text{grad}_T\, f = f_{x_1}\underset{\sim}{e}_1 + f_{x_2}\underset{\sim}{e}_2 , \quad \text{div}_T\, \underset{\sim}{F} = \text{div}\, \underset{\sim}{J} = J^1_{x_1} + J^2_{x_2}$$

we observe that with $\chi = e^{\sqrt{-i}\,\beta x_3}$

$$\text{curl}\, \underset{\sim}{F} = \underset{\sim}{J}^{\perp}_{x_3} - (\text{grad}_T\, f)^{\perp} - (\text{div}\, \underset{\sim}{J}^{\perp})\underset{\sim}{e}_3 .$$

$$(3.5)$$

$$\text{curl}\, \chi\underset{\sim}{F} = \chi\{\sqrt{-i}\,\beta\, \underset{\sim}{J}^{\perp} + \underset{\sim}{J}^{\perp}_{x_3} - (\text{grad}_T\, f)^{\perp} - (\text{div}\, \underset{\sim}{J}^{\perp})\underset{\sim}{e}_3$$

Substitution of (3.3) and (3.4) into (3.1) and (3.2) yields

$$\text{curl}\, \underset{\sim}{E}_n = \underset{\sim}{H}_n , \quad \text{curl}\, \underset{\sim}{H}_n = \alpha^2\underset{\sim}{E}_n \quad \text{in} \quad x_3 > 0 . \tag{3.6}$$

Whereas for $x_3 < 0$ from $(3.5)_2$ and (3.4) we have:

$$\text{curl}\, \underset{\sim}{E} \sim \chi\{\sqrt{-i}\,\beta\, \underset{\sim}{E}^{\perp}_o + \sum_{n=0}^{\infty} [\sqrt{-i}\, \underset{\sim}{E}^{\perp}_{n+1} + \underset{\sim}{E}^{\perp}_{n,x_3} - (\text{grad}\, e_n)^{\perp} - (\text{div}\, \underset{\sim}{E}^{\perp}_n)\underset{\sim}{e}_3]\beta^{-n}\} \tag{3.7}$$

$$\text{curl}\, \underset{\sim}{H} \sim \chi\{\sqrt{-i}\,\beta H^{\perp}_o + \sum_{n=0}^{\infty} [\sqrt{-i}\, \underset{\sim}{H}^{\perp}_{n+1} + \underset{\sim}{H}^{\perp}_{n,x_3} - (\text{grad}\, h_n)^{\perp} - (\text{div}\, \underset{\sim}{H}^{\perp}_n)\underset{\sim}{e}_3]\beta^{-n}\} \tag{3.8}$$

where $\underset{\sim}{E}_n = \underset{\sim}{\mathcal{E}}_n + \underset{\sim}{e}_3\, e_n$, $H_n = \underset{\sim}{\mathcal{H}}_n + h_n\, \underset{\sim}{e}_3$ .

We equate (3.7) to,

$$\underset{\sim}{H} = \chi \sum_{n=0}^{\infty} (\underset{\sim}{\mathcal{H}}_n + h_n)\beta^{-n}$$

and (3.8) to

$$i\beta^2 \underset{\sim}{E} = \chi\{i\beta^2 \underset{\sim}{\mathcal{E}}_o + i\beta^2 e_o + i\beta \underset{\sim}{\mathcal{E}}_1 + i\beta e_1 + \sum_{n=0}^{\infty} (i\underset{\sim}{\mathcal{E}}_{n+2} + ie_{n+2})\beta^{-n}\} .$$

Then we equate tangential and normal components of coefficients of like powers of $\beta$ . This shows us first that:

$$\underset{\sim}{\mathcal{E}}_o = -\underset{\sim}{\mathcal{E}}^o = \underset{\sim}{E}^o_T \; ; \; e_o \equiv 0 \, , \; e_1 \equiv 0 \, . \tag{3.9}$$

The next powers yield the equations:

$$\sqrt{-i}\; \underset{\sim}{\mathcal{E}}^{\perp}_1 = \underset{\sim}{\mathcal{H}}_o \quad , \quad h_o = \operatorname{div} \underset{\sim}{\mathcal{E}}^{\perp}_o = 0 \, . \tag{3.10}$$

In order to start the recursion process we first use $(3.6)_o$ and (3.9) to conclude that,

$$\operatorname{curl} \underset{\sim}{E}_o = \underset{\sim}{H}_o , \; \operatorname{curl} \underset{\sim}{H}_o = \alpha^2 \underset{\sim}{E}_o \text{ in } x_3 > 0, \; \underset{\sim}{E}^+_o = -\underset{\sim}{E}^o_T \text{ on } x_3 = 0 \, .$$

Thus $(\underset{\sim}{E}_o , \underset{\sim}{H}_o)$ is just the solution of $(P_{\alpha\infty})$ which we can solve. Bit from (3.2) we obtain,

$$\underset{\sim}{H}_o^- = \underset{\sim}{H}_o^+ = (\underset{\sim}{H}_o)_T^+ = \underset{\sim}{n} \times \text{curl}\ \underset{\sim}{E}_o \quad \text{on} \quad x_3 = 0 \ . \tag{3.11}$$

The right side of (3.11) is known (and easily computed with our process). Then (3.10), (3.2), and (3.11) yields,

$$(\underset{\sim}{E}_1)_T^+ = (\underset{\sim}{E}_1)_T^- = \underset{\sim}{E}_T^- = - \sqrt{i}\ (\underset{\sim}{H}_o^{\perp})^- = - \sqrt{i}\ ((\underset{\sim}{H}_o)_T^+)^{\perp} \ . \tag{3.12}$$

Then by $(3.6)_1$ we have a new problem for $(\underset{\sim}{E}_1, \underset{\sim}{H}_1)$ which is just like $(P_{\alpha\infty})$ but with new boundary values for $\underset{\sim}{E}_T$ as given by (3.11). Again this is solvable and thus we have found completely the first two terms in the expansion for $\underset{\sim}{E}$ and $\underset{\sim}{H}$ in $x_3 > 0$ . In $x_3 < 0$ $(E_n, H_n)$ can be obtained by integration (see [14]).

This process can be continued recursively, at each step we have to solve only exterior problems $(P_{\alpha\infty})$ with new data obtained from the inner expansion.

We note that to obtain the first order connection to the exterior field, that is $(\underset{\sim}{E}_1, \underset{\sim}{H}_1)$ it is not necessary to calculate any of the terms in the inner expansion, since by (3.12) the boundary values for $(\underset{\sim}{E}_1)_T$ are determined solely from $\underset{\sim}{H}_o$ , the infinite conductivity approximation.

Recently we obtained in [16] an analysis to establish the validity of the asymptotic procedure at least for the two dimensional situation of transverse-magnetic fields. Here we give the main results from [16].

Let $\Omega'$ be a region in $\mathbb{R}^2$ bounded by a smooth closed curve $\Gamma$ and let $\Omega = \mathbb{R}^2 - \bar{\Omega}'$. The problem to be studied is that of finding $u(x)$, $x = (x_1, x_2) \in \mathbb{R}^2$, such that,

$$\Delta u = -i\beta^2 u \quad \text{in} \quad \Omega' \,, \quad \Delta u = 0 \quad \text{in} \quad \Omega$$

$$u^- = u^+ + U^+ \,, \quad u_n^- = u_n^+ + U_n^+ \quad \text{on} \quad \Gamma \qquad (P_\beta)$$

$$u = O(1) \quad \text{as} \quad |x| \to \infty$$

Here $U$ is a given function with $\Delta U = 0$ in $\Omega$, save possibly for isolated singularities at $x = x^1, \ldots, x^k$. $n$ is the outer normal to $\Gamma$ and the plus and minus signs denote limits from $\Omega$ and $\Omega'$. $\beta$ is a positive constant.

We associate with $(P_\beta)$ a limit problem $(P_o)$ which is,

$$\Delta u = 0 \quad \text{in} \quad \Omega \quad , \quad u^+ = -U^+ \quad \text{on} \quad \Gamma$$

$$(P_o)$$

$$u = O(1) \quad \text{as} \quad |x| \to \infty$$

Thus $(P_o)$ is simply an exterior Dirichlet problem. A formal asymptotic solution procedure for large $\beta$, described in detail in [6] consists of two formal series of the form,

$$u \sim \sum_{m=0}^{\infty} v_m \beta^{-m} \quad \text{in} \quad \Omega$$

$$(3.13)$$

$$u \sim \psi \sum_{m=1}^{\infty} w_m \beta^{-m} \quad \text{in} \quad \Omega', \quad \psi = e^{i\sqrt{i}\,\beta\,\eta} \,,$$

Here $v_m$ and $w_m$ are independent of $\beta$ and $\psi$ is a function which decays exponentially as one moves from $\Gamma$ into $\Omega'$. The $v_m$ and $w_m$ can be computed recursively, $v_o$ being the solution of $(P_o)$ and $v_m$, $m \geq 1$, solutions of problems like $(P_o)$. The $w_m$ are determined by solving some ordinary differential equations (see [16]).

A solution procedure for the full problem $(P_\beta)$ is given in [6] and is quite complicated. The asymptotic procedure reduces one to solve only the exterior problem $(P_o)$. The physical situation leading to $(P_\beta)$ is a cylinder, of cross section $\Omega'$ and non-ferromagnetic metal, lying parallel to the $x_3$ - axis, the exterior is air and there is an incident electromagnetic field $(\underset{\sim}{E}^o, \underset{\sim}{H}^o)$ having the transverse magnetic form that is there are functions $E^o$, $H_1^o$ and $H_2^o$ of $x_1$ and $x_2$ only such that,

$$\underset{\sim}{E}^o = E^o \underset{\sim}{e}_3 \quad , \quad \underset{\sim}{H}^o = H_1^o \underset{\sim}{e}_1 + H_2^o \underset{\sim}{e}_2$$

The resulting total fields $(\underset{\sim}{E}, \underset{\sim}{H})$ will then have the same form. If one assumes that the conductivity of air as well as displacement currents in both air and metal can all be neglected and the magnetic permeabilities of air and metal have same value $\mu$ then Maxwell's equations can be scaled so as to read,

$$\text{curl } \underset{\sim}{E} = \underset{\sim}{H} \quad , \quad \text{curl } \underset{\sim}{H} = i\beta^2 \underset{\sim}{E} \quad \text{in } \Omega'$$

$$\text{curl } \underset{\sim}{E} = \underset{\sim}{H} \quad , \quad \text{curl } \underset{\sim}{H} = 0 \quad \text{in } \Omega$$

For fields of the above form ((T-M)case) there are scalar functions u and U , such that, the different sets of Maxwell equations inside and outside of the cylinder reduce to $(P_\beta)$. The transition conditions across $\Gamma$ in $(P_\beta)$ represent continuity of tangential components of $\underset{\sim}{E}$ and $\underset{\sim}{H}$ . For <u>infinite conductivity</u> $\beta = \infty$ one assumes that the exterior field can have no tangential component at $\Gamma$ ; this yields $(P_o)$ with $U = E^o$ .

Our asymptotic procedure reflects the <u>skin effect</u> in which the interior fields concentrate near the surface and decay exponentially into the cylinder. It gives the infinite conductivity approximation as first terms and then calculates corrections.

Under the assumptions - $\Gamma$ is a $C^\infty$ curve, U is $C^\infty$ on $\bar{\Omega}^+ \setminus \{x^1,\ldots x^k\}$ - we proof in [16] the following regularity for the solution u of the original problem $(P_\beta)$ .

THEOREM 3.1: *Given any constant* $|\bar{\gamma}| \geq 0$ *there is a constant* $K(\bar{\beta}\ ;\ \gamma) > 0$ *such that for any* $\beta \geq \bar{\beta}$ *and any index* $\gamma = (\gamma_1, \gamma_2)$ , $|\gamma| \leq |\bar{\gamma}|$ *we have for the solution* u *of* $(P_\beta)$

$$\| u^- \|_{1/2}(\Gamma) + \| u_n^- \|_{-1/2}(\Gamma) \leq k(\bar{\beta}) \ (\| U^+ \|_{1/2}(\Gamma) + \| U_n^+ \|_{-1/2}(\Gamma))$$

$$\| D^\gamma u \|_{L_\infty(\bar{\Omega}^*_{\bar{\eta}/2})} \leq K(\bar{\beta}) e^{-\beta\bar{\eta}/2\sqrt{2}} \ (\| U^+ \|_{1/2}(\Gamma) + \| U_n^+ \|_{-1/2}(\Gamma))$$

*where* $\bar{\Omega}^*_{\bar{\eta}/2} = \Omega' \setminus \bar{\Omega}'_{\bar{\eta}/2}$ *and* $\Omega'_{\bar{\eta}/2}$ *is the strip* $0 < \eta < \bar{\eta}/2$ *in* $\Omega'_{\bar{\eta}}$

If one deontes by $\Omega'_{\bar{\eta}}$ a boundary strip in $\eta'$ with $\eta = 0$ giving $\Gamma$ and if one sets

$$u_N = \begin{cases} v_o + \ldots + v_N \beta^{-N} & \text{in } \Omega \\ \psi(w_1 a^{-1} + \ldots + w_{N+1} \beta^{-N-1}) & \text{in } \Omega'_{\overline{\eta}} \end{cases} \tag{3.14}$$

then one finds $R_N = u - u_N$ is of order $\beta^{-(N+1)}$. In [16] we established the following convergence result for the asymptotic procedure.

THEOREM 3.2: (i) *Given any* $N \geq 0$ *there are constants* $\overline{\beta}$, $L_N$ *such that if* $u$ *is the solution of* $(P_\beta)$ *and* $u_N$ *is as in* (3.14) *then, for all* $\beta \geq \overline{\beta}$

$$\| u-u_N\|_1 (\Omega_{\overline{\eta}/2}) + \int_{\Omega^+} |\nabla(u-u_N)|^2 \, dx + \| u-u_N\|_{1/2}(\Gamma) + \| u_n-u_{N,n})\|_{-1/2}(\Gamma)$$

$$\leq L_N \beta^{-N-1}$$

(ii) *Given any* $\rho > 0$ *there is* $L_N(\rho)$ *such that if* $\Omega_\rho$ *is any subset of* $\Omega$ *with* $d(\Omega_\rho, \Gamma) > \rho$ *and* $\beta \geq \overline{\beta}$

$$\| u-u_N\|_{L_\infty} (\Omega_\rho) \leq L_N(\rho) \beta^{-N-1}$$

## 4. TWO-DIMENSIONAL INTERFACE PROBLEMS WITH CORNERS

First, let us again consider the example of a metallic cylinder in a transverse-magnetic field which was introduced at the end of section 2, that is we consider in the two-dimensional cross-section $\Omega'$ and its exterior $\Omega$.

$$\begin{aligned} &\operatorname{curl} \underset{\sim}{E} = \underset{\sim}{H}\,, \quad \operatorname{curl} \underset{\sim}{H} = \alpha^2 \underset{\sim}{E} \quad \text{in} \quad \Omega \\ &\operatorname{curl} \underset{\sim}{E} = \underset{\sim}{H}\,, \quad \operatorname{curl} \underset{\sim}{H} = i\beta \underset{\sim}{E} \quad \text{in} \quad \Omega \end{aligned} \tag{4.1}$$

with continuous tangential components $\underset{\sim}{n} \times \underset{\sim}{E}$, $\underset{\sim}{n} \times \underset{\sim}{H}$ across the interface $\Gamma$. The incident fields due to a wire, carrying a periodic current $I(t) = \mathrm{Re}\,\{I_o e^{-i\omega t}\}$, $I_o \in \mathbb{R}$, are

$$\begin{aligned} &\underset{\sim}{E}^o(x,y,z,t) = \mathrm{Re}\,\{I_o\, \phi_\alpha(x,y)\underset{\sim}{e}_3 e^{-i\omega t}\} \\ &\underset{\sim}{H}^o(x,y,z,t) = \mathrm{Re}\,\{I_o\, (\phi_{\alpha,y}\underset{\sim}{e}_1 - \phi_{\alpha,x}\underset{\sim}{e}_2) e^{-i\omega t}\} \end{aligned} \tag{4.2}$$

with the Hankel function of first kind and order zero

$$\phi_\alpha(\underset{\sim}{x}) = -\frac{i}{4} H_o^{(1)}(\alpha|\underset{\sim}{x} - \underset{\sim}{x}_o|), \quad \underset{\sim}{x} = (x,y,o) \in \Omega \tag{4.3}$$

The incident fields $\underset{\sim}{E}^o$, $\underset{\sim}{H}^o$ satisfy Maxwell's equations in $\Omega \backslash \{\underset{\sim}{x}_o\}$ and the solution $\underset{\sim}{E}$, $\underset{\sim}{H}$ of (4.1) has the same form (4.2) if $\phi_\alpha$ is replaced by the solution $u$ of the interface problem (4.4) below. Since $\underset{\sim}{E}$, $\underset{\sim}{H}$ are automatically divergence free and the scattered fields $\underset{\sim}{E} - \underset{\sim}{E}^o$, $\underset{\sim}{H} - \underset{\sim}{H}^o$ satisfy a radiation condition, the eddy current problem (4.1) is reduced to

$$\Delta u + \alpha^2 u = 0 \quad \text{in} \quad \Omega \,, \quad \Delta u + i\beta^2 u = 0 \quad \text{in} \quad \Omega'$$
$$u^- = u^+ + \phi_\alpha \,, \quad u_n^- = u_n^+ + \frac{\partial}{\partial n}\phi_\alpha \quad \text{on} \quad \Gamma \tag{4.4}$$

For household currents the frequency $\alpha$ is very small and instead of (4.4) one considers the Laplace-Helmholtz interface problem $(P_\beta)$ of section 3.

If the metallic cylinder is made of ferromagnetic material, the constant $\beta$ is usually large. This leads to the perfect conductor approximation $(P_{\alpha\infty})$ of section 2 and therefore in our two-dimensional situation here it leads to the exterior Dirichlet problem.

$$\Delta u + \alpha^2 u = 0 \quad \text{in} \quad \Omega \,, \quad u^+ = -\phi_\alpha \quad \text{on} \quad \Gamma \tag{4.5}$$

where $u$ satisfies the Sommerfeld radiation condition.

In [22] we give for the above problems several formulations with boundary integral equations. As an illustration of our results in the preceding sections we present here only the direct method. With the Hankel function (4.3) the operators of the simple layer and of the double layer and their normal derivatives become (see [8])

$$\begin{aligned}
\mathcal{V}_\gamma\psi(x) &= \frac{i}{2}\int_\Gamma \psi(y)\, H_o^{(1)}(\gamma|x-y|)\, ds_y \\
\mathcal{N}_\gamma\psi(x) &= \frac{i}{2}\int_\Gamma \psi(y)\, \frac{\partial}{\partial n_y} H_o^{(1)}(\gamma|x-y|)\, ds_y \\
\mathcal{K}_\gamma\psi(x) &= \frac{i}{2}\int_\Gamma \psi(y)\, \frac{\partial}{\partial n_x} H_o^{(1)}(\gamma|x-y|)\, ds_y \\
\mathcal{D}_\gamma\psi(x) &= \frac{i}{2}\int_\Gamma \psi(y)\, \frac{\partial}{\partial n_x}\frac{\partial}{\partial n_y} H_o^{(1)}(\gamma|x-y|)\, ds_y
\end{aligned} \tag{4.6}$$

The Green's second identiy together with the jump relations yields on $\Gamma$ an Agmon-Douglis-Nirenberg system on $H^{1/2}(\Gamma) \times H^{-1/2}(\Gamma)$ similar to (1.48), (1.49).

$$- (\mathcal{D}_\alpha + \mathcal{D}_{\sqrt{i}\,\beta})\, u - (K_\alpha + K_{\sqrt{i}\,\beta}) \frac{\partial u}{\partial n} = \frac{\partial \phi_\alpha}{\partial n} \tag{4.7}$$

$$- (N_\alpha + N_{\sqrt{i}\,\beta})\, u + (V_\alpha + V_{\sqrt{i}\,\beta}) \frac{\partial u}{\partial n} = \phi_\alpha$$

For a smooth interface $\Gamma$ the coupled system is strongly elliptic satisfying the Garding inequality (1.52) in Theorem 1.11 (see also [22]).

In the following we are interested in the solution of (4.4) for small $\alpha$, that is in the <u>low frequency</u> case. The asymptotic expansion for small $z$

$$- \frac{i}{4} H_0^{(1)}(z) = \frac{1}{2\pi} \log z + m + O(z^2 \log z), \quad m \text{ a constant}, \tag{4.8}$$

shows for small $\alpha$

$$V_\alpha \psi = V_0 \psi - \frac{1}{\pi} (\log \alpha + m) \int_\Gamma \psi ds_y + O(\alpha^2 \log \alpha)$$

where

$$V_0 \psi(x) = - \frac{1}{\pi} \int_\Gamma \log|x-y| \quad \psi(y)\, ds_y \tag{4.9}$$

Here the term $O(\alpha^2 \log \alpha)$ contains smoothing pseudodifferential operators of order $-3$. Neglecting terms $O(\alpha^2 \log \alpha)$ in (4.7) we obtain with $v_0 = \frac{1}{\pi} \log |\cdot - y|$

$$
\begin{aligned}
&- (\mathcal{D}_o + \mathcal{D}_{\sqrt{i}\,\beta})u - (K_o + K_{\sqrt{i}\,\beta})\frac{\partial u}{\partial n} = \frac{\partial}{\partial n} v_o \\
&- (N_o + N_{\sqrt{i}\,\beta})u + (V_o + V_{\sqrt{i}\,\beta})\frac{\partial u}{\partial n} = v_o
\end{aligned}
\qquad (4.10)
$$

where $\mathcal{D}_o$, $N_o$, $K_o$ are obtained from (4.6) by substituting $v_o$ for the Hankel function. If $\beta$ is not too large, the system (4.10) changes by (4.8) to

$$
\begin{pmatrix} -\mathcal{D}_o & -K_o \\ -N_o & V_o \end{pmatrix}
\begin{pmatrix} u \\ \frac{\partial u}{\partial n} \end{pmatrix}
= \frac{1}{2}
\begin{pmatrix} \frac{\partial}{\partial n} v_o \\ v_o \end{pmatrix}
\qquad (4.11)
$$

For this system again the two-dimensional version of Theorem 1.11 holds if the interface $\Gamma$ is smooth.

The analysis above of pseudodifferential operators is restricted to the case of a sufficiently smooth boundary. For functions on polygons the Fourier transform is no longer directly applicable and has to be replaced by the Mellin transform which acts in weighted Sobolev spaces [1], [3].

For polygonal $\Gamma$ with a localization technique the mapping properties of the system (4.11) are derived by investigating the operators on a reference angle $\Gamma^{\omega}$ and posing compatibility conditions [5] via the spaces $(0 \leq s < 3/2)$ in [22]

$$H^2(\Gamma^\omega) = \{(u_-,u_+) \in H^s(\mathbb{R}_+)^2 : u_- - u_+ \in \tilde{H}^s(\mathbb{R}_+)\} ,$$

$$H^{-s}(\Gamma^\omega) = \{(u_-,u_+) \in H^{-s}(\mathbb{R}_+)^2 : u_- + u_+ \in H^{-s}(\mathbb{R}_+)\} \tag{4.12}$$

The identification of functions u on $\Gamma^\omega$ pairs $(u_-,u_+)$ of functions on $\mathbb{R}_+$ induces an identification $\hat{=}$ of integral operators on $\Gamma^\omega$ with (2x2) matrices of integral operators on $\mathbb{R}_+$ , namely

$$V \hat{=} \begin{pmatrix} \ell & \ell \\ \ell & \ell \end{pmatrix} + \begin{pmatrix} V_o & V_\omega \\ V_\omega & V_o \end{pmatrix}, \quad N_o \hat{=} \begin{pmatrix} 0 & N_\omega \\ N_\omega & 0 \end{pmatrix} \quad D \hat{=} \begin{pmatrix} D_o & D_\omega \\ D_\omega & D_o \end{pmatrix} \tag{4.13}$$

where for $g \in C_o^\infty [0,\infty)$

$$\ell g := -\frac{1}{\pi}\int_0^\infty \log y \; g(y)dy, \quad V_o g = \lim_{\omega\to o} V_\omega g \cdot$$

$$V_\omega g(x) := -\frac{1}{\pi}\int_0^\infty \log\left|1 - \frac{x}{y} e^{-i\omega}\right| g(y)dy ,$$

$$N_\omega g(x) := \frac{1}{\pi}\int_0^\infty \operatorname{Im}\left(\frac{1}{xe^{i\omega}-y}\right) g(y)dy, \tag{4.14}$$

$$D_\omega g(x) := \frac{1}{\pi}\int_0^\infty \frac{1}{x}\frac{\partial}{\partial\omega} \operatorname{Im}\left(\frac{1}{xe^{i\omega}-y}\right) g(y)dy, \qquad D_o g = \lim_{\omega\to o} D_\omega g$$

The kernely $N_\omega$ as well as $V_o$, $V_\omega$ (after the separation of a finite-dimensional part) are homogeneous of degree $-1$ and $0$ , respectively.

Therefore these operators can be algebraized by Mellin transformation. To this end one has to extend the domain of definition of the kernels and the functions on which the operators act from $[0,1)$ to the half-line $[0,\infty)$. Then one obtains for $g \in C_o^\infty(0,\infty)$

$$\widehat{V_o g}(\lambda) = \frac{\cosh \pi\lambda}{\lambda \sinh \pi\lambda} \hat{g}(\lambda - i)$$

$$(\mathrm{Im}\,\lambda \in (0,1)); \tag{4.15}$$

$$\widehat{V_\omega g}(\lambda) = \frac{\cosh(\pi-\omega)\lambda}{\lambda \sinh \pi\lambda} \hat{g}(\lambda - i)$$

$$\widehat{N_\omega g}(\lambda) = -\frac{\sinh(\lambda-\omega)\lambda}{\sinh \pi\lambda} \hat{g}(\lambda) \qquad (\mathrm{Im}\,\lambda \in (-1,1)) .$$

$$\widehat{D_\omega g}(\lambda) = (\lambda+i) \frac{\cosh(\pi-\omega)(\lambda+i)}{\sinh\pi(\lambda+i)} \hat{g}(\lambda+i)$$

$$\tag{4.16}$$

$$\widehat{K_\omega g}(\lambda) = -\frac{\sinh(\pi-\omega)(\lambda+i)}{\sinh\pi(\lambda+i)} \hat{g}(\lambda) , \qquad \mathrm{Im}\,\lambda \in (-1,1)$$

Here the Mellin transform $\hat{g}(\lambda)$ of a function $g \in C_o^\infty(0,\infty)$ is defined by

$$\hat{g}(\lambda) = \int_0^\infty x^{i\lambda-1} g(x)dx . \tag{4.17}$$

The explicit Mellin symbols in (4.15) are proved in [1 ], the representations in (4.16) are derived from (4.15) via (4.14) since there holds

$$\widehat{Xg}(\lambda) = \int_0^\infty x^{i\lambda-1} \, x\, g(x) \, d\lambda = \hat{g}(\lambda-i) \, .$$

Now, the special form (4.15), (4.16) and the local equivalence of norms of weighted and usual Sobolev spaces on $\mathbb{R}$ as well as interpolation yield with (4.12) and a suitable partition of unity the following continuity result for (4.11).

THEOREM 4.1 [22]: *For a polygonal interface* $\Gamma$ *and* $s \in (-1/2, 3/2)$ , *the mapping*

$$A_o : H^s(\Gamma) \times \tilde{H}^{s-1}(\Gamma) \times \mathbb{R} \to \tilde{H}^{s-1}(\Gamma) \times H^s(\Gamma) \times \mathbb{R}$$

*defined by* (4.11) *is continuous and a Garding's inequality holds for* $A_o$ *with* $s = \frac{1}{2}$ . (ii) *For* $s = 1/2$ *the system* (4.11) *is uniquely solvable and equivalent to* $(P_\beta)$ *of section* 3.

*Proof:* (i) With the decompositions (4.13) the system (4.11) decomposes as

$$A_o \mathrel{\hat{=}} \begin{pmatrix} D_o & D_\omega & 0 & K_\omega \\ D_\omega & D_o & K_\omega & 0 \\ 0 & N_\omega & V_o & V_\omega \\ N_\omega & 0 & V_\omega & V_o \end{pmatrix}$$

If we define on $\Gamma^\omega$ for $g \mathrel{\hat{=}} (g_-, g_+)$ , $g_+(x) = g(x)$ , $g_-(x) = g(xe^{i\omega})$ for $x \geq 0$ , the even part $g^e := g_+(x) + g_-(x)$ and the odd part $g^o(x) := g_-(x) - g_+(x)$

then the system becomes uncoupled in even and odd parts. We show in [3] that there exists a $\gamma = \gamma(\omega,\chi) > 0$ such that for any smooth cut-off-function $\chi$

$$(V\chi g, g)_{L_2(\Gamma^\omega)} \geq \gamma \| \chi g \|^2_{H^{-1/2}(\Gamma^\omega)}$$

and in [22]

$$(\mathcal{D}\chi g, g)_{L_2(\Gamma^\omega)} \geq \gamma \| \chi g \|^2_{H^{1/2}(\Gamma^\omega)} \quad .$$

A tedious analysis as in [1], [3] shows in [22] by means of the above inequalitites that $A_o$ satisfies a Garding inequality. The claimed continuity of $A_o$ follows from the explicit forms (4.15), (4.16) of the Mellin symbols by application of the method in [1], [3]. The assertion (ii) follows by standard arguments as used in the preceding sections. □

In the case of a curvilinear interface, $\Gamma$, that is piecewise $C^3$, the continuity and coerciveness result in Theorem 4.1 remains valid by combining the arguments for a smooth boundary $\Gamma$ with those for a polygonal one. We omit the details and refer to [2], [22].

Actually $A_o$ is bijective in a whole scale of function spaces for $s \in (-1/2, 3/2)$ if the spaces are augmented by special singularity functions which describe the local behaviour of the solution near corners (see [22]). In [22] we demonstrate this phenomena at the example of forward scattering (4.5). As indicated in [3] the Mellin transform together with the Cauchy integral theorem for analytic functions gives an expansion of the solution in terms of singularity functions as in [ 1 ]. If one augments the space of test and trial functions in the Galerkin procedure by these special singularity functions one obtains higher convergence rates of the numerical approximation [22 ].

*References:*


[1] Costabel, M., Stephan E., Boundary integral equations for mixed boundary value problems in polygonal domains and Galerkin approximation, THD-preprint 593, Darmstadt 1981. To appear at Banach Center Publications.

[2] Costabel, M. and Stephan E., Curvature terms in the asymptotic expansions for solutions of boundary integral equations on curved polygons, THD-preprint 673 Darmstadt (1982).

[3] Costabel, M., Stephan E., Wendland W.L., On boundary integral equations of the first kind for the bi-Laplacian in a ploygonal plane domain, THD-preprint 670, Darmstadt 1982.

[4] Folland, G.B., Introduction to Partial Differential Equations, Mathematical Notes 17, Princeton University Press, Princeton, N.J. 1976.

[5] Grisvard, P., Boundary value problems in non-smooth domains, Univ. of Maryland, MD20742 Lecture Notes 19(1980).

[6] Hariharan, S.I. and MacCamy, R.C., Integral equation procedures for eddy current problems, Journ. of Comp. Physics 45 (1982) 80-99.

[7] Hsiao, G. and MacCamy, R.C., Solutions of boundary value problems by integral equations of the first kind, SIAM Review 15 (1973), 687-705.

[8] Hsiao, G. and Wendland, W.L., A finite element method for some integral equations of first kind, Journ. Math. Anal. and Appl. 58 (1977) 449-481.

[9] Hörmander, L. Linear Partial Differential Operators, in Die Grundlehren der mathematischen Wissenschaften in Einzeldarstellungen 116, Springer-Verlag, Berlin, Heidelberg, 1963.

[10] Knauff, W. and Kress, R., On the exterior boundary value problem for the time harmonic Maxwell equations, Journ. Math. Anal. Appl. 72 (1979) 215-235.

[11] Koshlyakov, N.S., Smirnov, M.M. and Gliner, E.B., Differential Equations of Mathematical Physics, North-Holland Publishing Company, Amsterdam, 1964.

[12] Kress, R. and Roach, G.F., Transmission problems for the Helmholtz equation, J. Math. Phys. 19 (1978) 1433-1437.

[13] Kupradse, W.D., Randwertaufgaben der Schwingungstheorie und Integralgleichungen, Deutscher Verlag der Wissenschaften, Berlin, 1956.

[14] MacCamy, R.C. and Stephan E., Solution procedure for three-dimensional eddy current problems. Carnegie-Mellon University, Techn. Report (1981), to appear.

[15] MacCamy, R.C., and Stephan, E., A single layer potential method for three-dimensional eddy current problems, in Ordinary and Partial Differential Equations, Dundee (1982) (ed. by W.N. Everitt and B.D. Sleeman), Lecture Notes in Mathematics, Springer-Verlag, Berlin (1982).

[16] MacCamy, R.C. and Stephan, E., A skin effect approximation for eddy current problems, THD-preprint 679 Darmstadt (1982), to appear.

[17] MacCamy, R.C. and Stephan, E., A boundary element method for an exterior problem for three-dimensional Maxwell's equations, THD-preprint 681 Darmstadt (1982), to appear.

[18] Magnus, W., Oberhettinger, F. and Soni, R.P., Formulas and Theorems for the Special Functions of Mathematical Physics, in Die Grundlehren der mathematischen Wissenschaften in Einzeldarstellungen 52, Springer-Verlag, New York (1966).

[19] Müller, C., Foundations of the Mathematical Theory of Electromagnetic Waves, Springer-Verlag, Berlin-Heidelberg-New York, 1969.

[20] Rellich, F., Über das asymptotische Verhalten der Lösungen von $\Delta u + \lambda u = 0$ in unendlichen Gebieten, Jahresbericht d. DMV 53 (1943) 57-65.

[21] Seeley, R., Pseudo-Differential Operators, CIME, Cremonese, Rome (1969) (coordinated by L. Nirenberg).

[22] Stephan, E., A boundary element method for two-dimensional eddy current and scattering problems, to appear.

[23] Stratton, I.A., Electromagnetic Theory, McGraw-Hill, New York, N.Y. 1941.

[24] Wendland, W.L., Boundary element methods and their asymptotic convergence, in these Lecture Notes.

[25] Werner, P., Zur mathematischen Theorie akustischer Wellenfelder, Arch. Rat. Mech. Anal. 6 (1960) 231-260.

[26] Werner, P., Beugungsprobleme der mathematischen Akustik, Arch. Rat. Mech. Anal. 12 (1963) 115-184.